EXPLORATIONS IN COMPUTING

An Introduction to Computer Science

CHAPMAN & HALL/CRC
TEXTBOOKS IN COMPUTING

Series Editors

John Impagliazzo
ICT Endowed Chair
Computer Science and Engineering
Qatar University

Professor Emeritus, Hofstra University

Andrew McGettrick
Department of Computer
and Information Sciences
University of Strathclyde

Aims and Scope

This series covers traditional areas of computing, as well as related technical areas, such as software engineering, artificial intelligence, computer engineering, information systems, and information technology. The series will accommodate textbooks for undergraduate and graduate students, generally adhering to worldwide curriculum standards from professional societies. The editors wish to encourage new and imaginative ideas and proposals, and are keen to help and encourage new authors. The editors welcome proposals that: provide groundbreaking and imaginative perspectives on aspects of computing; present topics in a new and exciting context; open up opportunities for emerging areas, such as multi-media, security, and mobile systems; capture new developments and applications in emerging fields of computing; and address topics that provide support for computing, such as mathematics, statistics, life and physical sciences, and business.

Published Titles

Pascal Hitzler, Markus Krötzsch, and Sebastian Rudolph,
Foundations of Semantic Web Technologies

Uvais Qidwai and C.H. Chen, Digital Image Processing: An Algorithmic
Approach with MATLAB®

Henrik Bærbak Christensen, Flexible, Reliable Software: Using Patterns
and Agile Development

John S. Conery, Explorations in Computing: An Introduction to
Computer Science

CHAPMAN & HALL/CRC
TEXTBOOKS IN COMPUTING

EXPLORATIONS IN COMPUTING

An Introduction to Computer Science

John S. Conery

With Illustrations
by Phil Foglio

CRC Press
Taylor & Francis Group
Boca Raton   London   New York

CRC Press is an imprint of the
Taylor & Francis Group an **informa** business

A CHAPMAN & HALL BOOK

CRC Press
Taylor & Francis Group
6000 Broken Sound Parkway NW, Suite 300
Boca Raton, FL 33487-2742

Printed in the United States of America on acid-free paper
10 9 8 7 6 5 4 3

International Standard Book Number: 978-1-4398-1262-4 (Hardback)

Library of Congress Cataloging-in-Publication Data

Conery, John S.
    Explorations in computing : an introduction to computer science / John S. Conery.
        p. cm. -- (Chapman & Hall/CRC textbooks in computing)
    Includes index.
    Summary: "Based on a course taught by the author, this textbook provides an introduction to computer science for non-majors or pre-majors in beginning courses. The book uses an active learning and problem-solving approach to present key topics in computer science, along with real-world examples and practical algorithms. The author explains how to read, rather than write, programs and how to solve problems. To provide a good understanding of computing without covering everything in the field, short focused chapters include tutorial projects and interactive labs that use Ruby, a simple open-source language."-- Provided by publisher.
    Summary: "The title of this book, Explorations in Computing, conveys the idea of how we will use a similar active learning approach to study computation. Each chapter is organized around a single project that introduces an important concept or application in computer science. To complete the project, students type commands in Ruby, an interactive programming language,following a detailed script set out in the text. The aim is for the students to immerse themselves in the interactive environment, and experience first-hand what goes on inside a computer as it solves some interesting problems. Many parts of the projects are open-ended, and students are encouraged to continue exploring on their own, after using the exercises in the book as a starting point"-- Provided by publisher.
    ISBN 978-1-4398-1262-4 (hardback)
    I. Title. II. Series.

QA76.C58594 2010
004--dc22   2010032399

Visit the Taylor & Francis Web site at
http://www.taylorandfrancis.com

and the CRC Press Web site at
http://www.crcpress.com

To my friend Steve Sall, who introduced me to mountaineering and Oregon microbrews, and who (evidently without success) has tried to convince me to write shorter sentences.

Preface

In the context of human history, computers are a fairly recent invention. But the idea of *computation*—of solving a complex problem by repeated, systematic execution of a series of simple and straightforward operations—is thousands of years old. Ancient Greek, Egyptian, and Chinese philosophers discovered many important facts about numbers and their relationships, and developed computational methods that are still used today. In post-Renaissance Europe, mathematicians and scientists used computational techniques to fill books of mathematical tables. Numeric integration, carried out painstakingly by hand, was used to calculate the future positions the Moon and planets, and the nautical almanacs produced by "human computers" were essential for navigation well into the twentieth century.

Computation is now an essential part of modern life. Every day we write mail, share photographs, play music, read the news, and pay bills using our personal computers. Engineers use computers to design cars and airplanes, meteorologists use computers to predict the weather, pharmaceutical companies use computers to design new drugs, and investment firms use computational models to predict whether complex transactions are likely to succeed. Modern astronomers also rely heavily on computation. Computers perform calculations that track the movement of planets, asteroids, and comets, keeping an eye out for bodies that pose a potential hazard to the Earth. Astrophysicists use computation to investigate theories on how black holes are formed and how planets coalesce from clouds of interstellar dust. Telescopes today gather massive amounts of data, requiring sophisticated new methods to sift through the information and catalog objects.

Computer science is the study of computation. Given the name "computer science" one might think the field could be characterized as "the study of computers," but as the discussion above showed, the idea of computing has been around a lot longer than there have been machines to do the computations. In the words of one influential computer scientist, Edsger Dijkstra (1930–2002), "computer science is no more about computers than astronomy is about telescopes." Computer hardware plays a huge role, of course, since much of the motivation for studying computer science comes from the fact that computations are run on machines that perform a wide variety of essential tasks. For many people, a large part of the satisfaction of working in computer science derives from the fact that abstract ideas can be turned into programs that run on real computer systems and address important real-world problems.

Another misconception is that computer science is the same as "programming" and computer science courses are all about teaching students to write programs. Computer science is much more than simply writing software. Computer science is a rich intellectual field where practitioners apply a computational approach to address a wide variety of interesting and challenging problems. Computer scientists are engaged in research in core areas of theoretical computer science, computer systems design, algorithms, and programming languages, as well as more application-oriented areas such as databases, networking, and informatics.

This book is a textbook, intended for courses that are an introduction to computer science. The emphasis is on how computation helps people solve problems. Computer science is a huge field, and entire books have been written about algorithms, theory, programming languages, databases, networks, and other areas. Rather than trying to survey the entire field and give a brief introduction to each important area, the goals in this book are to focus on the fundamental idea of computation itself and to give readers some insight into how computation can be used to solve a variety of interesting and important real-world problems.

Active Learning

The distinguishing feature of this book is its active learning approach. Each chapter includes a tutorial project that guides students as they use an interactive environment to explore important ideas in computing by running programs, modifying them, and trying them out on different inputs.

One of the inspirations for this approach was the active learning embodied in a role-playing game called *The Oregon Trail*. Students who played this game learned about the great westward migration of the nineteenth century by making decisions for their character as they traveled from Missouri to Oregon in 1848, trying to manage their resources and avoid hazards along the way. By actively engaging with the material in a virtual environment, and making decisions that would affect the outcome of the game, students gained a much deeper appreciation of what life was like for the people who set out on that journey. One of the factors often cited for the success of this game is that students were able to try many different variations. Students could play the game several times, assuming several different roles, and often seeing a different outcome, even when they made the same decisions.

The title of this book, *Explorations in Computing*, conveys the idea of how we will use a similar active learning approach to study computation. Each chapter is organized around a single project that introduces an important concept or application in computer science. To complete the project, students type commands in Ruby, an interactive programming language, following a detailed script set out in the text. The aim is for the students to immerse themselves in the interactive environment and experience first-hand what goes on inside a computer as it solves some interesting problems. Many parts of the projects are open-ended, and students are encouraged to continue exploring on their own, after using the exercises in the book as a starting point.

An example of this approach is the project in a chapter on the *N*-body problem, where students set up an experiment that simulates the motion of the planets in the solar system. After running the basic exercises, which lead up to a simulation that shows the planets moving in elliptical orbits around the Sun, students can explore on their own, to see what happens in a chaotic system when the initial conditions change just slightly, or when the mass of one of the planets is increased to the point where the system has two bodies the size of the Sun. Another example is a project on the traveling salesman problem, where students run experiments based on a genetic algorithm. After running a set of preliminary experiments to learn how the algorithm evolves an efficient tour through random mutations, students can run more simulations, varying the simulation parameters to see what effect each has on the outcome.

Intended Audience

This book was written for students who want to learn what computer science is about. It was written with two different audiences in mind: students who intend to continue on to major in computer science, and would like a general overview of the field, and students who have chosen to major in a different field, but who would like to take a computer science course as a science elective.

Although the projects in this book are set up to run in an interactive programming environment, no prior experience with programming is necessary. To complete a project, students follow detailed instructions from the text, much like working on a tutorial to learn how to use a new software application. By working through the projects in each chapter, readers will build up a working knowledge of the concepts and terminology of Ruby programming, but the projects do not require students to write any of their own programs.

Although students do not need programming experience, they should be proficient computer users. Students who want to do the projects on their own computers will need to install some software (explained in more detail below), so they should be comfortable with the process of connecting to the Internet and downloading and installing applications. Several projects also require students to create folders and navigate through a file system, and to open, edit, and save text files.

The projects are based on a variety of different subjects, but none of the exercises assume any detailed knowledge of the subject area. The introduction to each section should provide the necessary background to work on the project in that chapter.

As might be expected for a science class, many projects do require a basic proficiency in math. Students should be comfortable with logarithms, exponents, square roots, and other basic mathematical functions. Projects do not require students to solve equations or to work through proofs, as they would in a math course, but students should understand what the functions are and how they are used. Students will gain a deeper appreciation for scalability and other important concepts in computer science when they have the necessary math background.

♦ Advanced Topics

Some chapters include more challenging ideas or exercises. In some cases there will be an entire section devoted to an advanced topic, and in other cases there may be recommended projects for students who want to learn more about a topic or do some more exploration on their own. These sections and projects are indicated by a ♦ symbol, to indicate material that requires more advanced skills, like the trails marked with a "black diamond" at a downhill ski area.

Notes for Students

You have no doubt heard the adage, "What you get out of a course depends on what you put into it." That saying is especially true for learning about computation with this book. Each chapter is built around a project that helps you explore a particular problem and ways of solving it computationally. If you work through the project and spend some time thinking about what your computer is doing as it runs a computation, you will be rewarded by gaining a deeper insight into how computers can help solve a wide variety of important problems.

A useful analogy for these projects are the lab projects that go along with an introductory chemistry course. An instructor selects a set of concepts the students should learn and then develops a set of lab projects to help students gain some experience and reinforce their learning of the concepts. The materials and methods are all spelled out in great detail, and students follow a set of well-specified steps. Those who continue on in chemistry will later learn to design their own experiments, but for beginners everything is set up by the instructor.

That same approach is taken here in this book. The "computational experiments" in each chapter are tutorials that contain detailed instructions for how to start a piece of software and then what to type in order to run the experiment. As you interact with the software you will see how the computation unfolds. The tutorials are designed so that you should be able to complete them in about the same amount of time you would spend on a lab project in a chemistry class. You could run through the tutorials in less time—about as fast as you can type, or if you get examples from web pages, as fast as you can cut and paste—but you should take the time to make sure you understand what your computer is doing as you carry out each step in the tutorial.

At the end of each chapter you will find a set of exercises. These are similar to the questions you would find in a more traditional textbook and are designed to test your understanding of the material in that chapter. If you have completed the tutorial and understood what happened at each step along the way you should be able to answer these questions.

Notes for Instructors

This book is an introduction to computer science for premajors and nonmajors, a course commonly called CS0 in the computer science literature. As part of a program for premajors, the book would be a suitable text for a first course in an introductory computer science sequence, or as part of a "great ideas" course. The book would also be a good text for a stand-alone science elective, or for a course on computational thinking. When augmented with programming assignments, it could also be used in a programming-first or objects-first CS1 course.

As mentioned above, the book is organized around a set of projects that give students an opportunity to experiment with important ideas in computer science. In most cases, the important concepts are algorithms, and the projects are examples of how algorithms provide computational solutions to important problems. An interactive programming language provides a "computational workbench" where students can experiment with algorithms by typing expressions and seeing the results. The interactive language sets up an environment where students can run computations and explore the effects of changing parameters or modifying operations performed at key steps of the computation.

But using an interactive programming language raises a difficult issue: won't students have to learn to write programs? The approach taken in this book is to base the experiments on a set of scripted tutorials. Each project in the book has a detailed set of instructions for how to perform an experiment by loading software that has been written already. The expressions students enter create and manipulate objects and call methods that implement the algorithm being studied.

The viability of the scripted tutorial approach is based on the fact that is is much easier to learn to read existing programs than to write new ones. Anyone who has tried to learn a foreign language knows how much easier it is to read phrases in the new language than it is to speak or write a sentence. A similar effect applies to programming languages as well. Beginning students can reach a surprising level of literacy by just learning a few fundamental concepts of object-oriented programming—objects, classes, methods, variables, and control flow—with the view that they are learning a language that is a notation for describing algorithms. Since students are only expected to understand programs, they do not need to learn how to design, implement, test, and debug their own code, and several messy details covered in introductory programming courses, like scope rules, call by reference, variable lifetimes, *etc.*, can safely be ignored.

I chose to use Ruby for these projects for several reasons, the foremost being the interactive programming environment that supports experimentation. Ruby has a very clean syntax, and for most operations it provides an intuitive notation. Ruby is open source and is easily downloaded and installed on a wide variety of systems.

An important question was whether to try to make the book a comprehensive introduction to the entire field of computer science, or whether to focus on fewer topics and go into them in more depth. I chose the latter. I think the projects will be much more interesting, and students will gain a better overall understanding of what computer science is about and how computer scientists think about problems, if the book has a few well-chosen examples, even if it means leaving out several important topics.

The topics presented in the book are outlined below. The general pattern for each chapter will be to first introduce the concept presented in that chapter; this introductory section will essentially be an essay that tries to make the case that the idea is interesting and worth understanding in more detail. The main part of the chapter will be the development, through a series of projects, of one or more algorithms that illustrate the idea and provide the student with a chance to experiment.

1 Introduction

The book starts with a general introduction to computation, expanding on the themes mentioned in the first section of this preface: computer science is not just about computers and is not just programming.

2 The Ruby Workbench

The second chapter is a practical introduction to Ruby and how it can be used as a "computational workbench" to set up experiments with computations. The tutorial takes the students through the construction of a simple program to convert temperature from Celsius to Fahrenheit, and introduces the ideas of variables, objects, and methods.

3 The Sieve of Eratosthenes

This chapter introduces the first real algorithm studied in the book. It also introduces a few more practical techniques used later in the book: making lists of numbers and iterating over a list. The tutorial starts with simple expressions involving integers, shows how to make a list of numbers, then how to selectively remove composite numbers, and leads finally to an algorithm that creates a complete list of prime numbers.

4 A Journey of a Thousand Miles

This chapter builds on the basic idea of iteration presented in the previous chapter. The project shows how iteration can be used to solve two common problems, searching and sorting, using linear search and insertion sort. An important idea in computing in this chapter is scalability, and students are introduced to $\mathcal{O}$ notation.

5 Divide and Conquer

The important idea in this chapter is that a more sophisticated strategy for solving a problem can lead to a more efficient computation. The tutorial shows how binary search takes up to $\log_2 n$ steps instead of n, and merge sort takes at most $n \log_2 n$ steps instead of n^2.

6 When Words Collide

The new concept in this chapter is that our ability to solve a problem computationally depends not only on the sequence of steps defined by an algorithm, but also on how the data is organized. The tutorial project is based on a data structure that implements an index for a large collection of data. Students learn about hash functions and eventually do experiments with a hash table that resolves collisions with buckets.

7 Bit by Bit

The projects in this chapter are related to encoding data: using patterns of binary digits to encode numbers and letters, the number of bits required to encode a set of items, text compression with Huffman trees, and error correction with parity bits.

8 The War of the Words

This chapter introduces the important ideas that functions can also be encoded as a string of bits and that instructions (bit patterns representing steps that implement functions) are stored in a computer's memory along with data. The tutorial uses the game of Corewar, which is a contest between two programs running in the same virtual machine; a program wins if it can write a halt instruction over the opponent's code. The tutorial leads the student through the phases of a processor's fetch-decode-execute cycle and emphasizes how a word that is a piece of data (the constant 0) for one program becomes an instruction (halt) for the other program.

9 Now for Something Completely Different

The big idea in this chapter is randomness, and how random numbers can be used in a variety of algorithms, from games to scientific applications. There is an interesting paradox here: can we really generate random outputs from an algorithm? Isn't a method in Ruby supposed to carry out exactly the same calculations and produce the same result each time it is called? The answer is that random numbers generated by an algorithm are pseudorandom, and the project takes students through the steps in the development and testing of a pseudorandom number generator.

10 Ask Dr. Ruby

The tutorial project in this chapter is based on a Ruby implementation of Joseph Weizenbaum's ELIZA program, and shows how very simple pattern matching rules can be used to transform input sentences, giving the illusion that the computer is carrying on a conversation. By the end of the chapter students will see how difficult natural language processing is, and how semantics and real-world knowledge are required for effective natural language understanding.

11 The Music of the Spheres

The big idea in this chapter is computer simulation. The project leads to an *ab initio* simulation of the motion of planets in the solar system. The chapter introduces issues related to verification and other topics in computer simulation.

12 The Traveling Salesman

The last chapter introduces the idea of intractable problems, building on ideas of scalability from earlier chapters. The project is based on a genetic algorithm, and gives students the opportunity to explore probabilistic solutions. The tutorial has students use predefined code for Map and Tour classes to create random tours, so they can see how tours can be mutated and how collections of tours evolve until an optimal or near-optimal solution is obtained.

Pedagogical Considerations

The chapters and projects described above have been used in a course at the University of Oregon (*CIS 170: The Science of Computing*). We cover the first two chapters during the first week, but after that we spend between one and two weeks on the remaining topics chosen for that term. Lectures emphasize material from the first sections of a chapter, describing the problem and how it might be solved computationally, and explaining how that week's lab project gives some experience with the computation. Students have an option of attending a lab session, where an instructor is available to help them work through the material, but many students do the tutorials on their own. Live demonstrations of the tutorial projects, both in lecture and in lab sessions, have proved to be very effective.

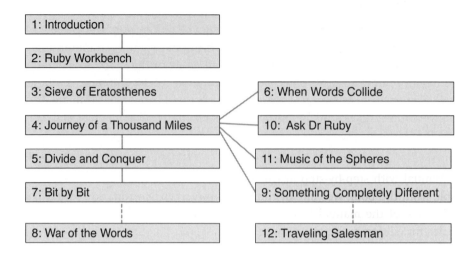

At the end of each chapter there is a set of exercises that ask questions about issues raised in the chapter. After the students have completed the tutorial, they are asked to answer a selected set of questions and submit them as a "lab report" that gives them a chance to explain what they learned. Similar questions are given on exams.

When selecting topics to use in a course, Chapters 1 through 4 should be used every term, since they introduce key concepts (algorithms, scalability) and practical lab skills (instantiating objects, calling methods, creating and iterating over containers) used in other projects. The remaining chapters are mostly independent, and can be selected according to the interests of the students. The chapters on data representations (Bit by Bit) and machine language programming (War of the Words) are both based on the idea of encoding information, but students will have no trouble completing the Corewar project without having done the data representation projects. Similarly, The Traveling Salesman uses random numbers, but students will get a lot out of this project even if they haven't seen how random numbers are generated.

It is also possible to organize a course that includes additional topics and activities beyond those described in this book. In Spring 2008, a few months before the national elections in the U.S., we used electronic voting as a theme for CIS 170. There were additional units on the history of elections and the need for privacy and security in voting, and the computer science topics included networking, encryption, and software engineering, all of which play a role in the design of electronic voting machines.

Software, Documentation, and Lab Manuals

All of the software used for the tutorial projects is written in Ruby. Students can do the projects on computers in an instructional lab, or they can install Ruby on their own computers. Ruby is open-source, and it is a straightforward process to install the Ruby interpreter and associated applications:

- Users of Microsoft Windows XP can download a "one-click installer" that automates all the installation steps.

- A single command typed in a terminal window will install Ruby on a Linux system.

- Ruby is already installed on Mac OS X 10.4 and later (although users running Mac OS X 10.6 may have to reinstall Ruby to work on labs that have interactive visualizations).

The software students will use for the projects is named RubyLabs. RubyLabs is written exclusively in Ruby, using only libraries and modules that are part of the standard Ruby distribution. There is one Ruby module for each lab project. All of the modules have been collected into a single "Ruby gem," which makes it easy to install all the lab software in one step at the beginning of the term. The RubyLabs gem also includes data files and sample Ruby code that students can copy and modify.

A Lab Manual with step-by-step instructions for installing Ruby and the RubyLabs gem is available from the book web site at `http://www.cs.uoregon.edu/eic`. There is a separate version of the manual for Windows XP, Mac OS X, and Linux. The manual also includes tips for editing programs and running commands in a terminal emulator.

The web site also has on-line documentation of all the modules in the RubyLabs gem. After the gem has been installed, this documentation can be read locally by a web browser, without having to connect to the Internet.

Web Site

The web site for this book is

    `http://www.cs.uoregon.edu/eic`

The web site will have
- copies of the lab manual (PDF documents that can be downloaded for free)
- links to the latest versions of the RubyLabs software and documentation
- errata and other news

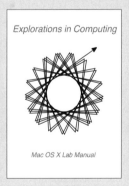

Explorations in Computing

Mac OS X Lab Manual

Acknowledgments

The most important group of people who influenced the development of this book are, without a doubt, the students at the University of Oregon who have been involved in one way or another with CIS 170, the *Science of Computing* course I have taught each year since the 2005-06 academic year. Students who took the course and provided invaluable suggestions (often without waiting for me to ask) include: Joyce Corrao-Clanon, Alex Forbes, Emily Hayes, Isla Globus-Harris, Peri Moritz, Paul Russ, Charles Sheinin, Richard Suhr, and Ace Taylor. Jeff Blakeslee developed an early version of the binary tree visualizations used in Chapter 7, and Michael Maag made a key contribution to the RubyLabs canvas that is used in all the visualizations.

Anyone who has taught a college or university level course knows how important it is to have motivated and engaged teaching assistants, and I have been lucky to work with the best: Megen Brittell, Tom Bulatewicz, Victor Hanson-Smith, and Shad Stafford. Shad also used an early draft of Chapters 2 and 3 in a course he taught at Pacific University in Forest Grove, Oregon, and I received many helpful comments from Shad and his students.

I was thrilled when Phil Foglio responded to my e-mail and said he would be willing to make a set of illustrations for the book. Many thanks to Phil, Kaja Foglio (self-proclaimed "scanning bot"), and the other folks at Studio Foglio, LLC for making it happen.

John Impagliazzo and Andrew McGettrick, the series editors for Chapman & Hall/CRC Press, provided several constructive suggestions during the early phase of the development of the book. From the rough draft I initially submitted, they were able to help me focus on a better choice of topics and more readable presentation. I am also indebted to the external reviewers, especially Jessen Havill, of Denison University, and Andrew Neel, from the University of Memphis.

While the students, reviewers, and editors all had a tremendous influence on the contents of the book, it would have remained just another interesting idea instead of a real book if not for the efforts of Randi Cohen, my editor at CRC Press. Randi somehow knew exactly when I needed encouragement and positive feedback, to keep me going when it seemed like there was no end in sight, and when to set a firm deadline, when it looked like I was going to keep exploring forever.

Finally, I am grateful for the support of my wife Leslie and my daughter Kathleen. Kathleen looked over my lecture notes, read early versions of some of the chapters, and, thankfully, let me know how some of my attempts at humor would have been received by others of her generation. I love you both.

John Conery
Eugene, Oregon

Contents

Chapter 1

Introduction

In the summer of 1821, an English mathematician named Charles Babbage was working with his friend, the astronomer John Herschel, to create a book of mathematical tables. Before computers and calculators were available, people who needed to solve mathematical equations would find the value of a function by searching a table in a reference book like the one shown in Figure 1.1. These books were essential for navigators, architects, merchants, bankers, and anyone else who used math in their profession. There were tables for interest rates, currency conversion, liquid and dry measures, and just about any other quantity that could be expressed with numbers. Not surprisingly, there was a strong demand for accurate tables, and a tremendous amount of time and effort went into creating and checking values printed in reference books.

Babbage and Herschel were working on a nautical almanac, a book of tables containing the positions of the Moon, planets, and stars. These almanacs were used by navigators to determine their location at sea. More than any others, these tables needed to be accurate, as there were concerns that errors in tables could lead to longer routes than necessary, or even shipwrecks. Babbage and Herschel would meet periodically to check the tables being made by a group of people they had hired to work through the tedious steps required to fill the rows. At one of their meetings, when reviewing the latest results, Babbage showed his frustration with the large number of errors by exclaiming, "I wish to God these calculations had been executed by steam!"

Of course Babbage wasn't attributing any special powers to water vapor. The year 1821 was the height of the Industrial Revolution, when machines powered by steam engines were beginning to automate tasks previously carried out by humans. Babbage was simply using the terminology of his time to express his wish that the calculations should be done automatically, by a machine, so the results would be more accurate and reliable.

The quote from Babbage brings up an interesting question. Steam power was helping transform physical labor, and machines were beginning to be widely used to augment, or even replace, human effort. But what about mental labor? What made Babbage think steam engines could help him solve mathematical problems?

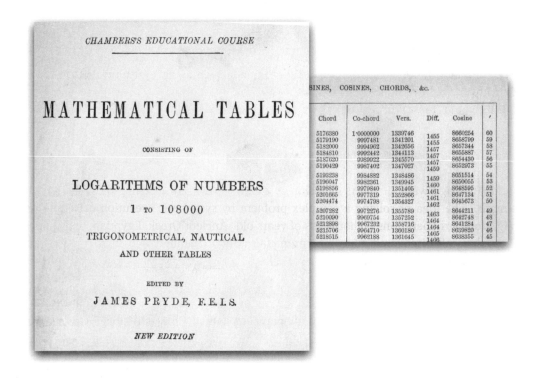

Figure 1.1: Chambers's Mathematical Tables, New Edition, *London, 1901. Before there were computers to calculate mathematical functions, if a person wanted to know the value of a trigonometric function, they would look in a table of sines and cosines. For example, to find the value of* $\cos 30°20'$, *the person would find the page for* $30°$, *then scan down to the row for* $20'$ *and look in the column labeled "cosine."*

The answer is that the idea of **computation**—solving a complex problem by repeated, systematic execution of a series of simple and straightforward operations—was already well established by the nineteenth century. Mathematicians had developed techniques for calculating entries in tables using using only the most basic operations of arithmetic, such as addition and subtraction, where the value in one row of the table can be determined using values from rows filled in previously. Many tables, including the ones in the book being prepared by Babbage and Herschel, were produced by groups of people who had no advanced mathematical skills, but were hired and trained to fill in a table by doing a specific sequence of additions and subtractions. Prior to the middle of the twentieth century, the word "computer" was a job title, referring to any person who was engaged in systematic calculation of values like those found in mathematical tables. Babbage realized that the simple operations carried out by human computers were mechanical in nature, and he dreamed of one day building a machine that would be able to carry out the steps in a computation automatically.

Today computation is so familiar we take it for granted. A navigator, surveyor, architect, or anyone else who needs to know the value of a mathematical function simply enters a number in a calculator and presses a button labeled with the name of the function. Computation is also at the heart of computer applications that help us with common tasks that seemingly have little or nothing to do with mathematics, such as using a word processor to write an essay, organizing a music library, or playing recorded music.

This is a book about computation. The focus is on the nature of computation, with the goal of showing how executing a series of simple steps can eventually lead to the solution of a complex problem. Computers are widely used, not just in math and science, but in every facet of modern life. The aims of this book are to explore the computations that take place in a wide variety of applications and to learn how this powerful idea helps people solve interesting and important problems.

1.1 Computation

Charles Babbage was not the first person to dream of using a machine to help him with his math. The idea of solving a complex problem by systematic execution of simple and straightforward operations is thousands of years old. Ancient Greek, Egyptian, and Chinese philosophers all discovered many important facts about numbers and their relationships. These mathematicians developed methods that are still used today to determine whether a number is prime or how to find the largest common denominator of a pair of numbers. They also devised many different tools to help them perform their calculations; in fact the word "calculate" comes from the Latin for "pebble," since small stones or beads were used in an abacus or similar device.

Many famous mathematicians in post-Renaissance Europe, including Johannes Kepler (1571–1630), Blaise Pascal (1623–1662), and Gottfried Leibniz (1646–1716), dreamed that one day there would be machines to carry out the steps in a computation automatically. As Leibniz once wrote,

> Astronomers surely will not have to continue to exercise the patience which is required for computation. It is this that deters them from ... working on hypotheses and from discussions of observations with each other. For it is unworthy of excellent men to lose hours like slaves in the labor of calculation which could safely be relegated to anyone else if machines were used.

Pascal invented a mechanical calculator that was able to perform additions and subtractions of numbers up to six digits long. Leibniz designed a calculator that could also do multiplications and divisions. These early machines were able to assist human computers, in much the way an abacus or other device might. It wasn't until the nineteenth century that designs for fully automated machines started to appear. Babbage himself played a key role. He designed a machine he called the Difference Engine, coming up with ingenious ideas that later influenced the builders of some of the first successful mechanical computers in the middle 1900s.

The first electronic computers were developed during World War II. Soon after the end of the war the idea of using machines to automate the calculations in science and business began to spread, and several companies started manufacturing computing machines. The machines were very big and very expensive, though, and were found only in the largest corporations, government bureaus, or university research labs. They were used for such diverse tasks as computing the trajectories of rockets and missiles, predicting the weather, and business data processing applications for payroll and accounting.

Fast forward fifty years, and computation has now become an essential part of modern life. Every day we write mail, share photographs, play music, read the news, and pay bills using our personal computers. Engineers use computers to design cars and airplanes, pharmaceutical companies use computers to develop new drugs, movie producers generate special effects and in some cases entire animated films using computers, and investment firms use computational models to decide whether complex transactions are likely to succeed. Not surprisingly, given the role astronomy has played in the history of computing, modern astronomers also rely heavily on computation. Organizations like the Jet Propulsion Laboratory use computers to carry out the calculations that track the locations of planets, asteroids, and comets with the goal of keeping an eye out for potential threats from impacts, such as the one involving the comet Shoemaker-Levy and Jupiter in 1994.

Computation plays a much more extensive role in modern science than the straightforward "number crunching" involved in the calculation of orbits. The phrase *computational science* refers to the use of computation to help answer fundamental scientific questions. Here the word "computational" is an adjective that describes how the science is done. Computational science is, like the more traditional approaches of theoretical science and experimental science, a way of trying to solve important scientific problems. Computational physicists use computers to study the formation of black holes, investigate theories of how planets form, and simulate the predicted collision, three billion years from now, of our Milky Way galaxy with Andromeda.

As we look at the wide variety of problems that are being solved with the help of computation, several questions naturally arise. Do these problems have anything in common? Is there some aspect of a problem that would lead one to believe it can be solved computationally? For that matter, what does it mean to "compute" something?

The generally accepted definition of a **computation** is that it is a sequence of simple, well-defined steps that lead to the solution of a problem. The problem itself must be defined exactly and unambiguously, and each step in the computation that solves the problem must be described in very specific terms.

Squaring without Multiplying

You might think that to compute the value of $f(n) = n^2$ you need to know how to multiply two numbers. But there is an easy way to square a number using only additions.

The diagram at right shows how to compute 5^2: at the top of a piece of paper write the first 5 odd numbers. Add the first two, and write the sum below them. Then add the next odd number to the sum computed on the previous step. Keep doing this until you have added the last odd number and at the end you'll have the value of 5^2.

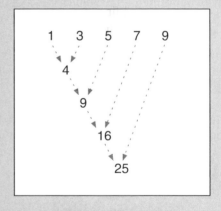

This technique is known as the "method of differences." The Difference Engine, the machine designed by Charles Babbage, was designed to employ this technique to compute the value of any polynomial.

Length (cm)	Average (Mean) Length:
Stream A	
40, 39, 69, 57, 50	$(40 + 39 + 69 + 57 + 50) \div 5 = 51$
Stream B	
44, 62, 59, 58	$(44 + 62 + 59 + 58) \div 4 = 56$

Figure 1.2: *To compare the average length of fish in two different streams, an ecologist would compute the mean length of fish seen in each stream. The mean length is defined as the sum of lengths divided by the number of fish.*

As a simple example of a computation, suppose an ecologist wants to compare the lengths of fish found in two different streams. One way to do this is to calculate the average length of the fish samples from each stream. To solve this problem the ecologist first needs to specify exactly what is meant by "average" length. The most common definition is the arithmetic mean. The steps in the computation are to add up the lengths of the fish found in the first stream, and compute the mean by dividing the sum of lengths by the number of fish (Figure 1.2). Next find the mean for the second stream by adding the lengths of those fish and dividing the sum by the number of fish in that stream. This is an example of a computation because each step (counting, addition, division) is very straightforward, and there is a clear specification of the starting point and the end result.

Note that there is nothing in the description of computation that involves the word "computer." A computation is a process, a sequence of simple operations that leads from an initial state to the desired final result. The process can be carried out entirely by a person, or by a person using the help of mechanical or electronic calculators, or completely automatically by a computer. The choice of which technology would be most effective depends on the situation. For a very small number of fish, an ecologist could write all the numbers in a single column on a sheet of paper, add them up, and then do the division using paper and pencil. For a larger sample, a person would likely want to use a calculator or abacus; the state of the device will show the running total, so that after the last length is added the final step is to divide the total by the number of fish. For very large samples—for example, using information in a database made by storing videotaped observations of fish passing through a ladder next to a dam—it would be most efficient to use a computer and program it to add the lengths and do the division. No matter what technology is used, these situations all involve the same basic computation: in each case the average length is determined by calculating the sum of lengths and then dividing by the number of fish. Note also that the process of computing the mean of a set of numbers applies to other situations as well. This same basic computation, of finding the sum of values in the set, and then dividing by the size of the set, can be applied to (almost) any set of numbers.

1.2 The Limits of Computation

In looking to the future, a natural question is whether the astonishing growth in the use of computation will continue. Most people who work in fields related to computing expect technology to continue to improve, with the components used to build computers getting smaller, faster, cheaper, and more energy efficient. The phrase "ubiquitous computing" refers to the prediction that computers will be so small and so inexpensive they will be everywhere, perhaps even in the clothes we wear. With this sort of explosion in the availability of computing power, it's natural to wonder what sorts of tasks these computers might be asked to perform. One way to address the question is to take a closer look at some of the areas where computers are being used today and ask whether there are any limitations on what might be done in the future.

Often the limits to what we can do with a computer are **technological barriers** that could be overcome by using a more advanced piece of hardware. In principle one could type an essay for a literature class on a cell phone or create a full-length animated feature using a laptop. But it would be more productive to write a paper on a computer with a mouse, a keyboard, and a screen large enough to show a full page of text. Since each frame in an animation is the result of hours of computing on a high-speed supercomputer, a personal computer is not a practical choice for making an animated movie.

There are many important problems that could be solved by computers but are limited by current technology. To take just one example, equations that model changes in weather are very well understood by meteorologists, but the accuracy of weather predictions is currently limited by computing power. With more powerful computer systems (and more detailed measurement of current conditions) it may be possible to make short-term weather forecasts with almost perfect accuracy.

In many cases, however, the limits of what a computer can do are things that cannot be overcome by moving to a more advanced piece of hardware. These barriers might be characterized as **computational limits**, since the difficulty lies in specifying how to solve a problem computationally rather than in the hardware used to carry out the computation. Here are some problems that would be very difficult, if not impossible, to solve by a computer:

- When we want to send a text or e-mail to someone, we need to know their contact information. If the person is in our address book it's trivial to look up their e-mail address, but if we don't know the person we're stuck. In some cases we can do a web search, *e.g.*, to find the name or e-mail address at a company we want to correspond with, but we can't ask a computer to find the e-mail address of someone we just met at a coffee shop.

- When using a computer to manage finances we can connect to a bank or credit card company to download a list of transactions, and we might be able to have a program do some calculations for different investment options, based on interest rates or projected earnings. But we know the computer can't choose the perfect investment because it can't predict which companies will succeed or future interest rates.

- A computer can help find mileage estimates for different types of cars, or admission and enrollment statistics for different colleges. But a computer can't determine whether the fun of driving a flashy convertible outweighs the practicality of an all-wheel drive station wagon, and people usually find it difficult to quantify all the attributes of different colleges and universities that would allow a computer to make a perfect decision on the best school to attend.

- It is common to talk with a computer via telephone to make a train reservation or arrange air travel. But these "conversations" are very limited, and generally we are restricted to one-word sentences that relate to a specific reservation. We would not get very far if we asked the computer for general travel advice, *e.g.*, "What's the weather like in Los Angeles this time of year? Should I go to Hawaii instead?"

In the previous section, a computation was defined to be a series of well-defined steps that lead from an unambiguous starting state to an equally well-defined final result. With this definition in hand, let's take a closer look at the situations described above and some other difficult problems. We will begin to see some interesting and important differences.

Ambiguous Problems

The problem of sending a message to a person we meet at a coffee shop is simply too vague to be solved computationally. It lacks a well-defined starting point and there is no sequence of steps a computer can carry out to solve it. It's worth noting this is a problem that cannot be solved by a human, either—a friend won't be able to help any more than a computer unless we supply a lot more information.

Some attributes students use when deciding which college to attend are well defined. The cost of tuition, living expenses, and the average GPA and SAT scores of entering freshmen are important factors. If these easily quantified items are the only criteria that are important, a computation could lead to a decision of which college is best. For most people, however, intangible qualities like geographical location and quality of life are important, and these are hard to put into terms that could be used in a computation (Figure 1.3). People do solve problems that involve intangibles, and it might seem like this is the sort of thing a person could do better than a computer. But the "solutions" obtained by people aren't like the solutions to the problem of computing the mean length of fish: they are recommendations, not unambiguous and reliably correct answers.

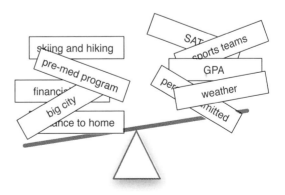

Figure 1.3: *In order to use a computer to choose a college to attend, each attribute would have to be well-defined and quantifiable.*

Natural Language

It would be nice if we could open a cell phone and say "send a message to Aleah, Katie, and Erica to see if they want to come over to study calculus." We could then work on something else while the phone composes a text message, sends it to our friends' phones, and negotiates with the other phones to find a time when everyone wants to meet. This sort of interaction is not possible with current computers, but researchers in a field known as artificial intelligence are trying to understand what is involved in these types of communications and developing computational methods to carry them out. A human personal assistant can accomplish this task, so this is an example of the kind of problem that humans solve but computers (currently) do not. It is an open question whether this problem is beyond the limits of computation. It very well might be possible for some future personal digital assistant to accomplish this task.

Intractable Problems

It's tempting to think a computer could easily win a chess tournament by considering all possible moves starting from a given board configuration, then examining each possible response its opponent could make, and eventually choosing the move that leads to a certain victory. This is an example of a problem that should be solvable by a computer. The problem has a well-defined input (the starting configuration of a chessboard), output (configurations where each move leaves the opponent's king in "check"), and very specific and well-defined set of operations (the legal moves for each chess piece).

Before we try to write a program that uses this strategy, however, we should do some back of the envelope calculations first. The number of possible chess games has been estimated to be more than 10^{43}. Even if this computation were run on a supercomputer far more powerful than the fastest computers available today, on a hypothetical machine that could somehow compute 1 trillion (10^{12}) alternative board combinations per second, it would require $10^{43}/10^{12} = 10^{31}$ seconds, or roughly 10^{21} years. To put this in perspective, the universe is only 10^{13} years old. So even though this is a well-defined computation, it is one no computer will ever complete, and in that sense it is well beyond the limits of computation. It goes without saying that no human will ever do this computation, either. When humans

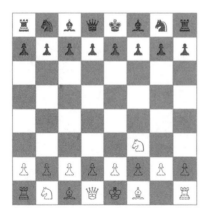

Figure 1.4: *If a computer tries to analyze every possible sequence of moves in response to this opening in a game of chess, it will have to consider over 10^{43} different games.*

and computers play chess successfully they are using other strategies than simple "brute force" exploration of all possible moves.

Computer scientists refer to this sort of problem as **intractable**. Small portions of the problem, such as different opening strategies or endgames, can be analyzed by considering every option, but evaluating every move in a full game is beyond reach. Chapter 12 explores another intractable problem, one that is often faced by organizations that need to do a substantial amount of scheduling or planning.

♦ Unsolvable Problems

Mathematicians in the 1930s made a startling discovery that some problems are simply unsolvable. In logic, these problems are called *undecidable*. The group of unsolvable problems includes determining whether paradoxes, like the familiar statement "this sentence is false," are true or false. The statement can't be true, because that would imply it is false, and likewise it can't be false, because that would make it a true statement about itself. In mathematical terms it is simply undecidable.

The computer science equivalent of the undecidable functions are called *noncomputable problems*. The most well-known, the Halting Problem, asks whether it is possible to examine a running program and decide whether that program will ever terminate. Imagine a situation where a an application is running on a laptop, and an icon appears on the screen that says the program is busy, so it will not respond to any keystrokes or mouse clicks. After five minutes we might start to wonder whether the application is progressing very slowly or has crashed. It would be nice to be able to run another program that would examine the first one and say "be patient, it will terminate" or "it crashed, you need to kill it and restart it." A fundamental result in theoretical computer science tells us that this problem is logically equivalent to an undecidable function and that it is impossible to write such a "halt-checking" program.

Unsolvability is a different type of limitation than intractability, the limitation encountered by the chess-playing program. The chess player will compute the perfect game of chess if we are patient enough to wait 10^{21} years, so it is only unsolvable in a practical sense. The halt-checker requires us to evaluate an undecidable function, so it is beyond the limits of computation in that we know it is impossible to write a general purpose program that could carry out a sequence of steps that will let it determine whether another program will terminate.

1.3 Algorithms

Let's return to the simple example of a computation presented in Section 1.1, where an ecologist wants to know the average length of fish observed in a stream. We now know what the computation involves: sum up the lengths and divide by the number of fish. The next question is, how do we describe the computation in sufficient detail so the steps can be carried out by a machine?

First consider how the ecologist might enlist the aid of a human research assistant. If the assistant has taken a statistics class, the ecologist can just give the assistant the data and expect them to compute the mean lengths. But if the assistant does not know how to compute a mean, the ecologist needs to describe the operation in detail: write the list of numbers on a piece of paper, and then cross them off one by one as they are added to a running sum, and after adding the last piece of data, divide by the number of fish.

A detailed description of how to solve a problem by first specifying the precise starting conditions and then how to follow a set of simple steps that lead to the final solution is known as an **algorithm**. An algorithm is characterized by

- a precise statement of the starting conditions, which are the inputs to the algorithm;

- a specification of the final state of the algorithm, which is used to decide when the algorithm will terminate;

- a detailed description of the individual steps, each of which is a simple and straight-forward operation that will help move the algorithm toward its final state.

In short, an algorithm is a specification for how to carry out a computation. Although the word "algorithm" can be used to refer to any method for systematically solving a problem, and algorithms were widely used long before anyone thought of building a machine to perform the steps in a computation, today the term generally refers to a method that will be carried out automatically by a computer.

In the description of an algorithm, the steps have to be simple enough to be "understood" by a machine. One way to think of what a machine is capable of doing is to think in terms of symbols, such as numbers or letters. The steps in an algorithm are basically **symbol manipulations** like simple arithmetic operations or comparisons that determine which words

---

A Brief History of Algorithms

The world *algorithm* comes from the name of a Persian mathematician, Mohammed ibn Mûsâ al-Khwârizmî (*ca.* 780–850), whose book on the use of Indian numerals introduced Europeans to the numeral 0. The book was translated into Latin with the title *Algoritmi de numero Indorum* ("al-Khwârizmî Concerning the Hindu Art of Reckoning").

The earliest algorithm, now known as Euclid's Algorithm, dates from at least 300 BC. This algorithm is still used today to find the lowest common denominator of two numbers. Other ancient algorithms include the Sieve of Eratosthenes, a method for making lists of prime numbers (and the basis for one of the projects in this book), and methods used by Sun Tzu and other Chinese mathematicians around AD 200.

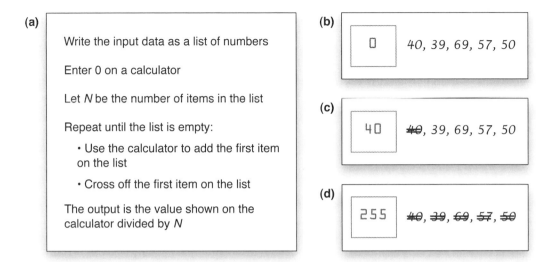

(a)

Write the input data as a list of numbers

Enter 0 on a calculator

Let *N* be the number of items in the list

Repeat until the list is empty:

   • Use the calculator to add the first item on the list

   • Cross off the first item on the list

The output is the value shown on the calculator divided by *N*

(b)   0   *40, 39, 69, 57, 50*

(c)   40   ~~*40*~~, *39, 69, 57, 50*

(d)   255   ~~*40*~~, ~~*39*~~, ~~*69*~~, ~~*57*~~, ~~*50*~~

Figure 1.5: *(a) An algorithm for computing the mean of a set of numbers. (b) The initial state of the computation defined by this algorithm, after entering a 0 into a calculator and writing the input data on a sheet of paper. (c) The state of the computation after adding the first number and crossing it off the list. (d) After the last number has been removed from the list, the calculator holds the sum of all the numbers. For this data set, the mean is 51, the result of dividing 255 by 5.*

come before others in the alphabet. By putting together a large number of simple symbolic operations, a machine can do very complex tasks, such as sorting long lists of names, counting millions of votes cast in an election, or using words to build an index of web pages gathered from the Internet.

As an example of how a task can be described as a sequence of symbol manipulations, the process followed by the research assistant to compute the mean length of a group of fish is shown in Figure 1.5. On the left is the algorithm, showing what needs to be done at each step. Three different stages of the computation are shown on the right. To initialize the computation, the number 0 is entered into a calculator, and all the values in the data set are written out on a sheet of paper. Each time the assistant removes a number from the list, it is added to the running total on the calculator and crossed off the list. By the time the last number has been crossed off, the value displayed on the calculator is the sum of all the lengths. Each step in the algorithm is a symbol manipulation, where the list becomes shorter by one item and the sum has been updated through simple arithmetic operations (which could easily be done by hand as well as a calculator). The final output is a single number, again the result of a simple arithmetic operation, in this case a division.

Often the descriptions of the steps of an algorithm are given in English or another human language, as in the algorithm for computing the mean value of a set of numbers shown in Figure 1.5. This notation, which is sometimes called **pseudocode**, is sufficient for talking about the algorithm, for describing the process to another person, or for trying to understand whether or not the algorithm works. But in order to run the algorithm on a computer, the steps have to be written more precisely as statements in a programming language. If the ecologists find they are spending too much of their time computing means by hand, they might decide to invest some time in implementing their algorithm in the form of a program and running it on their computer.

The idea that one can solve a problem through the use of an algorithm is the central concept in computer science. Computer scientists analyze the theoretical properties of algorithms, develop programming languages used to implement algorithms, and design computer systems to automatically execute the steps in algorithms. As technology opens the doors for new applications, computer scientists work with researchers in other fields to find ways to solve important real-world problems, either by inventing new algorithms or improving and adapting existing algorithms.

1.4 A Laboratory for Computational Experiments

The main goals for this book are to focus on the fundamental idea of computation and to show readers how computation can be used to solve some interesting and important real-world problems. A typical chapter will introduce a problem, explain why it is important, and give an overview of one or more algorithms that have been used to solve the problem.

The heart of each chapter is a tutorial project that has been designed to explore the algorithms described in that chapter. The tutorials in this book are comparable to projects in an introductory chemistry lab for nonmajors. Chemistry instructors for these courses design the projects by selecting materials and methods that are accessible to beginners, and students follow detailed instructions to carry out experiments that help them learn fundamental concepts about chemical processes. The computational projects in this book are similar: the programs and data have been prepared ahead of time, and we will work through the tutorial project to gain some insight into how an algorithm works by watching it run and experimenting with it.

The laboratory for these computational experiments can be any small desktop or laptop computer. The software we will use is an interactive environment based on a programming language named **Ruby**. An interactive language works much like a calculator: users type in expressions, and the system performs a calculation and prints a result. We will start by becoming familiar with the building blocks provided by Ruby. We will then work on projects by putting the pieces together in interesting ways to carry out experiments on a wide variety of algorithms.

In the jargon of computer programming, Ruby belongs to the family of programming languages known as **object-oriented languages**. These languages use the word "object" not only to refer to pieces of data, such as numbers and text, but also to collections of data, such as lists of numbers, and to things like functions and rules that can be applied to other objects. In each chapter we will be using Ruby as our computational laboratory, setting up a small virtual world were we create objects and carry out operations designed to experiment with algorithms related to the main topic of that chapter.

Here are some examples of how an interactive environment based on an object-oriented programming language can be used to set up and carry out experiments with computations:

- The first nontrivial algorithm presented in this book is the Sieve of Eratosthenes, a very old algorithm that has been used since the time of the ancient Greeks to make lists of prime numbers. The name of the algorithm is a hint to the basic idea: create a list of numbers, and then sift out those that are not prime. It is easy to set up a

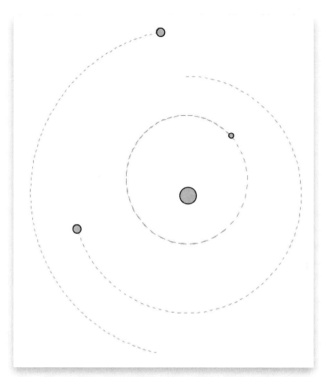

Figure 1.6: *The project in Chapter 11 (*The Music of the Spheres*) uses Ruby objects on our "computational workbench" to simulate the motions of the planets as they orbit the Sun. This picture is a snapshot of the interactive visualization displayed by Ruby during the simulation.*

straightforward program that repeatedly works its way through the list. After experimenting with the method, we will find that it is not necessary to do as many sifting operations as one might think, and the insight we gain from the initial experiments will be used to implement a more elegant version that does the minimal amount of work.

- One of the early milestones in artificial intelligence was a program named ELIZA, which gave the appearance of carrying out a conversation by playing the role of a psychiatrist. A user would type a sentence on a computer terminal, and ELIZA would respond. For example, if a person typed "I don't like computers" the program might print "Do computers worry you?" What was fascinating about ELIZA was how well it seemed to participate in a conversation, in spite of the fact that it only did very simple syntactic transformations on input sentences. For this project, we will create objects that represent the transformation rules and run experiments that apply the rules to test sentences.

- In a modern-day version of the computations supervised by Babbage and Herschel, we will use Ruby to simulate the motions of the planets as they orbit the Sun (Figure 1.6). The Sun and each of the planets will be represented as Ruby objects. We will see how to run the simulation and watch the motions of the planets in an interactive visualization. We will also be able to change the simulation parameters, for example, to see what would happen if there were two large objects the size of the Sun.

- A classic problem from the world of mathematics, known as the Traveling Salesman Problem, has the same basic structure as several important real-world problems that require efficient schedules. We will use what is known as a "genetic algorithm" to experiment with one way of solving the Traveling Salesman Problem. Each object in this project will represent a complete tour. We will create a set of tours and put them in a "virtual Petri dish," then sit back and watch as the tours mutate and evolve, eventually giving rise to an efficient solution to the problem.

The important point to keep in mind when working through these projects is that Ruby, like any programming language, is simply a notation for describing an algorithm. Any time one learns a foreign language there are myriad details to deal with, and that is certainly the case with a programming language. At times it will seem like this book is more concerned with teaching the ins and outs of Ruby programming than it is with the ideas of computation. But remember the goal here is "literacy." Readers will need to learn this new notation well enough to understand the basic steps of an algorithm, but will not have to memorize all the details that would be required to write their own new programs. Anyone who has ever tried to learn a foreign language knows it is easier to read sentences in a new language than it is to write a sentence. A similar thing will happen with Ruby. After a bit of practice it will be possible to understand the basic steps of an algorithm when they are written as statements in Ruby, even though it might be difficult to write a new Ruby program from scratch.

The reason we are using Ruby is that there is a tremendous benefit from using a real programming language as the notation for describing an algorithm. After an algorithm has been implemented in the form of a program, we can run it on a computer: we can apply it to different inputs, modify it, extend it, and carry out any number of experiments that will help lead to a deeper understanding of the algorithm.

Important Concepts Introduced in This Chapter

computation	A sequence of simple, well-defined steps carried out to solve a problem
algorithm	A description of how to solve a problem computationally; an algorithm includes a precise statement of the problem (the input), the desired solution (the output), and the order in which the steps will be executed during the computation
limitations	A problem might not be solvable by computation because it is ambiguous, it requires too many steps to complete, or it is mathematically impossible

To Learn More

There are several good books on the history of computing. Three that provided much of the historical background for this chapter are:

- *The Computer from Pascal to von Neumann* was written by Herman H. Goldstine, one of the pioneers of the field who worked with John von Neumann on one of the first electronic computers.

- *When Computers Were Human*, by David Alan Grier, tells the story behind several large scale computing projects in the days before computing machines.

- *The Difference Engine: Charles Babbage and the Quest to Build the First Computer*, by Doron Swade, describes the nautical almanac project supervised by Babbage and Herschel.

The Computer History Museum (http://www.computerhistory.org) has a variety of materials on early computers, including an on-line exhibit of Babbage's Difference Engine with a video of a complete reproduction of the machine in action.

Two books that provide introductions to the ideas of logical paradoxes and mathematically unsolvable problems are:

- *Gödel, Escher, Bach: An Eternal Golden Braid*, by Douglas Hofstadter, shows how the "self-referential" nature of the statement "this sentence is false" also appears in drawings by M. C. Escher and the music of J. S. Bach.

- *Logicomix: An Epic Search for Truth*, by Apostolos Doxiadis and Christos H. Papadimitriou, is a graphic novel that tells the story of Bertrand Russell, an English philosopher and mathematician whose work paved the way toward the discovery of unsolvable problems in the 1930s.

Exercises

1. What are some of the tasks you use a computer for? What are some of the limitations of how well the computer carries out these tasks? Explain some of the technological limitations you encounter, *e.g.*, how you could do a better job if you used a more powerful machine. Are there computational limits, *i.e.*, aspects of the problem that make it difficult or impossible to define an algorithm for part of the task?

2. Below is a list of fields where computers have been used. In some cases computers are well established, and people who work in that area rely heavily on computation, but in other cases the use of computers is still very tentative. Pick an area that interests you and write a short paper on how computers help solve problems in that area. Start by writing down some initial impressions and then do some research on the Internet to see what progress is being made by computer scientists and their colleagues from the problem domain. Questions to ask yourself as you do your research might include "What are the barriers to the use of computers in this field? Are those limits technological or computational? What are some of the social impacts and ethical issues arising from the use of computers in this area?"

    a) Medicine: Does your doctor use a computer in his or her practice? Can computers diagnose illnesses or prescribe medicines?

    b) Pharmacology: What is "rational drug design"? What role does computation play in the development of new drug treatments?

    c) Engineering: What role do computers play in the design and construction of new cars? airplanes? bridges? How has computing changed the way engineers work?

    d) Architecture: How are computers used to plan new buildings? How do they help architects come up with energy-efficient designs?

    e) Meteorology: Are computer models being used to generate weather forecasts? Do they track hurricanes and other storms? How well do these models predict weather 24, 48, or 96 hours in advance?

    f) Art and Entertainment: How have computers had an effect on music, video, or other artistic endeavors?

    g) Libraries: What impacts are computers having on your school library or local community public library?

    h) Banking and Finance: Do you do your banking on-line? Do you purchase and pay for any items using the Internet? What is the field of "computational finance" about?

    i) Journalism: How does your local paper or school paper use computer technology? How are blogs and social networking sites changing journalism?

    j) Government: What role does computer and information technology play in local government in your area? Do you live in a place where electronic voting technology is used?

3. Which of the following methods for finding a book at a library could be considered an algorithm? In each case you can assume you have a precise specification of the book you want to find, *i.e.*, you know the title, author, and date of publication, and you also know the library owns the book. The desired outcome of your search is that you either find the book or you learn the book has been checked out. Which of the following methods provide an effective set of steps for obtaining the book?

    a) Walk up to the first person you see, ask them where the book is.

    b) Find a librarian, ask them where the book is.

    c) Wait by the book return until the book you want is returned.

    d) Use an electronic catalog to find where the book is shelved, then use a map to find the shelf.

    e) Start at the shelf nearest the door, then look systematically, shelf by shelf, through all shelves in the library.

    f) Pick a shelf at random, see if your book is there; if not, pick another shelf at random and repeat.

    g) Recruit ten friends; divide the library into ten regions; assign each friend to a different region; ask them to search every shelf in their region and report back to you.

4. If you're a college football fan, you know that "computer polls" are used to rank teams at the end of the season based on the scores of the games played earlier in the year. There are often major differences between the rankings, so organizations that rate teams often throw out the highest and lowest computer rankings and take the average of the rest. Based on what you now know about algorithms, can you explain why different computers would give different rankings when they all use the same set of scores?

5. Use a spreadsheet to make a table of squares of numbers using only addition, following the method shown in the sidebar on page 4. Start by putting the number 1 in cells A1 and B1. Then use the "fill down" command to tell the spreadsheet that new values in column A should be the result of adding 2 to the value above; after you do this, the value in row i, column A should be the ith odd number. Next fill each cell in column B with the sum of the value in the cell to the left and the cell above, as shown in the picture on the next page. Is the value in row i, column B the value of i^2?

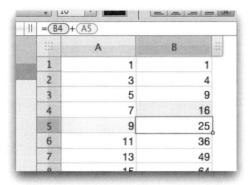

6. Suppose you need to make a table of squares of numbers, using the same technique as the previous problem, but you don't have a spreadsheet application on your computer. Can you describe the method for computing n^2 in enough detail that a friend can make the table? Use a format similar to that in Figure 1.5 to describe the input and the sequence of operations.

Chapter 2

The Ruby Workbench

Introducing Ruby and the RubyLabs environment for computational experiments

This chapter is a brief introduction to Ruby, the programming language we will use for the projects in the rest of the book. The tutorial project for this chapter is a simple program to convert temperature values from Fahrenheit to Celsius. The algorithm is trivial—the temperature conversion only requires Ruby to compute the value of a single equation—but going through the steps of implementing the program in Ruby is a good way to learn about the language and how we can use it for computational experiments. Once we get through the basics of how to set up and run experiments with Ruby we'll be ready to tackle the more challenging projects in later chapters.

Lab Manual

The instructions for setting up a "computational workbench" on your computer are in a Lab Manual that can be downloaded from the Explorations in Computing web site:

> http://www.cs.uoregon.edu/eic

There are versions of the manual for Microsoft Windows, Linux, and Mac OS X. Each has detailed instructions for installing and running the software you will use for the tutorial projects.

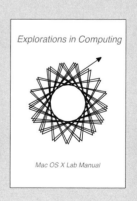

Explorations in Computing

Mac OS X Lab Manual

2.1 Interactive Ruby

A simple application for converting temperatures from Fahrenheit to Celsius is shown in Figure 2.1. The program has a window with two places (usually called "text boxes") for entering numbers. To convert 80°F to Celsius, the user types the number 80 in the Fahrenheit box. Then, when the button is clicked, the program will read the number from the Fahrenheit box, do the conversion, and display the result in the Celsius box.

More complicated programs might have more controls. For example, the application could have a menu to let the user select degrees Kelvin or some other scale, or a slider to control the number of digits of accuracy for the output. The parts of a program that get input data and present results define the program's **user interface**. Most programs that run on laptops or small desktop machines have a graphical user interface, or **GUI**, that consists of a set of windows, menus, buttons, and other controls.

If we try to use a graphical interface for the projects in this book we will encounter two problems. The first is that there would have to be a new interface for each project. In Chapter 10, the project is based on a program that carries on a conversation with the computer, so we will need a way to enter sentences and view the computer's responses. The experiments with the solar system simulation in Chapter 11 require a panel for 2D graphics to show the motion of the planets and controls to start and stop the simulation. A different interface for each project would force us to learn a new set of controls for each experiment.

A more serious drawback is that by its very nature a GUI tries to hide all the details of what happens inside a computer as a program is running. Users don't want to see what's going on inside the application when they click a button or select a menu item, they just want the computer to perform some function. For the projects in this book, however, we are interested in the computations that go on behind the scenes. We want to be able to monitor the progress being made by an algorithm as a computation progresses.

We can solve both of these problems by adopting an older style of interacting with a computer. Before mice and other pointing devices, and before large high definition color monitors, most computers were connected to a teletype or an electric typewriter. Users typed a command on the keyboard, to tell the computer what to do, and any output from the command was printed on paper. Teletypes and typewriters were later replaced by video display terminals (Figure 2.2), but the method for interacting with the computer remained the same: users typed commands, and the computer displayed results.

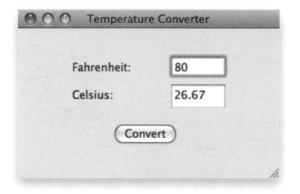

Figure 2.1: *A graphical user interface, or GUI, for a program that converts temperature values. Users enter a number in one text box and click the button. The application will convert the temperature and display the result in the other box.*

Figure 2.2: *Video display terminals, like these from Tektronix, allowed users to type commands on a keyboard and see the output on the display.*

This style of interacting with a computer is known as a **command line interface**. On a modern computer, we can run software that has a command line interface by using an application called a **terminal emulator**. The application displays a single window on the computer's screen. When the terminal emulator is active, it acts just like an old-fashioned video terminal. Users type a command, and any output generated when the computer executes the command is displayed in the window.

A terminal emulator with its command line interface is just what we need for our "computational workbench." At the beginning of a project we will type commands that run small parts of a computation. Then we will combine the pieces, and we will have Ruby display information in the terminal window so we can watch the complete algorithm in action. With this sort of control over the Ruby programs, will be able to make small changes to see what effect they have, or to run the algorithms several times, with different inputs each time. For the tutorial project in this chapter, we will start by typing simple arithmetic expressions to see how to do arithmetic with Ruby. By the end of the project we will be able to convert temperatures from Fahrenheit to Celsius by entering a single Ruby statement (Figure 2.3).

Figure 2.3: *A terminal emulator application allows us to run software with a command line interface. The first line shown here asks Ruby to compute the Celsius equivalent of 80°, and the second line is the output.*

The system we use to run Ruby is called **Interactive Ruby**, often abbreviated **IRB**. When IRB is ready for us to type a Ruby statement, it displays a **prompt** in the terminal emulator window. IRB's prompt is two greater than signs, so this is what you will see in your terminal window as soon as IRB is waiting for your input:

```
>>
```

Using an interactive programming language is similar to using a calculator, where we enter an equation and the computer displays the result. With IRB, we type a statement in Ruby, which could be a command to load some software, or a request to evaluate an arithmetic expression, or a number of other operations. Ruby will then execute the statement, and any output generated will be displayed in the terminal window.

To have Ruby evaluate an arithmetic expression, simply enter the expression in the terminal emulator window and hit the return key. Ruby will print => and the value of the expression. For example, to ask IRB to print the sum of 5 and 6, just type 5 + 6, hit return, and IRB will print the result. This is what you will see in the terminal emulator window:

```
>> 5 + 6
=> 11
```

Expressions with addition and subtraction use the familiar plus and minus signs. The people who designed the first programming languages had to choose symbols that were on the keyboards available at that time, and the conventions they chose are still used today. Ruby and other programming languages use an asterisk as the symbol for multiplication and a slash (/) for division:

```
>> 5 * 10
=> 50
>> 6 / 3
=> 2
```

A Note about Displayed Text

In this book interactions with Ruby are typeset in what is known as a fixed-width or "typewriter" font.

Tutorial projects will have Ruby statements you should type into your terminal emulator application. Right below the statement you will see the expected output from Ruby. To help distinguish between text you type and responses from Ruby, everything you type will be shown in slanted blue letters, and everything printed by Ruby will be shown in black. For example, in the temperature conversion project, you will type an expression to ask Ruby to convert 212°F to Celsius:

```
>> celsius(212)
=> 100
```

The >> on the first line is the prompt from IRB. The string following that is the expression you type to ask Ruby to calculate the temperature conversion. The text on the second line is what Ruby is expected to print as the output of the program.

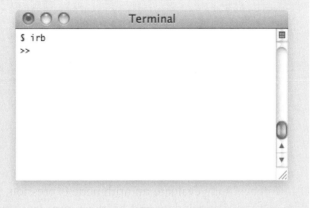

Interactive Ruby on Your Computer

Instructions for starting a terminal emulator application and running Interactive Ruby (IRB) vary from one operating system to another.

The Lab Manual that accompanies this book has detailed instructions for Windows XP, Linux, and Mac OS X.

In Ruby terminology the arithmetic symbols such as + and * are known as **operators**, and the numbers are **operands**. Ruby has several other operators, but we will postpone our discussion of these other symbols and what they mean until we need them for a project.

When an expression has more than one operator, Ruby applies the operators according to their **precedence**. Since multiplication should be performed before addition, the result of $3 + 4 \times 5$ is 23:

```
>> 3 + 4 * 5
=> 23
```

If we want Ruby to evaluate the operations in a different order we can use parentheses:

```
>> (3 + 4) * 5
=> 35
```

At the end of the session just type `quit` and IRB will exit:

```
>> quit
```

Tutorial Project

The first step in the tutorial project for this chapter is to make sure you can start the Interactive Ruby application. Open up a terminal window and start IRB (detailed instructions specific to your type of operating system can be found in the Lab Manual). Once you have started IRB and see the >> prompt in your terminal emulator window you are ready to start the tutorial project.

T1. Type a simple expression, *e.g.*, `13 + 2` and hit the return key. Ruby should print the result:
```
>> 13 + 2
=> 15
```

T2. Try some simple expressions involving other operators:
```
>> 6 - 3
=> 3
>> 3 * 7
=> 21
>> 8 / 4
=> 2
```

T3. Try some expressions with and without parentheses:

```
>> 3 + 4 * 5
=> 23
>> (3 + 4) * 5
=> 35
>> 8 - 4 / 2
=> 6
>> (8 - 4) / 2
=> 2
```

Make sure you understand the output in each of these examples. Is Ruby printing what you expected?

T4. Does Ruby care if you include spaces in the middle of your expressions? For example, do you get the same result when you type 3+4 with no spaces as you get when you type 3 + 4 with spaces before and after the plus sign?

T5. What happens if you leave out an operand, *e.g.*, if you type 3 + * 5 instead of 3 + 4 * 5?

T6. Type an expression that mistakenly uses a symbol instead of a number, *e.g.*, 3 + x:

```
>> 3 + x
NameError: undefined local variable or method 'x' for main:Object
```

The error message has some unfamiliar terminology, but at this point you can glean some information: the message has the word "undefined," and "x" is enclosed in quotes, so there's a good chance Ruby was complaining about the x in that expression. We'll see what "method" and "object" mean later in this chapter.

In grade school you might have learned a mnemonic like "my dear aunt sally" to remember the precedence of arithmetic operators (multiplication, division, addition, subtraction). Ruby and other programming languages have a slightly different rule: multiplication and division have the same precedence, as do addition and subtraction. If an expression has two operators with the same precedence, the one on the left is applied first.

T7. Evaluate the following expressions with IRB to see how Ruby applies these precedence rules:

```
6 / 3 * 4
8 * 3 / 4
5 - 4 + 2
```

Would any of these expressions have a different value using the "dear aunt sally" rules?

2.2 Numbers

All of the expressions in the previous section involved **integers**, *i.e.*, whole numbers with no fractional parts. We can write expressions that use real numbers as well. The general rule is that if we want Ruby to treat a number as a real number we have to include a decimal point when we write the number.

To be more precise, in programming languages, a number like 5.0 is a **floating point** number, not a real number. Real numbers are bothersome things like $1/3$ and $\sqrt{2}$ that have an infinite number of digits. Because numbers are stored in a finite amount of space inside a computer, we have to use an approximate value. Floating point numbers are approximations of real numbers, usually accurate to around 12 digits.

The distinction between integers and floating point numbers can be important. Consider what happens if we ask Ruby to divide 10 by 3. If we have Ruby do the operation using integers this is what we get:

```
>> 10 / 3
=> 3
```

Ruby printed 3 because the two operands are integers, and the result is the largest integer below the actual result of 3.3333... If we do the same calculation using floating point numbers, the result is more accurate:

```
>> 10.0 / 3.0
=> 3.33333333333333
```

An important thing to remember is that when Ruby does arithmetic with integers, it does not round off to the nearest integer, it *truncates*, as shown in this example:

```
>> 5.0 / 3.0
=> 1.66666666666667
>> 5 / 3
=> 1
```

The equation for converting Fahrenheit to Celsius is $C = (F - 32) \times 5/9$. We can use Ruby to help us convert 80°F to Celsius by typing the expression in IRB:

```
>> (80 - 32) * 5 / 9
=> 26
```

Note that all the numbers in this example are integers and that Ruby prints an integer result. A more precise result (which you would get if you used a temperature conversion program on the Internet or if you retype the expression using floating point values) is 26.67°C.

It may seem at this point like Ruby is being rather "pedantic" in the way it forces us to write 80.0 instead of 80 if we want a more accurate conversion. Why can't it simply recognize that we want the result to be 26.67 and not simply 26? The answer is that Ruby cannot always figure out what we are looking for and give us the result in a format we might prefer. There may be other situations where the integer value is in fact the correct value, even if it is truncated and not rounded. We will see many examples of algorithms

---

Floating Point Numbers

The term "floating point" comes from the fact that the internal representation of these numbers is based on scientific notation. The number 141.42 is stored internally as something like 1.4142×10^2.

This format makes it easier to represent very large numbers (1.4142×10^{23}) or very small numbers (1.1412×10^{-10}) by letting the decimal point "float" to the left or right as the exponent changes.

```
1.4142          141.42
```

Multiplying by 100 shifts the decimal point two places to the right

throughout the book that rely on the fact that operands and results are all integers. So the bottom line is that we need to be aware of the fact that algorithms rely on two basic types of numbers, integers and floating point, and be careful to specify the correct types of operands when we enter expressions.

Tutorial Project

T8. Use IRB to evaluate a simple expression such as 3 * 5 where both operands are integers:
```
>> 3 * 5
=> 15
```

T9. Repeat the previous expression, but use floating point numbers:
```
>> 3.0 * 5.0
=> 15.0
```

T10. Use Ruby to evaluate the expression that converts 100°F to Celsius:
```
>> (100 - 32) * 5 / 9
=> 37
```
Is this result what you expected, given the rules Ruby uses for evaluating arithmetic expressions? Is it an accurate calculation of the temperature in Celsius? (To answer this second question you can use a calculator that has temperature conversions built in, or use a program from the Internet to do the conversion.)

T11. Repeat the previous exercise, using Ruby to convert these temperature values to Celsius: 90°F, 70°F, 212°F, 32°F.

T12. The formula for converting from Celsius to Fahrenheit is $F = C \times 9/5 + 32$. Use this formula to convert the following temperatures to Fahrenheit: 0°C, 10°C, 20°C, 30°C, 100°C.

You now have enough experience with numbers in Ruby to complete the tutorial project in the remaining sections of this chapter, and you can skip ahead to the next section. The problems below look at expressions involving numbers in a little more depth, and are recommended if you will be doing any of the optional projects later in the book.

The first question: What happens if you try to mix integers and floating point numbers in an expression?

◆ Ask Ruby to evaluate 3.0 * 5.

◆ Next ask it to evaluate 3 * 5.0.

◆ Do you get an error message? If not, is the result an integer or a floating point value?

◆ Try a few more expressions that mix integers and floating point, *e.g.*, (80 - 32) * 5 / 9 or (80 - 32) * 5.0 / 9.

What Ruby did for these cases turns out to be a general rule: if Ruby is asked to apply an operation where one of the operands is an integer value, the integer is "promoted" to floating point, and then the operation is applied, generating a floating point result.

The next set of questions explore what can happen if we rearrange the order of evaluation in the temperature conversion equation.

◆ One way of describing the method for converting Fahrenheit to Celsius is "subtract 32 and multiply by 5/9." If we follow this prescription exactly, we should tell Ruby to compute the difference and then multiply it by 5/9: (80 - 32) * (5 / 9). Note how the parentheses are needed to tell Ruby to calculate 5/9. What happens if you type that expression into IRB? Can you explain what went wrong?

♦ Whenever you find yourself in a situation where Ruby is not evaluating a complicated expression the way you think it should, the best way to figure out what is happening is to break the complicated expression into smaller parts. The two parts of this expression are the subexpressions inside parentheses. What does Ruby produce for `80 - 32`? How about `5 / 9`? Does this help explain how Ruby evaluated `(80 - 32) * (5 / 9)`?

♦ We can fix this problem by using floating point values instead of integers. What is the result of `5.0 / 9.0`? What is `(80 - 32) * (5.0 / 9.0)`?

2.3 Variables

Suppose we want to calculate the area of a countertop shaped like the one shown in Figure 2.4a. The counter is a square with one corner missing, and the sides of the missing triangular piece are half as big as the edge of the square. Figure 2.4b shows one strategy for computing the area. Since it's straightforward to calculate the area of a square and the area of a right triangle, we can figure out the size of the countertop by subtracting the area of the missing triangle from the area of the square. The formula for the area of the countertop is thus $x^2 - (x/2)^2/2$, where x is the length of one edge of the square.

To use Ruby to compute the area, the first step is to calculate x^2, the area of the square. If the length of the edge is 109 cm, we just have to type this expression:

```
>> 109 * 109
=> 11881
```

A slightly simpler form uses Ruby's exponentiation operator, which is written with two asterisks in a row, so 109^2 is written this way:

```
>> 109 ** 2
=> 11881
```

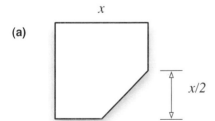

(a)

Figure 2.4: (a) A countertop with a corner cut out. Each edge of the missing triangle is half as big as the edge of the square. (b) The area of the counter is the area of the square minus the area of the triangle.

(b)

Layout Rules, Part I: Expressions

Ruby has very few hard and fast rules for how expressions can be written when we type them in IRB. We are allowed to use spaces almost anywhere we like.

Figuring out where to put spaces in an expression is mostly a matter of personal preference. Often a statement with a single operator is easier to read if there are spaces around the operator.

When an expression has several operators, it may be easier to understand if higher precedence operators are written without spaces.

```
>> x**2
=> 9
```

```
>> x ** 2        spaces here
=> 9             make it easier to
                 see the operator
```

```
>> x ** 2 + y ** 2
=> 25
```

```
>> x**2 + y**2       emphasize the
=> 25                fact that ** has a
                     high precedence
```

Similarly, we can calculate the area of the missing triangular piece, which is $(x/2)^2/2$:

```
>> ((109 / 2) ** 2) / 2
=> 1458
```

If we want to do the calculation in a single expression, we can put the two pieces together into one Ruby statement:

```
>> (109 ** 2) - (((109 / 2) ** 2) / 2)
=> 10423
```

This final expression is accurate, but it is rather complicated, which makes it hard to type correctly. It would be very easy to leave out a parenthesis, or enter 190 instead of 109, or type a single asterisk instead of a double asterisk. Complicated expressions are not only difficult to type, they are hard to read, and it takes much more effort to try to track down problems.

As a first step in simplifying this expression, we can use a **variable** to stand for the length of the edge. To introduce a variable named x to stand for the width of the countertop, we simply type

```
>> x = 109
=> 109
```

An expression like this is known as an **assignment statement**. An assignment has a variable name, an equal sign, and the value we want the variable to represent. Ruby printed 109 because an assignment statement is an expression, just like the other expressions we have been entering, and IRB always prints the value of any expression typed into the terminal window. Here is the expression that computes the area of the countertop, rewritten to use the new variable:

```
>> (x**2) - (((x / 2)**2) / 2)
=> 10423
```

The new version is more readable, but not by much, since it still has a lot of parentheses. We can simplify the expression even further by introducing two more variables, one to stand for the area of the square and the other for the area of the triangle. We could use simple names like y and z for the two new variables, but Ruby allows us to use complete words as variable names. In this example, the new variable names are square and triangle, which makes it easier to remember what we want each variable to represent:

```
>> square = x ** 2
=> 11881
>> triangle = ((x / 2)**2) / 2
=> 1458
```

Now we can use the new variables to compute the area of the countertop:

```
>> square - triangle
=> 10423
```

There is another advantage to using the expressions with variable names. Suppose, after computing the area the first time, we double-check the measurements and find the edge is actually 107 cm. To recompute the area we can type another assignment statement that has x on the left side:

```
>> x = 107
=> 107
```

The old value of x is erased and the new value replaces it. We can then ask Ruby to repeat the expressions that calculates the area, and the new value of x will be used to compute new values for square and triangle.

The tutorial project for this section will give you some experience with reevaluating expressions you entered earlier. A feature called "command line editing," which is implemented in most terminal emulators, will save you a lot of typing when you do these exercises. Be sure to read the section on terminal emulators in the Lab Manual when you are ready to start the tutorial.

What's in a Name?

In projects later in the book we will be creating several variables in each IRB session. In order to keep everything straight it's important to choose meaningful names.

Most mathematical equations use single letters like **x** and **y** for variables, but variable names in Ruby can be complete words like **square** and **triangle**.

There are some restrictions on what can be used for a variable name.
- All names must start with a letter.
- The remainder of the name can have a mix of upper and lowercase letters, digits, or an underscore, *e.g.,* **squareSide**, **square_side**, or **sq123**.
- Case is important: **a** and **A** are two different names in Ruby.
- Variable names must begin with a lowercase letter. We will see names that begin with uppercase letters later in the book, but when choosing a name for a variable make sure it begins with a lowercase letter.

(a)   (b)

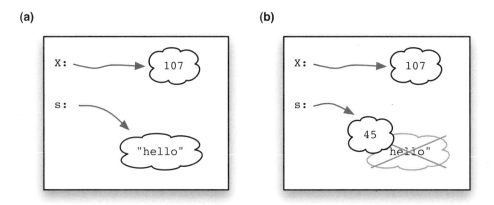

Figure 2.5: *(a) The computer's memory (the "object store") after defining variables named* x *and* s. *(b) If we decide to use* s *to hold an integer the old object referred to by* s *is discarded.*

So far all the expressions in this chapter have been arithmetic expressions based on numeric values. Ruby also lets us write expressions that use **strings** as values. Strings are used in programs that work with names, addresses, product descriptions, and a wide variety of other kinds of data. To write a string in Ruby, simply enclose a sequence of characters in double quotes:

```
>> s = "hello"
=> "hello"
```

We won't be using strings in the temperature conversion project, so we will put off the discussion of what we can do with strings until we need them in other projects. But knowing that variables can refer to a variety of different types of data will give us a better idea of what variables are and how Ruby evaluates expressions that contain variable names.

A useful way to visualize how Ruby manages variables is to imagine the computer's memory as a large whiteboard, initially empty. When a variable is defined, Ruby finds space for the value, and it writes the variable's name and a pointer to the value. Figure 2.5a shows what the whiteboard might look like during the IRB session after defining the variables x and s in the previous examples.

The generic name for a piece of data in Ruby is **object**. Integers, floating point numbers, and strings are simple kinds of objects, and there are dozens of other kinds of objects, as well. How Ruby chooses to represent objects inside the computer's memory is usually not important for the projects in this book, so objects are shown as abstract quantities inside clouds in drawings like the one in Figure 2.5. Programmers often refer to the memory that holds data items as "object storage," or in a abbreviated form, the **object store**.

We saw earlier that the value of a variable can change—we updated the value of the variable x by associating it with a new number. It turns out we can also change what kind of object is associated with a variable. Ruby doesn't complain if, after defining s to be a string, we then type an expression that makes s refer to a number:

```
>> s = 9 * 5
=> 45
```

Ruby is perfectly happy to throw away the string object that used to be the value of s, create a new object representing the number 45, and associate s with the new object (Figure 2.5b).

There are several important consequences of this flexible and dynamic behavior concerning variable names, and we will come back to this topic in future projects. For now the important concept to learn is that variables are simply names of objects. When we use a variable in an expression, Ruby looks in the object store to find the value of the variable, and it uses that value in its calculations.

Tutorial Project

T13. Type two expressions that define variables x and y:

```
>> x = 6
=> 6

>> y = 5
=> 5
```

T14. Try out a few expressions using these variables:

```
>> x + 3
=> 9

>> x * y
=> 30

>> (x + 3) * y
=> 45
```

Did you get what you expected?

T15. Change the value of x:

```
>> x = 2
=> 2
```

T16. Repeat the three expressions that used x. Did the values of these expressions change after you changed the value of x?

T17. Set x to the length of the long edge of the countertop:

```
>> x = 109
=> 109
```

T18. Define the area of a square that is x centimeters on each side:

```
>> square = x ** 2
=> 11881
```

Is the result what you expect, *i.e.*, is it x^2?

T19. Define the area of the missing right triangle that is $x/2$ centimeters on each leg:

```
>> triangle = ((x / 2) **2) / 2
=> 1458
```

Is this the correct value?

T20. Ask Ruby to compute the area of the countertop:

```
>> square - triangle
=> 10423
```

The next exercise is to recompute the area of the countertop after changing the value of the variable named x. You can either retype the expressions for square and triangle, or cut and paste the text from the previous exercises, or, if your terminal emulator supports it, use command line editing to reevaluate the previous expressions.

With command line editing you just need to hit the up-arrow key on your keyboard two times, and then hit the return key. This is a technique we will use in the next section, and often in projects later in the book, so it's worth learning how it works on your terminal emulator.

T21. Change the value of x by entering a new assignment statement:

```
>> x = 107
=> 107
```

T22. Recompute the values of `square` and `triangle` using the new value of x.

T23. For the next exercise define a variable with a string value:

```
>> s = "hello"
=> "hello"
```

T24. What do you think will happen if you try to add 3 to s?

```
>> s + 3
TypeError: can't convert Fixnum into String
```

That rather cryptic looking error message has the word `TypeError`, which is a hint that Ruby did not know how to add a string and a number, since they are two different kinds of objects.

T25. Assign s a new, numeric, value:

```
>> s = 7
=> 7
```

Now ask Ruby to evaluate `s + 3` again. What happened this time?

From this set of experiments with variable names, do you see how names are simply references to objects? When Ruby evaluates an expression containing a variable, it effectively replaces the name with the current value of a variable and then evaluates the resulting expression.

2.4 Methods

The examples in the previous sections demonstrated how to build expressions out of numbers and arithmetic operators like + and *. Ruby also has an extensive library of mathematical functions often found on scientific calculators, including logarithms and trigonometric functions.

To use these functions in Ruby, we write the name of the function, and then the values it operates on in parentheses following the name. For example, the name of the function that computes a square root is `sqrt`, so to ask Ruby to calculate the square root of 16 we would type this expression:

```
>> sqrt(16)
=> 4.0
```

Ruby also allows us to create our own functions. For the tutorial project in this section, we will see how to define a function named `celsius` that will convert temperature values, using the formula described in Section 2.2. Once we make the function, we can use it in an expression, just like those already defined in Ruby's library. To convert a temperature to Celsius we will simply type the value in parentheses right after the name `celsius`:

```
>> celsius(80)
=> 26
```

In Ruby a function is known as a **method**. Here is some more terminology associated with methods:

- When an expression contains a method name, we say Ruby **calls** the method as it evaluates the expression.

- A method is often defined in terms of **parameters**.

- When a method is called, we **pass** it **arguments**, which are objects used by the method.

- The value computed by a method is **returned** as the result of the method call.

The Ruby statements that define the `celsius` method are shown in Figure 2.6. The definition only requires three lines of text. The first line has the name of the method being defined and the names of parameters that will be used when the method when it is called. The `celsius` method has one parameter, but in general methods can have any number of parameters (including zero).

The parameter is a variable we can use in the statements inside the method. In this example we're telling Ruby we want to use a variable named `f` to stand for the temperature to convert. When the method is called, the value passed as an argument is stored in the variable. For example, when we type `celsius(50)`, Ruby will set `f` to `50` before it starts to evaluate any of the expressions inside the method.

The remaining lines in the definition, before the word `end` on the last line, make up the **body** of the method. There can be any number of Ruby statements in the body, typically written one per line. The other method shown in Figure 2.6 has three lines in its body to compute the area of the countertop in the examples in the last section. The statements in the body are executed in order when the method is called. The value of the expression that follows the word `return` will be the result of the method.

After the new methods have been defined, we can call them by typing their names and values we want to pass as arguments. To ask Ruby to find the Celsius equivalent of 212°F, simply type in an expression which calls the `celsius` method, passing it the number 212:

```
>> celsius(212)
=> 100
```

```
# Convert temperature f to Celsius

def celsius(f)
  return (f - 32) * 5 / 9
end
```

```
# Compute the area of the countertop
# when the long edge has length x

def countertop(x)
  square = x**2
  triangle = ((x/2)**2) / 2
  return square - triangle
end
```

Figure 2.6: *Examples of how to define new methods in Ruby. The first line specifies the name of a method and the names of any parameters that should be passed when the method is called.*

Layout Rules, Part II: Methods

It is a good idea to indent statements in the body of a method, as shown here, otherwise it can be difficult to tell where the method definition starts and where it ends.

There is one place where Ruby is strict when it comes to writing spaces in expressions: there cannot be a space between a method name and the opening parenthesis in a call to a method.

We can, however, put spaces inside the parentheses if it helps make the parameters easier to read.

```
def countertop(x)
  square = x**2
  triangle = ((x/2)**2) / 2
  return square - triangle
end
```

Ruby will print an error message
if there is a space here

```
>> a = countertop (100)

>> a = countertop( 10 * 10 )
=> 8750
```

We can create a variable to hold the area of the counter by calling the `countertop` method and saving the return value:

```
>> area = countertop(107)
=> 10045
```

Note our new methods work like other methods in Ruby: we pass a value, the method performs a calculation, and we get back a result that can be used like any other value.

One way to enter the method into IRB is to simply type the lines in the definition, one right after the other. This is what the terminal window would look like if we typed in the definition of the `celsius` method:

```
>> def celsius(f)
   return (f - 32) * 5 / 9
   end
=> nil
```

The word `nil` is a special value that means "nothing." A method definition, like an assignment and every other statement in Ruby, is an expression, and IRB has to print something for every expression we type. For historical reasons IRB prints `nil` after we enter a method definition, but the method has been defined. Note that IRB does not print a prompt when it is in the middle of reading the definition.

Typing a method definition in IRB is not very practical. For one thing, if we discover a mistake on one line, we would have to retype the entire definition, and that can be very tedious when a method has more than one or two lines in the body. Another drawback is that all the definitions we enter into IRB are lost when we exit the session. As soon as we type `quit` to terminate the session, all the variables and methods we created are thrown away. If we want to use the same methods in the next IRB session we have to type them all in again.

Instead of typing the method definition one line at a time in IRB, we can put the definition in a text file, and then have IRB read the definition from the file. The main advantage to this approach is that if we want to change the definition it's much easier to edit the file and reread it than it is to retype the entire definition in IRB.

To tell IRB to read the text in a file, use a method named `load`. If the definition of the method is in a file named `celsius.rb`, this is how to tell IRB to read the definition from the file:

```
>> load "celsius.rb"
=> true
```

Note that the argument to `load` is a string: it's the name of the file surrounded by double quotes. If Ruby encounters any problems, *e.g.*, if the file doesn't exist or if there are "syntax errors" like missing parentheses, IRB will print an error message, otherwise it prints `true`.

The `.rb` at the end of the file name is a common way to indicate a file is a plain text file containing a Ruby program, similar to the convention that word processor documents have names ending in `.doc` or PDF files have names ending with `.pdf`. Ruby doesn't care whether or not the file name ends in `.rb`, but it's a good idea to follow this convention so you can quickly tell which of your files contain Ruby programs.

An important thing to remember is that if you use your text editor to modify a program, you need to use the editor's "save" command to update the file on your hard drive. After saving the changes, you need to let IRB know the method has been modified. To do that, simply type the `load` command again, and IRB will reread the file and replace the old version of the method with the new one.

When a method definition is entered into a text file, it is common to include extra lines that explain what the program does. These lines are called **comments**. In Ruby, a comment begins with a # symbol. The two methods shown in Figure 2.6 both have comments at the beginning of the file.

Reserved Words

The words `def`, `return`, and `end` used in method definitions have a special meaning to Ruby. Words like these are known as **reserved words** or **keywords** in a programming language. They are reserved because we are not allowed to use them as the names of variables or new methods.

For example, if you try to define a variable named end IRB will print an error message:

```
>> end = 5
SyntaxError: compile error
(irb):65: syntax error, unexpected kEND
```

Unfortunately these error messages are not very helpful, and the message you get will vary depending on the context and the word you try to redefine.

Tutorial Project

The first few exercises in this part of the tutorial will give you some experience writing expressions that call methods.

T26. Use Ruby's `sqrt` method to compute the square root of 25:

```
>> sqrt(25)
=> 5.0
```

If you get an error message saying `sqrt` is not defined, it means Ruby's Math library was not loaded; refer to the section in your Lab Manual that has instructions for setting up the RubyLabs software on your computer so the Math module is loaded automatically each time you start IRB.

T27. Ruby has a method named `rand` that will return a random value between 0 and a value you pass as an argument. Type this expression to get a number between 0 and 99:

```
>> rand(100)
=> 18
```

You will probably see a different result than the number shown above. Use the command line editing feature of your terminal emulator to repeat this expression a few times; you should see a different result each time you call `rand`.

Next we'll experiment with a method we define, instead of methods already built into Ruby. Use your text editor to create a file named `celsius.rb` containing the lines shown in Figure 2.6.

T28. Load your `celsius` method into IRB:

```
>> load "celsius.rb"
=> true
```

If you get a message that says "no such file to load" double-check the spelling of the file name. Also, make sure you are running IRB in the same directory where the file was stored. If you get a message that says "syntax error" make sure the lines in the file are exactly as shown in the definition of `celsius` in Figure 2.6.

T29. Call the new method:

```
>> celsius(50)
=> 10
>> celsius(95)
=> 35
```

Go back to your text editor and modify the method so it uses floating point numbers. Change `32` to `32.0`, `9` to `9.0`, and `5` to `5.0`, and then save the file.

T30. Load the new version of the method:

```
>> load "celsius.rb"
=> true
```

T31. Test your new version:

```
>> celsius(100)
=> 37.7777777777778
```

If the result is a floating point value (as shown above) then you successfully updated the version of the method used by IRB.

♦ Did you notice that in the call to the new version of `celsius` the argument was an integer, but the result was printed as a floating point number? Can you explain why? (Hint: what did you discover by doing one of the earlier optional exercises with mixed integer and floating point expressions?)

♦ What do you suppose would happen if you pass a string value to `celsius`, *e.g.*, if you ask Ruby to evaluate `celsius("forty")`? Try it. Was your hypothesis correct?

2.5 RubyLabs

In the Ruby community software that has been written to be shared with other users is distributed in the form of "Ruby gems." The software you will be using for the projects in this book is available as a gem called RubyLabs. One of the steps described in the Lab Manual, in the section on installing Ruby and setting up your environment, explained how to run a command line application named `gem` to download and install the RubyLabs gem. If you were able to complete this command and configure your environment, the RubyLabs software will be included automatically at the start of each IRB session.

RubyLabs is a collection of **modules**. Each module defines objects, methods, and data files that will be used for the tutorial projects in one chapter. At the start of each IRB session, you will type a command that loads the module for the project. For example, the `countertop` method described in this chapter is part of a module named IntroLab. Including this module in your IRB session makes the `countertop` method available for that session:

```
>> include IntroLab
=> Object

>> countertop(100)
=> 8750
```

Don't forget that Ruby is very particular about upper and lower case letters, so make sure you type the name of the module exactly as it's shown, with an upper case `I` and an upper case `L`. If you spelled "include" correctly and typed the name exactly as it's shown, but still see an error message instead of the word `Object`, it means the RubyLabs software is missing or not configured properly, and you will have to refer to the section on "troubleshooting" in your Lab Manual.

Programmers refer to the file that contains a Ruby method as the **source file**, and the text in the file is often called the **source code**. The RubyLabs module has several methods for working with the source code for methods we will be using in our experiments. To see the statements in a method, we just call `Source.listing`, passing it the name of a method, and it will print the source code on the terminal:

```
>> Source.listing("countertop")
  1:    def countertop(x)
  2:       square = x**2
  3:       triangle = ((x/2)**2) / 2
  4:       return square - triangle
  5:    end
=> true
```

A listing is simply a printout of every line in the program, along with a line number at the start of each line.

A method named `Source.checkout` will save a copy of a program in a file. The name "checkout" comes from software engineering. Programmers who work on large projects with other people usually save their code in a shared library or repository. When they want to work on a piece of the project, they check out a copy, make their changes, and check it back in. If you want to have your own copy of the `countertop` method, in order to make changes and to some further experiments on your own, simply call `Source.checkout`:

```
>> Source.checkout("countertop")
Saved a copy of source in countertop.rb
=> true
```

As you can tell by the message it printed, `Source.checkout` created a file that has the same name as the method, along with the extension `.rb`, which is the convention for text files containing Ruby programs.

There are several other methods that work with the programs in the RubyLabs modules. These methods will allow us to print the value of a variable each time it is updated, or count the number of times a statement is executed, or measure how long it takes a program to run. These other methods will be introduced in the next chapter, when they will be used in experiments with the first algorithm we will study.

Tutorial Project

T32. Start a new IRB session, and tell IRB to include the methods defined in the IntroLab module:
```
>> include IntroLab
=> Object
```
If you get an error message instead of the word `Object` you need to refer to your Lab Manual and follow the instructions for installing the RubyLabs gem.

T33. Now that the module has been included you can call the `countertop` method:
```
>> countertop(100)
=> 8750
```
Use a calculator to make sure this the right answer. Did `countertop` compute $100^2 - 50^2/2$?

T34. Use the `Source.listing` method to print the source code for `countertop` on your terminal window (don't forget to capitalize the `S` in `Source`):
```
>> Source.listing("countertop")
   1:    def countertop(x)
   2:      square = x**2
   3:      triangle = ((x/2)**2) / 2
   4:      return square - triangle
   5:    end
=> true
```

T35. Use `Source.checkout` to get your own copy of the method:
```
>> Source.checkout("countertop")
Saved a copy of source in countertop.rb
=> true
```
You should now have a file named `countertop.rb` in your project directory.

2.6 Summary

The main goal for this chapter was to introduce the programming language named Ruby and the interactive environment that will be used for "computational experiments" throughout the rest of the book. The tutorial project illustrated how Interactive Ruby (IRB) works by implementing a trivial program to convert temperatures from Fahrenheit to Celsius.

Using IRB is similar to using a calculator. We enter an expression, and then Ruby evaluates the expression and prints the result. Numeric expressions can have any of the basic operations from arithmetic. For historical reasons (dating to the time when programs were created with keypunch machines that had a limited number of characters) Ruby uses an asterisk as the symbol for multiplication and a forward slash for division. We will see other arithmetic operators later when there are projects that need them.

The project brought up the point that there are two types of numbers in Ruby: integers and floating point numbers. Floating point numbers are finite approximations of real numbers. Numbers, strings, and other types of data are represented inside a computer as objects. Since we typically don't need to know the details of how pieces of data are represented in memory, we can think of them as abstract objects, drawn as clouds in figures that show the state of the system. Up to this point we have been describing objects simply as data, but in computer science the term "object" has a richer connotation. For the first few chapters in this book, however, we will simply use the word to mean "a generic piece of data."

Variables are symbolic names associated with objects. When Ruby evaluates an expression that contains a variable name, it looks in the "object store" for the current value of the variable. The value is substituted in place of the variable so Ruby can compute the value of the expression. Variables are created by assignment statements that consist of a variable name, an equal sign, and an expression that defines the value for the variable. We can change the value of a variable at any time simply by reusing it in another assignment statement.

Interactive Computing

Using the word "interactive" to describe situations where we use a terminal emulator to run a program might seem confusing at first. A person using an application with a GUI is certainly interacting with the program, so why use "interactive" to describe a language that can be used like a calculator?

The term has a long and rich tradition in computer science, going all the way back to the early 1960s. At the time, most programs carried out calculations that ran for several hours. A user would start a program, often by loading a deck of punched cards through a card reader, and then come back later to pick up the results from a printer.

The developers of LISP, one of the very first programming languages, took a different approach. Programmers typed a LISP expression on a typewriter, and the computer evaluated the expression and printed the result. Since most calculations took only a few seconds, the output was printed immediately, and this way of using the computer became known as "interactive computing."

Concepts and Terminology Introduced in This Chapter

IRB	An abbreviation for Interactive Ruby, a program with a command line interface for interactive computing with Ruby
terminal emulator	An application that mimics the actions of an old-fashioned computer terminal
integer	A whole number, a number with no fractional parts
floating point	A technique for storing approximate values of real numbers in a computer
string	A short piece of text used as data in a program
object	A generic piece of data; an object can be an integer, floating point number, string, or one of the more complex types of data introduced later in the book
variable	A name associated with an object
assignment statement	A Ruby expression that sets the value of a variable, creating a new variable if the name has not been used before
method	A function or complex operation; includes built-in methods like `sqrt` and other operations from Ruby's Math library, or user-defined methods like `celsius` that are collections of one or more Ruby expressions evaluated when the method is called
module	A collection of methods and data; a RubyLabs module contains objects and methods used for a computational experiment

---

While simple operations such as addition and multiplication are carried out by operators, more complicated functions are implemented in the form of methods. Ruby's Math library has methods for computing square roots and several other operations. In later chapters we will learn about methods that operate on strings and other kinds of objects.

We can enclose a set of Ruby statements between the words `def` and `end` in order to define our own methods. The method that converts temperature values from Fahrenheit to Celsius requires only three lines of text:

```
def celsius(f)
   return (f - 32) * 5 / 9
end
```

An example of a statement that uses this method is

```
>> t = celsius(90)
=> 32
```

In this call to `celsius`, Ruby creates a variable named `f`, assigns it the value 90, evaluates the expression in the body of the method, and returns the value of 32, which is then saved as the value of the variable `t`.

One way to understand what is going on inside the system when we type an expression in IRB is to think of how we simplify an expression in algebra. Suppose $x = 4$ and $y = x + 5$, and we are asked to find the value of $z = \sqrt{y}$. To solve this problem, we use the value of x to rewrite the second equation as $y = 4 + 5$, which is then simplified to $y = 9$. Now the third equation can be rewritten as $z = \sqrt{9}$, and one last rewrite gives $z = 3$.

Ruby operates in much the same way. If we type an expression like

```
>> a = countertop(x + 2)
```

Ruby will do a series of operations that are the equivalent of the equation rewriting steps you would do to perform the same calculation. To evaluate this expression, Ruby would look up the value of x, and use it to calculate x + 2. The result would be passed as an argument to the `countertop` method, and whatever value is returned by the method is stored in a.

The main thing to remember is that to evaluate an expression, Ruby makes sure it has all the values it needs. If it sees the name of a variable, it looks up the current value of the variable and substitutes that value in the expression. When it sees the name of a method, it calls the method and then substitutes the value returned by the method into the original expression. In later chapters we will encounter several examples of Ruby expressions that, at first, may seem very complex. The key to understanding what Ruby will do is to remember that Ruby will evaluate these expressions by calling methods and substituting values as they are needed.

Exercises

The exercises below ask you what Ruby would print as the value of an expression. To make sure you understand how Ruby works you should try evaluating the expressions yourself first, and then going to IRB to check your answers.

1. Suppose you start an IRB session and enter the following three assignment statements:

```
>> x = 4
>> y = 7
>> z = 3.5
```

Show what Ruby will print after it evaluates each of the following expressions:

```
>> x * 2
=>
>> x ** 2
=>
>> x * y
=>
>> x * z
=>
>> y / x
=>
>> y / z
=>
>> sqrt(x)
=>
```

2. Suppose you type the following three assignment statements in IRB:

```
>> a = 5
>> b = 7
>> c = 11
```

Show what Ruby will print after it evaluates each of the following expressions:

```
>> a * b + c
=>
>> a + b * c
=>
>> (a + b) * c
=>
>> sqrt(a + c)
=>
>> 2 * sqrt(a * b + 1)
=>
```

3. Show what Ruby will print when it evaluates this expression (assuming you have loaded your `celsius.rb`) file into IRB):

```
>> celsius(60)
=>
```

4. Do you think Ruby will print anything different if you pass a floating point number to `celsius`? What would Ruby print for this expression?

```
>> celsius(60.0)
=>
```

5. Do you think your `celsius` method can deal with negative numbers? What will Ruby do with this expression?

```
>> celsius(-3)
=>
```

6. As a quick test of the accuracy of your method suppose you want to see what `celsius` would print when you pass it well-known values. What expression would you enter to have Ruby compute the Celsius temperature of the freezing point of water? The boiling point?

7. One of the expressions in the body of the `countertop` method calculates the area of the missing triangular piece:

```
triangle = (( x/2 ) **2 ) / 2
```

Are all the parentheses required in this expression? Explain what would happen if the statement was rewritten in one of these forms:

```
triangle = (x/2) ** 2 / 2
triangle = x/2 ** 2 / 2
```

8. What will happen if you pass a negative number to the `countertop` method? If you try it in IRB, and you get back a number, it this an error, or do you think Ruby computed the correct result?

```
>> countertop(-10)
=>
```

♦ Use your text editor to create a file named `fahrenheit.rb`. Type in the definition of a method named `fahrenheit` that converts temperature values from Celsius to Fahrenheit, using the equation $F = C \times 9/5 + 32$. Load the file into IRB and test your method by calling `fahrenheit(0)`, `fahrenheit(100)`, and a few other temperatures.

♦ If you have driven on an interstate highway, you may have seen "speedometer test sections" that have a series of signs posted exactly one mile apart. The idea is for you to drive at a constant speed and measure the time it takes you to travel between signs so you can see if the speedometer on your car is accurate. To compute how fast you are going, in miles per hour, calculate $3600/t$, where t is the number of seconds it takes to drive between two signs. Write a Ruby method that will compute the speed of a car given the number of seconds it takes to travel one mile.

♦ Look on the Internet to find a formula for computing body mass index (BMI) as a function of height and weight. Define a Ruby method named `bmi` that will compute a body mass index using two parameters corresponding to height and weight values. Load the method into IRB and test it.

Chapter 3

The Sieve of Eratosthenes

An algorithm for finding prime numbers

Prime numbers have fascinated people for thousands of years. Mathematicians have long recognized that most integers are composite numbers, meaning they are the product of two smaller integers, but some special numbers, the prime numbers, are not evenly divisible by any smaller pair of numbers.

For most of their history prime numbers were only of theoretical interest, but today they are at the heart of a variety of important computer applications. The security of messages transmitted using public key cryptography, the most widely used method for transferring sensitive information on the Internet, relies heavily on properties of prime numbers that were discovered thousands of years ago.

The project in this chapter explores the Sieve of Eratosthenes, an algorithm that was invented over two thousand years ago to make lists of prime numbers. When the algorithm is implemented in Ruby, it will be in the form of a method named `sieve`. After completing the tutorial project, we will be able to use `sieve` to make a list of all the prime numbers less than a specified limit. For example, if we want to know all the prime numbers less than 1000, we just have to pass that number in a call to `sieve`:

```
>> sieve(1000)
=> [2, 3, 5, 7, 11, 13, ... 983, 991, 997]
```

The main goal for this chapter, as it is in the other chapters in this book, is to understand the algorithm and the computation it defines. We will start with an informal description of the process, and use the sieve to make a small list the way it was done for thousands of years, writing numbers on pieces of paper. Working through the example raises an interesting question about the algorithm that needs to be resolved before we can get a computer to carry out the steps of the computation.

Implementing the algorithm in Ruby and doing some experiments on our "computational workbench" provides an opportunity to introduce some new concepts in computing and some new features of the RubyLabs software that will be used in later projects. In the exercises up to this point, all the pieces of data were simple objects like numbers and strings. The steps in the Sieve of Eratosthenes algorithm work on lists of numbers, not just individual values. In the process of implementing the `sieve` method we will see how to create and manipulate lists as collections of objects in Ruby.

The new features of the RubyLabs module introduced in this chapter are methods that allow us to take a closer look at a Ruby program as it is running. In much the same way technicians use a variety of instruments to probe and measure a piece of electronic equipment, we will use Ruby methods that display the state of a computation, count the number of times key steps are executed, and measure the amount of time it takes to execute a program. These new methods are "tools" on our Ruby workbench that will be used in several other computational experiments throughout the book.

3.1 The Sieve Algorithm

To see how the Sieve of Eratosthenes got its name, imagine numbers are rocks, where the shape of each rock is determined by its factors. Even numbers, *i.e.*, all the multiples of 2, have a certain distinctive shape, multiples of 3 have a slightly different shape, and so on. Now suppose we have a magic bowl with adjustable holes in the bottom. As a first step in making a list of prime numbers, put a bunch of rocks in the bowl, and then adjust the holes to match the shape for multiples of 2. When we shake the bowl, all the rocks that are multiples of 2 will fall through the holes (Figure 3.1). Next adjust the holes so they match the shape for multiples of 3, and shake again so the multiples of 3 fall out. Keep repeating the adjusting and shaking steps until only prime numbers are left.

Instead of just starting with a random collection of numbers, and sifting out values in no particular order, the steps of the algorithm tell us how to proceed in a very precise manner. As a result of taking a more systematic approach, we are guaranteed that when we're done we will have every prime number within a specified range.

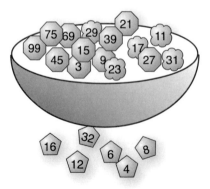

Figure 3.1: *The Sieve of Eratosthenes is a "magic bowl" that lets composite numbers fall through holes in the bottom. The first time the bowl is shaken, even numbers (multiples of two) fall out.*

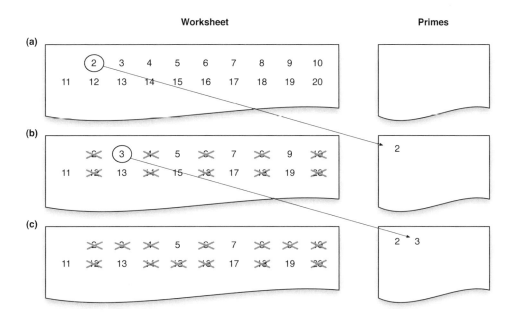

Figure 3.2: *(a) A list of numbers from 2 to n, where n = 20. (b) The state of the algorithm after the first scan. The number 2 has been identified as a prime, and the multiples of 2 have been removed. (c) After the second scan 3 has been identified as a prime and the multiples of 3 have been removed.*

The first step in the algorithm is to make a list containing all the integers between 2 and some upper limit n. Since we don't really have a magic bowl, we need to use paper and pencil or some other real technology. Start by writing down all the integers between 2 and n on a sheet of paper. We'll call this the "worksheet." Figure 3.2a shows the list for $n = 20$.

To keep things neat, we'll use a separate piece of paper for our list of prime numbers. Each time we discover a new prime we'll write it on this second piece of paper. The first number on the list, 2, is prime, so copy it to the primes sheet. Multiples of 2 are not prime, so scan through the worksheet and remove the multiples of 2.

After all the multiples of 2 have been removed, the worksheet will look like sheet in Figure 3.2b. The lowest remaining number on the worksheet is 3. Add 3 to the list of primes on the second sheet, and repeat the process of going through the worksheet and removing all the multiples, this time removing multiples of 3 (Figure 3.2c). Keep repeating this process, copying the lowest remaining number to the primes list and removing its multiples from the worksheet, until the only numbers left on the worksheet are prime numbers.

The method is fairly straightforward, but this description leaves out some important details. The first detail is to define what it means to "remove a number from the worksheet." If we're going to use paper and pencil to carry through the steps of the algorithm, we can do something like what's shown in Figure 3.2, and cross out numbers to show they are no longer part of the list. If we're using a whiteboard it might be neater and easier to erase multiples instead of crossing them out. As we work through the tutorial project, we will see how to use Ruby to organize the numbers in an object that represents a list, and we will use Ruby methods to remove numbers from the list when we discover they are multiples of some other number.

The most important thing missing from the description of the algorithm is a statement of when we can stop doing the copying and sifting operations. If you do a few more steps of the example in Figure 3.2, where $n = 20$, you'll soon notice you aren't crossing out any more numbers. For example, after you "shake the bowl" to cross out multiples of 11, the only remaining numbers are 13, 17, and 19. You could continue and search for multiples of 13, but you can tell at a glance that there aren't any. The smallest composite number that is a multiple of 13 is $2 \times 13 = 26$, and this list only goes up to 20, so clearly there aren't any more multiples of 13 in the list. There aren't multiples of 17 or 19, either, and all the numbers left are prime.

Simply saying "repeat until all the numbers on the worksheet are prime" might be sufficient if we're telling another person how to make a list of prime numbers, but it is not specific enough to include it as part of a Ruby method. If we want to implement this algorithm in a programming language, we need a more precise specification of when the algorithm is finished.

We'll return to this question later in the chapter, but the first order of business is to figure out how to make lists of numbers and how to scan lists to remove composite numbers.

Tutorial Project

Take some time to make sure you understand how the Sieve of Eratosthenes works. Use the algorithm to make a list of prime numbers less than 50. Make a worksheet like the one shown in Figure 3.2 on a piece of paper (or a whiteboard, if you have one), then carry out the copying and scanning steps until you are left with only prime numbers.

How many passes did you make over the full list before you were left with all prime numbers? Can you think of a formula based on the upper limit n that describes a general rule for when you can stop?

3.2 The mod Operator

If we are going to use Ruby to generate a list of prime numbers, one of the first questions we need to address is how to determine whether one number is divisible by another. One way to see if x is a multiple of y is to check the remainder after dividing x by y.

To see why the remainder is relevant, consider a number like 10, which is a composite number because it is the product 2×5. An important fact about composite numbers is that if we divide a number by one of its factors, the remainder will be 0. In the case of the number 10, $10 \div 5 = 2\ R\ 0$ and $10 \div 2 = 5\ R\ 0$.

Ruby has an operator that computes the remainder of a division operation. The operator is called the **mod operator**, and its symbol is a percent sign. To compute the remainder of x divided by y, we write

```
x % y
```

To try out this operator in IRB, we can just pick some some small values where we know the remainder. Since $10 \div 3 = 3\ R\ 1$, we expect the result of `10 % 3` to be 1:

```
>> 10 % 3
=> 1
```

To verify that the remainder of dividing 10 by either of its two factors is 0:

```
>> 10 % 2
=> 0
>> 10 % 5
=> 0
```

The mod operator will come in handy in a surprisingly large number of projects through-out this book. The name "mod" comes from number theory, which calls this function the *modulo operation*. We use modulo arithmetic every day, even though we might not call it by that name. It is sometimes called "clock arithmetic" since operations are similar to figuring out the time on a clock. If it is currently 5 o'clock, and you want to know what time it will be 8 hours from now on a 12-hour clock, you can get the answer with Ruby's mod operator:

```
>> (5 + 8) % 12
=> 1
```

Tutorial Project

Start a new session with IRB.

T1. Type this expression to ask Ruby to do a division operation using floating point numbers:
```
>> 18.0 / 3
=> 6.0
```
Since the value after the decimal point is 0, there is no fractional part, *i.e.*, 18 is evenly divided by 3.

T2. The result of dividing 19.0 by 3, however, does have a fractional part:
```
>> 19.0 / 3
=> 6.33333333333333
```
The result is (approximately) $6\frac{1}{3}$.

T3. As explained in the previous chapter, when Ruby divides two integers, it just throws away the fractional part. Type the two previous expressions, but this time use integers instead of floating point numbers:
```
>> 18 / 3
=> 6
```
```
>> 19 / 3
=> 6
```

T4. Ruby's mod operator tells us what the remainder of a division operation is. Type this expression to verify that the remainder of dividing 18 by 3 is 0:
```
>> 18 % 3
=> 0
```

T5. Type this expression to see that the remainder of dividing 19 by 3 is 1:
```
>> 19 % 3
=> 1
```

T6. Try some experiments on your own. Pick any pair of integers x and y such that x is a multiple of y (for example $x = 25$ and $y = 5$). Ask Ruby to compute the remainder of x divided by y using the % operator. Is the remainder always 0 when x is a multiple of y?

T7. Try some experiments involving pairs of numbers x and y where x is not a multiple of y. Is the remainder always nonzero?

3.3 Containers

Now that we've seen how to determine whether a number is prime or composite we are ready to tackle the problem of how to create the lists of numbers we need for the worksheet and the list of primes.

In everyday life a **list** is a collection of items. Some lists are ordered, but many are just random collections of information. Ingredients in recipes are typically ordered because cooks like to have them presented in the order in which they are used, but shopping lists are often just random or semirandom collections of items (even if you're one of those super-organized people who arranges grocery lists by sections, the items within a section can be in any order, *e.g.*, there's no reason to put radishes before cucumbers in the produce list).

Mathematicians also deal with collections. A *set* is one of the fundamental concepts in mathematics. Usually items in a set are not given in any particular order, but if order is important mathematicians say they have an "ordered set" or a "sequence."

Computer scientists have a wide variety of ways of organizing collections of data, including sets, sequences, graphs, trees, and many other structures. In programming language terminology, a structure that holds a collection of data items is known as a **container**.

In Ruby we are going to use a container called an **array** to represent a list of numbers. Ruby arrays are ordered collections of data. The items in an array can be numbers, strings, or any other type of data object we can create in Ruby.

The simplest way to make an array in Ruby is to write the items in the array between square brackets, separated by commas. To have a variable a refer to a list of the numbers from 1 to 5 we would write

```
>> a = [1, 2, 3, 4, 5]
=> [1, 2, 3, 4, 5]
```

The expression on the first line above is an assignment statement. Like the assignments in the previous chapter, it has the name of a variable on the left side; the only difference here is the value on the right side is an array of numbers. Ruby handles this assignment like it did the others: if the variable a does not exist it is created, and the name a is associated with an object that represents a list of elements (Figure 3.3).

Arrays *vs.* Lists

In the computer science and programming literature the words "array" and "list" have subtle differences. Both are names for ordered collections of data, but lists are more flexible than arrays. Arrays typically have a predetermined size, but lists can grow and shrink. What Ruby calls an array is really more like a list.

The designers of the Ruby language had historical reasons for choosing the term "array" as the name of the data structure we will use for this project, and the name has stuck. For the most part in this book we do not need to be concerned with the differences between arrays and lists, and we can simply use Ruby's arrays to implement lists of items.

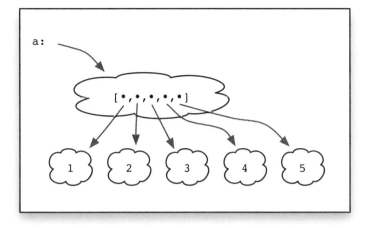

Figure 3.3: *The object store after creating an array of numbers. The variable* a *is a reference to an array object, which is a container that holds references to other objects.*

Note that Ruby responds to this assignment the same way it did the others, by printing the value of the new variable. To verify that a is an array, we can just ask Ruby to print its value:

```
>> a
=> [1, 2, 3, 4, 5]
```

There are all sorts of interesting things we can do with our new array. We can add items to the end, insert them in the middle, delete items, invert the order, make copies, and carry out dozens of other useful operations. For this project, though, we only need to know how to do a few of these things, and we'll put off investigating the other operations until we need them.

Most operations on arrays are performed by calling a method. In the previous chapter, we saw methods that are called by writing the method name followed by parameters enclosed in parentheses. Most operations on arrays, however, are specified using a different notation, where we write the name of an object, a period, and then the method name. For example, the method named length counts the number of items in an array. After making the array named a as shown above, we can ask Ruby how many items are in it by typing

```
>> a.length
=> 5
```

Here Ruby sees that a is the name of an array object, so the expression a.length means "call the method named length to operate on the object named a." As before, the output from IRB is the result returned by the method, in this case the number of items in the array a.

Two other examples of methods defined for arrays are first, which returns the value at the front of the array, and last, which returns the value at the end:

```
>> a.first
=> 1
>> a.last
=> 5
```

Note that, as was the case with the methods we saw in the previous chapter, the values returned by array methods are plugged into the surrounding expression, where they are used just like any other values:

```
>> n = a.length - 1
=> 4
>> a.first + a.last
=> 6
```

Array methods can have parameters, just like the parameters used to define `celsius` and other other methods described in the last chapter. An example of an array method with a parameter is `include?`, which returns `true` or `false`, depending on whether the value passed as an argument is contained somewhere in the array:

```
>> a.include?(2)
=> true
>> a.include?(7)
=> false
```

To attach an item to the end of an array we write an expression that uses the **append operator**, which is identified by the << symbol (two less than signs in a row, with no space between them). This expression uses the append operator to add 6 to the end of `a`:

```
>> a << 6
=> [1, 2, 3, 4, 5, 6]
```

In Ruby terminology, << is a method, just like `first` and `include?`. What distinguishes this method from the others is simply the syntax used to call it. Instead of writing an object name, a period, and a method name, we use the operator in an expression.

Before we start the project for this section, let's look at two special cases for making arrays. The first is the expression for making an array with nothing in it:

```
>> a = []
=> []
```

This array is known as an **empty array**. It's analogous to the empty set in math, or, in real-life terms, to a three-ring binder with no paper. The binder is still a binder, even if it's empty. An empty array is an object, just like any other array object, it just doesn't contain any references to other objects (Figure 3.4). Empty arrays are very common in a wide variety of algorithms. An algorithm will often create an initially empty array, and add items to it in future steps, or start with an array of items and repeatedly delete items until the array is empty.

The other special situation is a case where we want to make a list of all numbers in a specified range. Ruby has a notation for specifying ranges: write the lowest value, two periods, and then the highest value. For example, the range of numbers from 1 to 10 is written `1..10`. We can pass a range to a method named `Array`, and this method will create a new array containing every number in the range:

```
>> a = Array(1..10)
=> [1, 2, 3, 4, 5, 6, 7, 8, 9, 10]
```

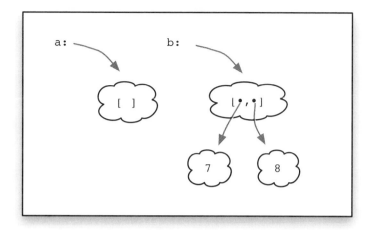

Figure 3.4: *In this example the object store has two arrays. The array named a is empty, since it does not contain references to any other objects, but b holds references to two integers.*

Looking ahead a little bit, these two techniques for creating arrays will be very useful when we implement the Sieve of Eratosthenes in Ruby. The initial worksheet will be an array with every number in the range from 2 up to the maximum value n, and the initial list of primes will be an empty array.

To summarize this brief introduction to arrays, we saw in this section that arrays are a new type of object that can be used in Ruby programs and in sessions with IRB. In all of the examples, IRB followed the same general set of rules it did in the previous chapter: Ruby waits for you to type an expression, then it evaluates the expression and prints the result. What we learned in this section is that

- objects in an expression can be arrays as well as numbers and strings

- operations on arrays are performed by methods

- when Ruby sees an array method in an expression it calls the method to perform an operation using the array, and the method returns a value that is plugged in to the expression

- most array methods are called by giving the name of an array, a period, and the name of the operation, but sometimes a method is called when an operator like << is used in an expression.

Tutorial Project

T8. Use IRB to create an array containing a few even numbers:
```
>> a = [2, 4, 6, 8]
=> [2, 4, 6, 8]
```

T9. Use the `length` operation to find out how many items are in the array:
```
>> a.length
=> 4
```

T10. Add a new number to the end of the array:
```
>> a << 10
=> [2, 4, 6, 8, 10]
```

T11. Did the previous expression change the length of the array? Call the `length` method to check your answer.

T12. Type this expression to make a new empty array named `b`:
```
>> b = []
=> []
```

T13. Use the `length` method to find out how many elements are in `b`. Did you get 0?

T14. A method named `empty?` will test whether an array is empty or not. If we invoke this operation on the new array `b` Ruby tells us the array is empty:
```
>> b.empty?
=> true
```
Note the question mark is part of the name of the operation; make sure you don't leave a space between the "y" and the question mark.

T15. Use the `empty?` method to see if `a` is empty.

T16. Make an array to hold the sequence of numbers between 1 and 20:
```
>> seq = Array(1..20)
=> [1, 2, 3, 4, 5, ... 19, 20]
```
Important: the "A" in this method name is capitalized. Ruby is very fussy about capitalization.

T17. Use the `length` method to count the items in this array. Are there 20 numbers?

What do you think will happen if we use a variable instead of an integer in a range expression? In other words, will Ruby allow us to write something like n..m, where n and m are variables we defined previously? Let's find out.

T18. Assign values for two new variables `n` and `m`:
```
>> n = 7
=> 7
>> m = 11
=> 11
```

T19. Create an array with all the numbers in the range from `n` to `m`:
```
>> seq = Array(n..m)
```
What did you get?

T20. At the beginning of this section there was a claim that arrays can hold any type of item. Let's test this by making an array of strings:
```
>> colors = ["green", "yellow", "black"]
=> ["green", "yellow", "black"]
```

T21. Use the << operator to add a new string to the end of `colors`:
```
>> colors << "steel"
=> ["green", "yellow", "black", "steel"]
```

T22. Use the `length` method to figure out how many items are now in `colors`.

T23. What do you think will happen if you try to add a string to the end of `a`, which is an array of numbers? Or if you try to add a number to the end of `colors`, which is an array of strings? Will Ruby complain, or will it make a mixed array? Use IRB to test your hypothesis. What did you learn?

3.4 Iterators

Recall from the introduction that the Sieve of Eratosthenes makes a list of integers, and then repeatedly goes through the list to discard composite numbers. We've just seen how to create an array to hold a list of numbers, and now we're ready to see how we can work through an array to check numbers to see if they are multiples of some other number.

The general technique for performing an operation on every item in a container is known as **iteration**. The word comes from the Latin word *iter*, which means "road" or "path." We often talk about iterating over an array, which means we start at the front and step through the array, one item at a time, in order to perform some operation.

Ruby and other object-oriented languages have several methods known as **iterators** that apply an operation to each item in the container. The simplest iterator for arrays is named each. This operation is invoked just like length and other methods, by giving the name of the array we want to work through, a period, and then the name of the operation. But we also need to tell the iterator what we want it to do with each element in the array. We do this by writing the operation we want to perform in braces {} following the word each.

As an example of how to call an iterator, let's use each to display every item in an array. First, let's make the array:

```
>> a = [1,2,3]
=> [1, 2, 3]
```

To display an object, we'll use a method named p, which prints objects on the terminal (see the sidebar on the next page, on Displaying Output). What we want to do is call p once for each item in the array. Here is the call to each that does it:

```
>> a.each { |x| p x }
1
2
3
=> [1, 2, 3]
```

In this expression, Ruby selects the items from a, one at a time. When an item is selected it is put in a variable named x, and then Ruby evaluates the expression p x. Since p is the method that prints a value on the terminal, evaluating this expression causes Ruby to print the value of x. There are three items in a, so the expression p x is evaluated three times, once for each item in a, and as a result three lines are printed on the terminal.

A useful mnemonic for the notation used with the each iterator is to think of how mathematicians specify the members of a set. If we want to describe the set of numbers between 1 and 10, we can write

$$\{ x \mid 1 \leq x \leq 10 \}$$

When read aloud, this expression is "the set of all x such that x is greater than or equal to 1 and less than or equal to 10." The syntax of the Ruby expression is very similar, except Ruby uses a pair of vertical bars instead of just one. In Ruby, we give the name of a variable between a pair of | symbols, followed by an expression that contains this variable. The role of the iterator is to store items from the array, one after another, in the variable, and for each item Ruby evaluates the expression to the right of the | symbols.

Displaying Output in the Terminal Window

Ruby has a method named p, which stands for "print." Whenever we call p, we can pass it one or more objects as arguments, and Ruby will display the objects in the terminal window.

Here is an example that shows how p prints a string on the terminal:

```
>> p "Hello"
"Hello"
=> nil
```

Notice that there are three lines here. The first line is the expression that has the call to p. The second line is the string printed by p. It's just a copy of the string we passed it—the argument passed to p was the string "Hello", and that's what p displayed. The third line was printed by IRB as the result of the call to p. Since p always returns nil you will always see this value when you call p from IRB.

This method isn't very useful in interactive Ruby because IRB prints the results of expressions anyway, but it's going to come in very handy when we start experimenting with more complicated algorithms. At various points in an algorithm we are going to want to call p to print the values of selected variables so we can keep track of how the algorithm is progressing.

Tutorial Project

T24. Make a small array of strings to experiment with expressions involving iterators:
```
>> colors = ["red", "yellow", "green", "blue", "white"]
=> ["red", "yellow", "green", "blue", "white"]
```

T25. Use the each iterator to print each item in a:
```
>> colors.each { |x| p x }
"red"
"yellow"
"green"
"blue"
"white"
=> ["red", "yellow", "green", "blue", "white"]
```

T26. Strings, like arrays, have a method named length. If s is a string, a call to s.length will tell us how many characters are in the string. Make a string and call length to have it count the number of characters:
```
>> s = "hello"
=> "hello"
>> s.length
=> 5
```

T27. Repeat the previous call to `a.each`, but this time, instead of printing a string, ask Ruby to print the length of the string:

```
>> colors.each { |x| p x.length }
3
6
5
4
5
=> ["red", "yellow", "green", "blue", "white"]
```

Make sure you understand what happened when Ruby evaluated these calls to `colors.each`: Ruby took the items one at a time from the array named `colors`, put them in the variable named `x`, and evaluated the expression between the braces. In the first example, Ruby printed the item from the array, and in the second it printed the length of the item.

3.5 Boolean Values and the `delete_if` Method

Earlier we saw arrays have a method named `include?` that will check to see if an array contains a specified item. Here are some more examples of this method, this time with an array of strings:

```
>> cars = ["bmw", "audi", "mini", "prius"]
=> ["bmw", "audi", "mini", "prius"]

>> cars.include?("mini")
=> true

>> cars.include?("alfa")
=> false
```

The `true` and `false` values returned by this method are known as **Boolean values**.

Another way to produce Boolean values is by writing an expression that compares items. In Ruby, the operator that compares two objects has the symbol ==, which is written as two equal signs with no space in between. This operator is called the **equality operator**. The result of a comparison is `true` if the two objects are the same, or `false` if they are different. For example, we can ask Ruby if the array of car names has four items:

```
>> cars.length == 4
=> true

>> cars.length == 6
=> false
```

The equality operator can also check to see if two strings are the same. Here are some examples, using the array of car names:

```
>> cars.first
=> "bmw"

>> cars.first == "bmw"
=> true

>> cars.first == "vw"
=> false
```

Here are some more examples, this time using integers:

```
>> n = 10
=> 10
>> n % 5
=> 0
>> (n % 5) == 0
=> true
>> (n % 5) == 1
=> false
```

Note how these last two expressions are evaluated: Ruby first computes the value of n % 5 (*i.e.*, it finds the remainder of dividing n by 5) and then it compares the result to the number on the right side of the operator. Since n % 5 is 0, the first expression using == has a value of true, but the second is false.

Expressions such as the ones shown above, where the value of the expression is either true or false, are known as **Boolean expressions**. One very useful place for Boolean expressions is in an iterator named delete_if. As its name implies, it is used to remove items from an array. The iterator walks its way through the array, putting each item into a Boolean expression. If the expression evaluates to true the item is deleted from the array.

Here is an example that uses the delete_if iterator to remove all the zeroes from an array of numbers:

```
>> a = [3, 0, 2, 5, 0, 1, 4, 0, 6]
=> [3, 0, 2, 5, 0, 1, 4, 0, 6]
>> a.delete_if { |n| n == 0 }
=> [3, 2, 5, 1, 4, 6]
```

This new iterator operates just like the each method introduced in the last section. It selects the items from a, one at a time, and stores them in the variable named n. Numbers that cause the Boolean expression to be true (all the zeroes) are removed from the array.

Here is a slightly more complex example. Since the value of n % 2 will be 0 only when n is an even number, this expression will remove the even numbers from a:

```
>> a
=> [3, 2, 5, 1, 4, 6]
>> a.delete_if { |n| (n % 2) == 0 }
=> [3, 5, 1]
```

The mnemonic for remembering how the each iterator works is also helpful for understanding an expression that involves delete_if. In essence, what the expression above says is "for all items *n* in the array *a*, delete *n* if *n* mod 2 is 0."

Before you start the projects in this section, a word of warning: The assignment operator, written with a single equal sign, and the operator that tests for equality, written with two equal signs, are very different things! Unfortunately, there are many situations where Ruby is happy to evaluate an expression that contains either operator. If you are not getting the results you expect, either in this section or later in the book, double-check to make sure you are typing the right number of equal signs.

Tutorial Project

T28. Type these expressions to make sure you understand the `==` operator:

```
>> x = 12
=> 12
>> 10 == 12
=> false
>> x % 2
=> 0
>> x % 2 == 0
=> true
>> x == 3 * 4
=> true
```

T29. One way to see if a number is even is to ask Ruby to check the remainder after dividing by 2; an even number has a remainder of 0:

```
>> 3 % 2
=> 1
>> 3 % 2 == 0
=> false
>> 4 % 2
=> 0
>> 4 % 2 == 0
=> true
```

T30. Make an array of numbers from 1 to 20:

```
>> a = Array(1..20)
=> [1, 2, 3, 4, 5, ... 18, 19, 20]
```

T31. Remove the even numbers by calling `delete_if` to delete the numbers that are evenly divisible by 2:

```
>> a.delete_if { |n| n % 2 == 0 }
=> [1, 3, 5, 7, 9, 11, 13, 15, 17, 19]
```

T32. Repeat the expression that creates the array, so you once again have an array of numbers from 1 to 20:

```
>> a = Array(1..20)
=> [1, 2, 3, 4, 5, ... 18, 19, 20]
```

T33. Repeat the expression that removes items, but change the 2 to a 3, so `delete_if` removes multiples of 3 (this would be a good place to use command line editing):

```
>> a.delete_if { |n| n % 3 == 0 }
=> [1, 2, 4, 5, 7, 8, 10, 11, 13, 14, 16, 17, 19, 20]
```

T34. Make an array of strings:

```
>> cars = ["bmw", "audi", "mini", "prius", "vw", "kia"]
=> ["bmw", "audi", "mini", "prius", "vw", "kia"]
```

T35. This expression will delete the strings that have more than three letters:

```
>> cars.delete_if { |x| x.length > 3 }
=> ["bmw", "vw", "kia"]
```

Do you see how Ruby evaluated this expression? It put each string in x, checked to see if the string had more than 3 characters, and deleted it if it did.

3.6 Exploring the Algorithm

The projects in the previous sections illustrated the basic building blocks for a Ruby implementation of the Sieve of Eratosthenes. We now know how to make lists of numbers, and we know how to iterate over lists to remove numbers that are evenly divisible by other numbers. The next step is to systematically apply these operations, so that after repeating the sifting operation implemented by `delete_if` a sufficient number of times we will be left with a list that contains only primes.

In this section we will use IRB as a "scatchpad." We'll make a list pretty much the same way we did it by hand in the introduction to this chapter, but instead of writing a list of numbers on a piece of paper and crossing them off one at a time, we'll let Ruby take care of all the clerical work. We will create Ruby arrays to keep track of the worksheet and the list of primes. In the next section we'll see how to collect all these pieces into a single program, so Ruby can do everything, not only the clerical work with the number lists but also the decision making about how many times to repeat the copying and sifting operations.

The steps of the algorithm are shown in Figure 3.5. The first step is to create the arrays that will represent the worksheet and the list of primes. To show how the algorithm works we will make a list of primes less than 100, so these two Ruby statements will initialize the arrays:

```
>> worksheet = Array(2..100)
=> [2, 3, 4, 5, ... 99, 100]
>> primes = []
=> []
```

Now we're ready to carry out Step 2 of the algorithm. Recall that to get the value of the first item in an array we can call the method named `first`, so to copy the first item from `worksheet` to the end of `primes` we just type

```
>> primes << worksheet.first
=> [2]
```

The Sieve of Eratosthenes

To make a list of every prime number less than n:

1. Create an array named `worksheet` with every integer from 2 up to n, and create an initially empty array named `primes`.

2. Copy the first number in `worksheet` to the end of `primes`.

3. Iterate over `worksheet` to remove every number that is a multiple of the most recently discovered prime.

4. Halt if every number in `worksheet` is prime, otherwise go back to step 2.

Figure 3.5: *The Sieve of Eratosthenes, using Ruby to manage the lists of numbers.*

Note that this step does not remove the 2 from the front of the worksheet; all it does is make a copy of the first item. We can check the contents of the arrays after this step:

```
>> primes
=> [2]

>> worksheet
=> [2, 3, 4, ... 99, 100]
```

For Step 3 of the algorithm, we clearly want to use the `delete_if` method to iterate over `worksheet`. The question is, what expression do we put between the braces? Can we write an expression that is `true` when an item `x` is a multiple of the most recently discovered prime? Recall from the tutorial project in the previous section that we can tell if a number `x` is even (*i.e.*, if it's a multiple of 2) by seeing if `x % 2` is 0. In general, to see if a number is a multiple of any value `n` check to see if `x % n` is 0. That means the expression between the braces should be `x % primes.last == 0`. If we try that expression on the `worksheet` array, this is what we should see:

```
>> worksheet.delete_if { |x| x % primes.last == 0 }
=> [3, 5, 7, 9, ... 97, 99]
```

Note that every even number has been deleted from `worksheet` because when this expression was evaluated `primes.last` was 2.

The worksheet still has several composite numbers, so we have to repeat the expressions that implement Steps 2 and 3 of the algorithm. This is the result:

```
>> primes << worksheet.first
=> [2, 3]

>> worksheet.delete_if { |x| x % primes.last == 0 }
=> [5, 7, 11, 13, ... 95, 97]
```

Note carefully what happened the second time we asked Ruby to execute these statements. The number at the front of the worksheet was 3, and it was copied to the end of `primes`. When `delete_if` was called the second time, `primes.last` was 3, so `delete_if` removed all the multiples of 3 from `worksheet`.

We're not done yet. Since 25, 35, and several other multiples of 5 are still in `worksheet`, we need to once again repeat the two statements that copy the first item in `worksheet` to the end of `primes` and remove multiples of the most recent prime.

After a few more repetitions we will have sifted out all the composite numbers. This is what the two arrays will look like after 10 rounds of copying and deleting:

```
>> primes
=> [2, 3, 5, 7, 11, 13, 17, 19, 23, 29]

>> worksheet
=> [31, 37, 41, 43, 47, 53, 59, 61, 67, 71, 73, 79, 83, 89, 97]
```

The array named `primes` has the small prime numbers that were copied from `worksheet`, and the numbers that are still left in `worksheet` are the rest of the prime numbers less than 100.

Tutorial Project

T36. Type the two expressions that initialize the arrays:
```
>> worksheet = Array(2..100)
=> [2, 3, 4, 5, ... 99, 100]

>> primes = []
=> []
```

T37. Copy the first number in `worksheet` to the end of `primes`:
```
>> primes << worksheet.first
=> [2]
```

T38. Since 2 is the last number in `primes`, this statement will delete the multiples of 2:
```
>> worksheet.delete_if { |x| x % primes.last == 0 }
=> [3, 5, 7, 9, ... 97, 99]
```

T39. Repeat the previous two expressions. This is a situation where command line editing will save a lot of time and prevent a lot of errors due to typing mistakes.

T40. Do you see how, after executing the copying and deleting steps a second time, the multiples of 3 are now gone from `worksheet`?

T41. Keep repeating these two expressions until you are convinced all the numbers left in `primes` and `worksheet` are all prime numbers.

T42. Since one number is added to the end of `primes` in each round of copying and deleting, asking Ruby to evaluate `primes.length` will tell you how many times you executed those steps. How many did you do before you decided there were no more composite numbers left?

3.7 The `sieve` Method

The discussion in the previous section described the key steps in the Sieve of Eratosthenes algorithm. After doing the exercises you should have a fairly good understanding of how each cycle of the algorithm removes all the multiples of the most recently discovered prime number.

What we would like to do next is to put these operations into a new method named `sieve`. The goal is to be able to call the method from an IRB session to have Ruby generate an array containing all the prime numbers less than a specified value. For example, to get the list of primes less than 50 we would type:
```
>> sieve(50)
=> [2, 3, 5, 7, 11, 13, 17, 19, 23, 29, 31, 37, 41, 43, 47]
```

The first step in making the `sieve` method is to create a file to hold the method definition. Ruby methods are commonly found in files that have the same name as the method, so the file to make for this project should be named `sieve.rb`. A good first step is to simply type in the outline of the method:
```
def sieve(n)

end
```

This definition tells Ruby that we are making a new method named `sieve`, and that when the method is called we will be passing a parameter.

Next we can type in the lines that create the two arrays and an expression that will return the final result. The only thing different about the assignment statement that initializes the worksheet is the upper bound, the number `n`, which will be set to the value passed as an argument when we call the method. Here is the new outline:

```
def sieve(n)
  worksheet = Array(2..n)
  primes = []
  return primes
end
```

At this point we don't really have an array of prime numbers to return, but don't worry, we'll be adding more statements.

A good habit to get into is to save this version and load it into IRB. It won't do much if we call it, but by going through the exercise of loading the outline and calling the method, we will discover if we accidentally made any typing mistakes. Several things can go wrong, even in this simple program. People leave out parentheses, or misspell "seive," or make any of a number of possible typing mistakes. The more often we check to make sure things are on the right track, the sooner we catch these sorts of errors.

If we tell Ruby to load the first version of the method, and there are no typing mistakes, this is what we will see:

```
>> load "sieve.rb"
=> true
```

The `true` means the file was loaded with no errors. If you see an error message, the message will contain the line number, and you should go to that line number in your text editor to correct the mistake. As one last test of the outline, we can call the method:

```
>> sieve(50)
=> []
```

This is what we expected, since `primes` is initialized to be the empty list. Once again, by running this simple test, we can be assured we didn't make a silly mistake, like leaving off the `end`, or typing in the wrong name for the method.

The next phase in the development of the `sieve` method is to figure out how to have Ruby repeat the two main operations in the algorithm, the step that copies the front of `worksheet` to `primes` and the step that uses `delete_if` to remove multiples of that prime. When we used IRB as a workbench to explore the algorithm we simply repeated the execution of two expressions until it was "obvious" there were no more composite numbers left in `worksheet`. It may be obvious to us, but not to the computer. We need to figure out how to specify a terminating condition that we can write as an expression in Ruby.

Recall from the previous exercise that every time we call `delete_if` the worksheet becomes shorter by at least one number. So one way to have Ruby control the algorithm is to have it keep transferring numbers from `worksheet` to `primes` until `worksheet` is empty. Ruby will do too many steps—it will keep repeating the two operations long after the last composite number has been sifted out—but let's go ahead and implement this version anyway, because it is a good way to introduce a new technique for describing iteration.

Earlier in the chapter, iteration was defined as a process that "walks" through a container to perform an operation with every object in the container. A more general definition is that iteration is any process that repeatedly executes a set of operations. One way to tell Ruby to repeat the copying and sifting operations is with a **while statement** that looks like this:

```
while worksheet.length > 0
  primes << worksheet.first
  worksheet.delete_if { |x| x % primes.last == 0 }
end
```

Although this may look complicated, the two lines in the middle are identical to what we've been using in our tests. When you type these two lines into your `sieve.rb` file, you can just copy and paste from your terminal emulator window and insert the lines into the method definition.

The four lines shown above tell Ruby to execute the statements between the line with the word `while` and the line with the word `end` as long as `worksheet.length` is greater than 0, *i.e.*, as long as there are numbers in the worksheet. When Ruby gets to the first line, it evaluates the Boolean expression next to the word `while`. If the value of this expression is `true`, Ruby evaluates the expressions on the following lines, up to the line that has the word `end`. Ruby then goes back to the line with `while` and checks the Boolean expression again. In this example, as long as there are items in `worksheet` Ruby will keep copying the first item from `worksheet` to the end of `primes` and sifting multiples of that value from `worksheet`.

An old term to describe iteration, dating from the first programming languages, is *loop*; the Ruby code above is an example of a **while loop**. An important fact about every `while` loop is that eventually the expression that controls the loop must evaluate to `false`, otherwise the program never terminates—it is caught in an infinite loop.

So does the `while` loop in our `sieve` method terminate? The answer is "yes," because the worksheet shrinks by at least one item on every iteration. To see why, note that the call to `delete_if` removes the first item from `worksheet` because every number is its own

```
# Make a list of all prime numbers between 2 and n, iterating until worksheet is empty

    def sieve(n)
      worksheet = Array(2..n)
      primes = []                              ** preliminary version **

      while worksheet.length > 0
        primes << worksheet.first
        worksheet.delete_if { |x| x % primes.last == 0 }
      end

      return primes
    end
```

Figure 3.6: *The first version of the* `sieve` *method, using a* `while` *loop that tests for an empty array.*

multiple, *i.e.*, x % x == 0 for any integer x. Eventually the array shrinks down to the empty array and the `while` loop will terminate.

After adding the `while` loop to the method definition in the file, the complete method should look like the definition in Figure 3.6. Save the file and reload the method, and we should be able to call `sieve` to have it make a list of primes:

```
>> load "sieve.rb"
=> true

>> sieve(50)
=> [2, 3, 5, 7, 11, 13, 17, 19, 23, 29, 31, 37, 41, 43, 47]
```

Success! We have a method that creates a list of prime numbers!

At this point you might want to try a few more calls, perhaps getting a list of primes up to 100 or even 1000. But don't get too carried away—if you ask for a list of primes up to 10,000 you'll start to see Ruby taking a lot longer to build the list. In the next section we'll improve the method so it requires far fewer iterations in the `while` expression.

Tutorial Project

If you have not already done so, use your text editor to create a file named `sieve.rb` and type in the method shown in Figure 3.6. You could simply type in the entire program, as shown, but it is a good idea to follow the procedure described in this section. First type in the outline, then load it and test it, and then continue with the next part of the code. This sort of "continual testing" is a helpful way to find errors soon after they occur. If you wait until you have typed the entire program it can be much more difficult to figure out which lines contain errors.

Load your new method into IRB and test it by calling it a few times. Once the method is producing lists of prime numbers you are ready to move on to the next section.

3.8 A Better Sieve

The challenge set out in the tutorial project at the end of Section 3.1 was to devise a formula for when to stop sifting numbers from the worksheet. It turns out we can stop as soon as the number at the front of the list is greater than $\sqrt{n}$.

To see why, consider a useful fact about multiplication. Any number that will be removed from the list is a composite number, which means it is the product of two smaller numbers *a* and *b*. It must be the case that either $a < \sqrt{n}$ or $b < \sqrt{n}$. If not, the product *ab* will be greater than *n*, because the product of two numbers that are both greater than $\sqrt{n}$ will be greater than *n*. Since our worksheet only goes up to *n* we can stop filtering when all the numbers in the worksheet are greater than $\sqrt{n}$, which will be the case as soon as we find a prime that large. In the case of $n = 20$ the algorithm can terminate after removing multiples of 3 from `primes`, because the new head of the worksheet is 5, and $5 > \sqrt{20} \approx 4.472$.

In Chapter 2 we saw Ruby has a method named `sqrt` that will compute square roots. It's easy to change the `while` statement so it checks to see whether the first number in the worksheet is greater than $\sqrt{n}$. Simply change the line with the `while` expression so it looks like this:

```
while worksheet.first < sqrt(n)
```

```
# Make a list of all prime numbers between 2 and n, iterating until the first number in
# worksheet is greater than the square root of n.

    def sieve(n)
      worksheet = Array(2..n)
      primes = []

★     while worksheet.first < sqrt(n)
        primes << worksheet.first
        worksheet.delete_if { |x|  x % primes.last == 0 }
      end

★     return primes + worksheet
    end
```

Figure 3.7: *The complete definition of the* sieve *method, which uses the Sieve of Eratosthenes algorithm to generate a list of prime numbers between 2 and an arbitrary upper limit n. The stars highlight the differences between this version and the preliminary version of Figure 3.6.*

This new Boolean expression still ensures the loop will terminate. Every time the loop body is executed, the worksheet decreases by at least one number (the number moved to the end of primes), so the value of the first number in worksheet is always increasing. Eventually a number greater than $\sqrt{n}$ will be seen at the head of the list.

If you edit your file to change the Boolean expression at the head of the while loop, and then reload the method and test it by calling sieve(100), you will notice that the new version has a small "bug." We seem to be missing a lot of numbers all of a sudden. The program will print a list of prime numbers, but they will only go up to 7. The problem is that at the end of the while loop the prime numbers are spread across the two arrays: all the primes less than $\sqrt{n}$ are in the primes array and all the remaining primes are still in worksheet.

To fix this bug we have to figure out how to splice together the primes array and the worksheet. In Ruby it's easy to splice together two arrays. If x and y are arrays, the expression x + y creates a new array that has all the items in a followed by all the items in b. So the final expression in the method is simply

```
    return primes + worksheet
```

Tutorial Project

The next section has a number of experiments based on the final version of the sieve method. These experiments will use a version of sieve that is included with RubyLabs. If you would like to do some explorations on your own, use your text editor to make the two changes discussed above and indicated by stars in Figure 3.7.

After making these changes you can reload your file and call the new version of sieve. This new version is much more efficient. A call to sieve(100000) to make a list of all prime numbers less than 100,000 should only take a second or two. The experiments in the next section should make it clear why the new version is so much more efficient.

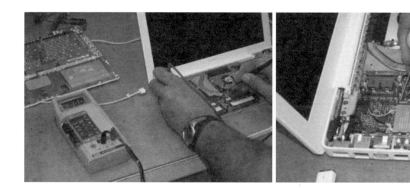

Figure 3.8: *One way to see if a piece of computer hardware is working is to connect a set of probes to a meter (left) and then use the probes to measure signals coming from the CPU or other parts of the system (right). The RubyLabs module has a number of "software probes" that will be used to monitor the execution of* `sieve` *and other methods.*

3.9 Experiments with the Sieve

This new version of the `sieve` method is clearly more efficient than the original version, since the new termination condition for the `while` loop means there are fewer calls to the `delete_if` method. But just how much more efficient is it?

One way to answer this question is to use a set of "probes" to monitor the progress of the `sieve` method. Hardware technicians use probes to measure electrical signals on the pins of CPUs, memory chips, and other components when they are tracking down hardware problems. One end of a pair of probes is connected to a meter. When the technician touches the other end of the probe to the hardware, electrical signals are measured and displayed on the meter (Figure 3.8).

For our projects we will be using "software probes." Each probe is a method defined in RubyLabs. We will call these methods to attach probes to `sieve` or any other method we want to watch. After attaching a probe, we can call our method, and the probe will display outputs in the terminal window that show us what the method is doing.

The tutorial project in this section will use probes to watch the worksheet get shorter and shorter as composite numbers are filtered out. We will use other probes to count the number of times the `while` loop is executed. If we let the loop run until the `worksheet` array is empty, the method needs almost 10,000 iterations to make a list of prime numbers between 2 and 100,000. What we are going to see is that with the new terminating condition the algorithm makes only 65 iterations. Our new version will be over 150 times more efficient!

The first RubyLabs probe to try is named `time`. Simply type the word `time` followed by a Ruby expression enclosed in braces. The `time` method will measure how long it takes Ruby to evaluate the expression between the braces. This is how to measure the time it takes `sieve` to make a list of prime numbers less than 1000:

```
>> time { sieve(1000) }
=> 0.015341
```

The result of this call is the number of seconds Ruby spent during the execution of the call to `sieve(1000)`. Note that `time` doesn't show us the list of prime numbers; all we see is the number of seconds it took Ruby to make the list.

```
>> include SieveLab
=> Object

>> Source.listing("sieve")
   1:    def sieve(n)
   2:       worksheet = Array(2..n)
   3:       primes = []
   4:
   5:       while worksheet.first < sqrt(n)
   6:          primes << worksheet.first
   7:          worksheet.delete_if { |x| x % primes.last == 0 }
   8:       end
   9:
  10:       return primes + worksheet
  11:    end
=> true
```

Figure 3.9: *For the projects in this section you need to use the version of the* sieve *method that comes with RubyLabs. Start a new IRB session, include SieveLab, and call the* listing *method to see the source code for the RubyLabs version of* sieve.

A method named trace allows us to watch variables as they change during the execution of a program. Before we call this method, we have to tell Ruby which variable to watch and when to print a new value of the variable. This is where the idea of the "software probe" comes into play.

To explain how probes work, we are going to attach a probe to the sieve method so Ruby shows us the value of worksheet at the start of each round of sifting. Start by getting a listing of the method, which will show the line numbers (Figure 3.9). The line we are most interested in is line 6, which has the statement that copies the first item in the worksheet to the primes array at the start of each iteration. If we tell Ruby to print the current value of worksheet just before line 6 is executed, we will be able to see the list getting shorter and shorter as the algorithm progresses. Here is the command that creates the probe:

```
>> Source.probe("sieve", 6, "p worksheet")
=> true
```

The method named Source.probe attaches a software probe. The first argument is a string that identifies the method we want to watch, the second is a line number within that method, and the third is a Ruby statement we want to have executed each time the specified line is executed.

After attaching this probe, we can call trace to monitor the progress of our sieve method:

```
>> trace { sieve(20) }
[2, 3, 4, 5, 6, 7, 8, 9, 10, 11, ... 19, 20]
[3, 5, 7, 9, 11, 13, 15, 17, 19]
=> [2, 3, 5, 7, 11, 13, 17, 19]
```

Notice how `trace` called `p worksheet`, the expression in the probe, to print the current version of the worksheet just before the statement on line 6 was executed. Since it it took two iterations to make a list of primes up to 20, the statement on line 6 was executed twice. The first output line shows the worksheet before any numbers were sifted out, and the second line shows the list after multiples of 2 were removed.

After we have used `trace` to monitor `sieve` for small values of n, we will want to run some more experiments to see how efficient the method is for larger values of n. For these experiments we don't really need to see the contents of the worksheet. In fact, if we use `trace` to monitor the execution of `sieve(1000)`, the screen will be filled with numbers and it will be hard to figure out how many rounds of sifting take place.

For experiments that simply need to count the number of times a line is executed, attach a new kind of probe, called a "counting probe." Call `Source.probe` and pass it a special argument named `:count`, as shown in this example:

```
>> Source.probe("sieve", 6, :count)
=> true
```

The word `:count`, with the colon at the front, is a special type of string called a **symbol**. Symbols are often used in Ruby programs to specify options. In this example, `:count` tells `Source.probe` to attach a counting probe instead of a probe that monitors the state of a variable.

After attaching a counting probe, using the expression shown above, we can use a method named `count` to call `sieve` and count how many times line 6 is executed:

```
>> count { sieve(20) }
=> 2
>> count { sieve(100) }
=> 4
```

According to the output, it only takes two rounds of copying and sifting to make a list of primes up to 20, and four rounds for the list of primes between 2 and 100.

Symbols in Ruby

Names that start with colons are called **symbols** in Ruby. Symbols are special types of strings: they are not enclosed in quotes, but are simply sequences of letters that start with a colon.

Symbols are commonly used to specify options in Ruby. For example, passing the symbol `:count` as an option to `Source.probe` tells Ruby to count the number of times a statement is executed.

We'll be using symbols in future labs, as well. The important thing to remember is that a symbol is just another type of string, but one that contains a single word and starts with a colon instead of being surrounded by quotes.

Tutorial Project

Important Note: To work on the projects in this section of the tutorial you need to start a new IRB session. Type `quit` to exit if your previous IRB session is still running.

The lab module for this project is named SieveLab. Start a new IRB session, and tell Ruby you want to use the SieveLab module:

```
>> include SieveLab
=> Object
```

Note: if you get an error message saying something about an "uninitialized constant" first make sure SieveLab is spelled properly (it must be capitalized exactly as you see it here). If you typed SieveLab correctly, this message means the RubyLabs package is not being loaded into IRB when you start a new session. Refer to the Lab Manual for suggestions on how to fix this problem.

T43. Try out the `time` method by using it to measure how long it takes to make some lists of prime numbers:

```
>> time { sieve(100) }
=> 0.001292

>> time { sieve(1000) }
=> 0.015364
```

The number you see will be different, since the execution time depends on processor speed and many other factors that vary from one computer to another.

T44. Call the `listing` method to get a printout of the `sieve` method:

```
>> Source.listing("sieve")
  1:     def sieve(n)
...
 11:        end
=> true
```

Your output should look like the code shown in Figure 3.9.

T45. Attach a probe to line 6 of the `sieve` method, so the worksheet is printed just before the statement on line 6 is executed:

```
>> Source.probe("sieve", 6, "p worksheet")
=> true
```

T46. To verify the probe is attached call `Source.probes` (note the "s" at the end of the name) to get a list of all probes:

```
>> Source.probes
sieve   6: p worksheet
=> true
```

The output confirms that the expression `p worksheet` will be evaluated each time line 6 in `sieve` is executed.

T47. The probe will be activated when we use the `trace` method. This expression tells Ruby to call `sieve(50)`, and whenever Ruby reaches a line in the program that has a probe attached the probe expression is executed:

```
>> trace { sieve(50) }
[2, 3, 4, 5, 6, 7, 8, 9, 10, 11, 12, 13, 14, 15, 16, ... 49, 50]
[3, 5, 7, 9, 11, 13, 15, 17, 19, 21, 23, 25, 27, 29, ... 47, 49]
[5, 7, 11, 13, 17, 19, 23, 25, 29, 31, 35, 37, 41, 43, 47, 49]
[7, 11, 13, 17, 19, 23, 29, 31, 37, 41, 43, 47, 49]
=> [2, 3, 5, 7, 11, 13, 17, 19, 23, 29, 31, 37, 41, 43, 47]
```

Can you see how the probe is working? And can you see why the `sieve` method terminated when it did? After removing multiples of 7 from `worksheet`, the smallest number left is 11. Since $11 > \sqrt{50}$ the loop terminates.

T48. Repeat the previous statement a few more times, passing different values to `sieve`.

T49. Attach a counting probe to line 6 of the `sieve` method:

```
>> Source.probe("sieve", 6, :count)
=> true
```

Note: there can be only one probe per line, so the new probe replaces the one set there previously.

T50. Use the `count` method to find out how many times that line is executed when making a list of primes between 2 and 50:

```
>> count { sieve(50) }
=> 4
```

Is this result correct? How many lines were printed by `trace` when it monitored the same method call?

T51. Try a few more experiments on your own, passing different values to `sieve`.

After using `count` and `trace` to monitor the execution of the `sieve` method, you should have a clear understanding of how the algorithm works and how the `while` loop terminates when the number at the front of the worksheet is greater than $\sqrt{n}$.

Next try some experiments with a much bigger list by asking `sieve` to make a list of primes up to 100,000. As a result of the way the probes are implemented in RubyLabs, having a probe attached does not interfere with the measurement of execution time when you call `time`, but `sieve` will run much more slowly when you call `count`.

T52. Measure how long it takes to make a list of primes between 2 and 100,000 (remember Ruby doesn't use commas in large numbers, so 100,000 is written as a 1 followed by five 0's):

```
>> time { sieve(100000) }
=> 2.720835
```

T53. Repeat the previous expression, but change `time` to `count`. Since the probe is active this will take a lot longer, but you will eventually see the number of iterations:

```
>> count { sieve(100000) }
=> 65
```

3.10 Summary

This chapter explored the Sieve of Eratosthenes, an algorithm that has been used for thousands of years to make lists of prime numbers. To use this algorithm to find all primes between 2 and some maximum value n, start by making a list of all numbers from 2 to n, and then systematically "sift out" the composite numbers.

The project introduced an important new idea in computing: programs often need to work with collections of objects. An array is a simple kind of container, where objects are stored in a linear order. We saw two different ways to to make arrays in Ruby. We can simply write a list of items to put in the array, for example a list of strings or numbers enclosed between square brackets, or we can make an array of numbers in a specified range. Arrays were used in the Ruby implementation of the sieve to hold the initial list of numbers and the growing list of prime numbers that will be the eventual output from the method.

Concepts and Terminology Introduced in This Chapter

array	An ordered collection of objects; in this chapter arrays were used to implement lists of numbers
iteration	A technique for solving a problem by repeating a set of steps
iterator	A method that applies an operation to each item in a collection, *e.g.*, to iterate over an array to print the elements or remove composite numbers
while loop	A Ruby statement that repeatedly executes a set of statements
mod operator	An operator that computes the remainder of a division; in Ruby, the remainder after dividing x by y is written `x % y`
Boolean expression	An expression that has a value of `true` or `false`

Another important idea in computing introduced in this chapter is iteration, which generally means "repetition." Two forms of iteration were used in this project: iterating over an array, to repeatedly perform some operation for each object in the array, and repeating an operation until a certain condition was met. The first form of iteration was used to filter out composite numbers: a method named `delete_if` examines each object in the worksheet and evaluates an expression to see if the object is a number divisible by the most recently discovered prime. This second form of iteration was carried out by a "`while` loop" that repeated the key steps of the algorithm until no more composite numbers were left in the worksheet.

Prime numbers play an important role in modern cryptography. Algorithms that encrypt messages, for example the credit card numbers you submit to a secure web site when you order something online, use "encryption keys" that are created by multiplying together two large prime numbers. Encrypted messages are safe as long as an intruder cannot break the key into its prime factors. The prime numbers used to make keys are huge. To make it very difficult for an intruder, keys should be around 600 digits long. The programs at secure web sites that choose prime numbers to make keys can't simply use the Sieve of Eratosthenes to generate their primes, since they would have to make an initial worksheet with every number from 1 to 10^{600} (to put this in perspective, there are only 10^{80} atoms in the entire universe). But the sieve is still used by these web sites. One of the steps in the algorithm that makes keys relies on a list of small primes, less than 10,000, and Eratosthenes' sieve is still used today to make these shorter lists.

Exercises

Try to answer questions 1 through 5 without using IRB. To answer questions 6 and 7 start IRB and load SieveLab or your version of `sieve.rb` so you can run the `sieve` method.

1. What does Ruby print for each of the following expressions?
   ```
   >> 12 % 5
   =>
   >> 12 % 4
   =>
   >> 97 % 7
   =>
   ```

2. Suppose we define an array with this expression:
   ```
   >> names = ["fred", "frodo", "fanny", "fonzie", "phil"]
   ```
   What would Ruby print for the following two expressions?
   ```
   >> names.each { |x| p x.length }
   =>
   >> names.delete_if { |x| x.length % 2 == 0 }
   =>
   ```

3. Briefly explain what Ruby did to evaluate that last expression (the one that contains a call to `delete_if`).

4. The improved `sieve` method shown at the end of the chapter stops removing numbers from `worksheet` as soon as it finds a prime number larger than a certain cutoff value. What is the cutoff value when we call the method to make a list of primes from 2 to 1000? That is, what is the value of the number n such that the method can stop filtering after it finds a prime larger than n?

5. To make a list of prime numbers between 2 and 100,000 we have to filter out multiples of numbers up to $\sqrt{100,000}$, which is 316. At first glance it might seem our new improved algorithm should have to do 316 iterations. But we just saw that `sieve` executes the loop only 65 times. Can you explain why it does not have to make 316 iterations?

Use IRB to answer the following questions about prime numbers.

6. What is the largest prime number less than 1000?

7. The table below shows a count of the number of primes less than x for selected values of x. The number in the column labeled $\pi(x)$ is the number of primes less than x. Can you figure out how to use the `sieve` method to count the number of primes less than 100? Less than 1000? Do the numbers agree with those shown in the table?

x	$\pi(x)$
10	4
100	25
1000	168
10,000	1229
100,000	9592

Chapter 4

A Journey of a Thousand Miles

Iteration as a strategy for solving computational problems

One of the things computers do best is manage information. Computers store everything from small collections of personal data, like address books, photo catalogs, and financial records, to huge databases containing scientific data, medical records, and corporate finances. People constantly use their computers to search for information, whether it's in a music library on a personal computer, or at a commercial music site, or even across the entire Internet. Computers also spend a lot of time rearranging data to make it easier for users to find what they need.

Two fundamental operations used in this wide variety of information management applications are *searching* and *sorting*. In their most basic forms, search algorithms scan through a collection to locate a particular item of interest, and sort algorithms reorganize collections so they are put in a particular order. In this chapter we will study two of the most common searching and sorting algorithms. The search algorithm, called *linear search*, scans an array from beginning to end to see if it contains a specific item. The sort algorithm, *insertion sort*, rearranges the items in an array to put them in order.

The two algorithms apply the same problem-solving strategy used by the Sieve of Eratosthenes. That algorithm makes a list of prime numbers by repeatedly executing the same operation, computing remainders, over and over. Instead of doing numerical calculations, however, searching and sorting simply compare items and move them around in memory. We will see how the same strategy for solving a problem, namely, setting up iterations that repeat a set of simple operations, will help applications effectively manage collections of information.

The experiments with the linear search and insertion sort algorithms will also introduce one of the most important concepts in computer science: the idea of **scalability**. These two algorithms are efficient and work well for small arrays of data, but as a data collection grows they become less and less effective. Computer scientists have developed a precise definition of what it means for an algorithm to be scalable, in order to compare algorithms on the basis of their ability to work on large problems, and we will use this definition to analyze the scalability of linear search and insertion sort.

After gaining experience with the basic searching and sorting algorithms in this chapter, we will look at two different algorithms in the next chapter. Through the use of a more sophisticated problem-solving strategy, these other algorithms are much more efficient on large arrays. But simplicity is a virtue, and for small arrays of data the two algorithms we will study in this chapter are still widely used.

4.1 Searching and Sorting

Searching and sorting are familiar operations in daily life, and their common meanings carry over into the world of computation. **Searching** is a process of looking for a particular piece of data in a large collection, and a **sorting** algorithm is used to arrange data in a specified order.

Searching and sorting commands are common features in applications we use every day. For example, an application that manages a personal music library (Figure 4.1) might sort songs by artist name, song name, genre, or other attributes. The application can also help locate a particular song by letting a user search for songs that have a specific string in the title. Some other examples of searching and sorting are found in spreadsheets and word processors. We can arrange the data in a spreadsheet by selecting a set of rows and columns and then invoking a sort operation, and we can search for words and phrases in a word processing application by invoking a "find" command.

As we begin our exploration of searching and sorting algorithms, we are going to need some new terminology that will let us be more specific about what we want the algorithms to do and what we expect to happen at each step. The two algorithms we will study both

"A Journey of a Thousand Miles
   Begins with a Single Step"

The title of this chapter is based on a quote from *Tao Te Ching*, by Lao-Tzu, a Chinese philosopher who lived in the 6th century BC.

The quote succinctly captures the nature of iteration as a strategy for solving computational problems. By repeating the same basic steps over and over, one can eventually solve some very large problems.

The Sieve of Eratosthenes uses this strategy when it repeatedly copies and sifts out multiples of prime numbers. In this chapter we will see how a similar strategy can be used to find an item in an array or to rearrange the array so it is sorted.

Song ▲	Artist	Album
Absolute Beginners	David Bowie	Absolute Beginners
Accidentally Like a Martyr	Warren Zevon	Excitable Boy
Across The Great Divide	The Band	The Night They Drove Old Dixie Down
Across The River	Peter Gabriel	Secret World Live
Across The Universe	The Beatles	Let It Be
Acute Schizophrenia Paranoia Blues	The Kinks	Muswell Hillbillies

iter ✕

Song	Artist	Album
Lady Writer	Dire Straits	Communiqué
Paperback Writer	The Beatles	Hey Jude
Tell Him	Exciters	The Big Chill
A Whiter Shade of Pale	Procol Harum	The Big Chill

Figure 4.1: *Examples of searching and sorting.* **Top:** *the user has clicked on the Song column name to sort the table entries according to the name of the song.* **Bottom:** *the user has typed "iter" in the search box, and the program is showing all the songs that have that string in the song name, artist name, or album title.*

work on lists, which are implemented in Ruby in the form of array objects. The first new concept is the idea of an **array index**, which is simply a number that identifies a position within the array. The first item in an array has the index 0, the next is index 1, and so on up to $n - 1$ for an array of n items. Note that the index of the first item is 0, not 1.

To access an item in the middle of an array in Ruby, simply write the name of the array followed by a number inside "square brackets." Here is an example, using an array of five strings:

```
>> vowels = ["a", "e", "i", "o", "u"]
=> ["a", "e", "i", "o", "u"]
>> vowels[0]
=> "a"
>> vowels[1]
=> "e"
>> vowels[4]
=> "u"
```

Not surprisingly, since searching and sorting are such important operations, Ruby has a number of different methods we can use to search through arrays or sort arrays according to various criteria. The method named `include?` looks through an array to see whether it contains a specified value, returning `true` if the search succeeds or `false` if it fails:

```
>> vowels.include?("e")
=> true
>> vowels.include?("x")
=> false
```

If we not only want to know *if* an item is in the array, but *where* it is, we can call a method named `index`. The most important thing to remember about `index` is that if we are searching an array that has *n* items, and the search succeeds, the result will be a number between 0 and $n - 1$:

```
>> vowels.index("e")
=> 1
>> vowels.index("u")
=> 4
```

If the search fails, `index` returns a special object named `nil`, which means "nothing" or "nowhere":

```
>> vowels.index("x")
=> nil
```

Array Indexes

Computer scientists are famous for beginning at 0 instead of 1 when they start assigning labels to things. Array indexes are a good example. If an array has 10 items, the locations in the array are labeled from 0 to 9.

If you type an expression with an array index operator and you see a result that doesn't look right, the first thing to check is to make sure you count starting at 0. If you type s[1] expecting to see the first item in an array s you'll get the wrong value. You need to remember to type s[0].

```
>> s = ["z", "w", "i", "t", "t", "e", "r", "i", "o", "n"]

>> s[1]
=> "w"
>> s[0]
=> "z"
>> s.length
=> 10
>> s[10]
=> nil
>> s[9]
=> "n"
```

z	w	i	t	t	e	r	i	o	n
0	1	2	3	4	5	6	7	8	9

To sort the items in an array simply call a method named `sort`:

```
>> a = ["a", "n", "a", "g", "r", "a", "m"]
=> ["a", "n", "a", "g", "r", "a", "m"]

>> a.sort
=> ["a", "a", "a", "g", "m", "n", "r"]
```

There are many other searching and sorting operations defines for arrays in Ruby. We can search for all the items that match a pattern (*e.g.*, all words that start with "a"), or sort in reverse order, and lots more, but we'll save the discussion of those methods for a later project when we need them.

Tutorial

T1. Make a small array of strings to use for testing index expressions:

```
>> notes = ["do", "re", "mi", "fa", "sol", "la", "ti"]
=> ["do", "re", "mi", "fa", "sol", "la", "ti"]
```

T2. The array method named `first` returns the item at the beginning of an array:

```
>> notes.first
=> "do"
```

T3. We can use an index expression to do the same thing. Since the first item in an array is at location 0, asking Ruby to evaluate `notes[0]` should give the same result as `notes.first`:

```
>> notes[0]
=> "do"
```

T4. Since there are 7 strings in this array we can find the last one at location 6:

```
>> n = notes.length
=> 7

>> notes[6]
=> "ti"
```

T5. Try asking Ruby for values at other locations, using any index between 0 and 6.

T6. What do you think will happen if you give an index that is past the end of the array? For example, if you ask for `notes[12]`?

```
>> notes[12]
=> nil
```

Recall that `nil` is the special object that stands for "nothing." This is simply Ruby's way of saying you have asked for something that does not exist.

T7. Next try the `include?` method. This method should return `true` or `false` depending on whether the array contains the specified item:

```
>> notes.include?("re")
=> true

>> notes.include?("bzzt")
=> false
```

T8. Use the `index` method to find out where the items are located:

```
>> notes.index("re")
=> 1

>> notes.index("bzzt")
=> nil
```

Do you see why the first expression above returned a value of 1?

T9. A call to `notes.sort` will generate a new list containing all the items in `notes`, but rearranged into ascending (alphabetical) order:

```
>> notes.sort
=> ["do", "fa", "la", "mi", "re", "sol", "ti"]
```

T10. Note that the call to `sort` did not change the array—it returned a *copy* of the array, with all the items arranged in ascending order. You can verify this by asking Ruby to print the original array again:

```
>> notes
=> ["do", "re", "mi", "fa", "sol", "la", "ti"]
```

4.2 The Linear Search Algorithm

The `include?` and `index` methods described in the previous section use the linear search algorithm to search through an array. To see how this algorithm works, we will pretend that these methods are not already part of Ruby and that we have to write our own methods for doing a search. The project for this section will be to write two new methods, which we'll call `contains?` and `search`.

As the name implies, a linear search is a simple process that starts at the front of an array and compares items one by one until it finds what it is looking for. The algorithm "walks through" an array to look at each successive item. The new twist for this algorithm is that if we find what we're looking for we can stop the iteration early—there is no need to continue on to the end of the array (Figure 4.2).

To stop the iteration early, we need a new programming construct known as **conditional execution**. The idea is to tell Ruby to execute a statement only under certain situations. In the search algorithm, we will have a return statement that will be executed only if Ruby finds the item it is looking for.

As an example of how to use conditional execution, let's look at a simple example first. Suppose we have a set of quiz results stored in an array named `scores`. If we want to print all the scores, we just need to use the `each` method to iterate over the array and print each item:

```
>> scores.each { |x| puts x }
86
97
70
. . .
```

If we want to see only the highest scores, we can change the statement between braces so the `puts` is executed only if the score is greater than a certain value. What we do is attach a **modifier** to the end of a statement. In this case the modifier is simply the word `if` followed by a Boolean expression. Ruby will execute the statement preceding the word `if` only when the expression is `true`. This statement tells Ruby to print only the scores higher than 90:

```
>> scores.each { |x| puts x if x > 90 }
97
95
. . .
```

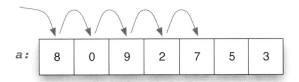

Figure 4.2: *Using a linear search to look for the number 7 in an array of numbers. The iteration stops as soon as 7 is found.*

As a second example, this statement tells Ruby to look through the array and print a message if it finds a perfect score:

```
>> scores.each { |x| puts "aced it!" if x == 100 }
```

Conditional execution is a new tool for our "computational workbench." Let's put it to use to create the `contains?` method. Here is a first "prototype." This version will scan the array and print a message when it finds a specified item.

```
def contains?(a, k)
  a.each { |x| puts "found it!" if x == k }
end
```

The name of the method is `contains?`, and it takes two parameters, the array to search and a value to look for. The value is named `k` because it's sometimes called a "key" in the literature on search algorithms. When the method is called, the iterator will look at each item in the array and print the message if the item matches the key. To see if anyone got a perfect score the call would be

```
>> contains?(scores, 100)
```

This method is almost what we want for the linear search algorithm. It does everything we need in terms of looking at the items in the array one by one, but the iteration always continues on to the end of the array, even after it finds the key. What we want to do is exit the loop as soon as we find a match. This is easy enough: all we need to do is return from the method as soon as we find the key. Since this method should return `true` when the key

Save Yourself Some Typing

The experiments in this chapter will use versions of `contains?` and other methods that are included as part of the RubyLabs module.

If you want to do some experiments on your own, you can type in the code shown in the figures in this chapter, or you can check out a copy of the RubyLabs methods.

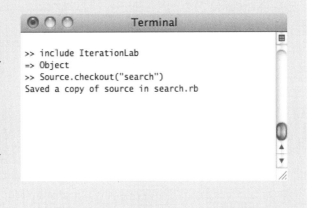

```
# Search array a to find item k

def contains?(a, k)
  a.each { |x| return true if x == k }
  return false
end
```

Figure 4.3: *The RubyLabs method named* contains? *is our version of the built-in method named* include?.

is in the array, all we have to do is change the expression inside the iterator so it looks like this:

```
a.each { |x| return true if x == k }
```

But now we have a new question: what do we do if the array does not contain the value k? In other words, what happens if the iteration runs all the way to the end, so the return statement is never executed? This problem also has an easy solution—just tell the method to return false if the iteration doesn't terminate early.

The final version of our new method is shown in Figure 4.3. Make sure you understand how this program works. When the method is called, it is passed an array and a value named k. The iterator is used to examine each item, starting from the front. If the iterator ever finds an item that matches k, the return statement is executed, and the result of the call is true. If the iterator does not find anything that matches k, the last statement in the method is executed, and the result of the call is false.

Next let's look at how we might implement the search method, which will be our own version of Ruby's index method. The idea here is to do a linear search, but when we find an item we want to return the location where it was found instead of just true. The easiest way to do this would be to change the each to a different iterator that is already part of Ruby and was designed for just this situation, but we're going to write out the loop using a while statement. This more verbose form will make it easier to attach a probe to monitor the execution of our method, and it will also be a piece of code that will form the basis of the sort method we'll look at later in this chapter.

Ruby's while statement was introduced in Section 3.7. It repeatedly executes the statements in the "body" of the loop as long as a Boolean expression is true. An outline of our new method is

```
def search(a, k)
  i = 0
  while i < a.length
    ...
    i += 1
  end
end
```

The variable i holds an index value. This variable is initially set to 0 since the index of the first item in the array is 0. The iteration continues while the index is less than the length of the array; recall that if an array has n items the index values range from 0 to $n - 1$.

Figure 4.4: *The RubyLabs method named* `search` *is our version of the built-in method named* `index`.

```
# Return the location of item k in array a

1:    def search(a, k)
2:      i = 0
3:      while i < a.length
4:        return i if a[i] == k
5:        i += 1
6:      end
7:      return nil
8:    end
```

The statement just before the end of the loop uses an operator we haven't seen yet. This expression is a combination of addition and assignment that tells Ruby to add 1 to the index. Adding 1 to a variable is very common in all sorts of algorithms, not just iterative algorithms. The phrase "increment the variable by 1" or simply "increment the variable" means a program should use an assignment to add 1 to the current value of a variable.

We can fill in this outline by adding the statements that compare an item in the array with the key value k and that return the desired values. On each iteration, i is the current index, so a[i] is the value stored in the array at location i. What we want to do is return i as soon as we find that a[i] matches k:

```
return i if a[i] == k
```

A listing of the `search` method is shown in Figure 4.4. We will refer to the line numbers in this figure when setting the software probes that trace the execution of the method. Note that if the iteration gets all the way to the end of the array, so that the `return` statement inside the loop is never executed, the statement at the end returns the value `nil`.

The computational experiments with our new linear search algorithms will use the Ruby-Labs `trace` method that was introduced in the last chapter. In these experiments, we will attach a "software probe" to a line in a method we want to observe, and then call `trace` so we can monitor the state of the computation.

The IterationLab module has a method that was designed to help us follow the progress of searching and sorting algorithms. The method is named `brackets`. It makes a string that has square brackets around some part of an array. The simplest way to use `brackets` is to pass it an array and an index. The string that comes back will have square brackets around the items from the index to the end of the array.

Here are some examples of how `brackets` works. Suppose we have a small array of strings:

```
>> a = ["H", "He", "Li", "Be", "B", "C", "N"]
=> ["H", "He", "Li", "Be", "B", "C", "N"]
```

If we call `brackets(a,2)` Ruby will make a string that has an opening bracket right before the string "Li," which is at a[2], and a closing bracket after the last item in a:

```
>> brackets(a, 2)
=> " H   He [Li   Be   B   C   N]"
```

If we call `brackets(a,3)` the brackets will enclose the items from `a[3]` to the end of the array:

```
>> brackets(a, 3)
=> " H  He  Li [Be  B  C  N]"
```

In the linear search experiments we will attach a probe to `search` to tell Ruby to print brackets around the part of the array that has not been searched yet. When we call `trace` we will see how the unsearched region gets shorter and shorter on each iteration.

Tutorial

The RubyLabs module with the definitions of `contains?`, `search`, and other methods used in this chapter is named IterationLab. When you start an IRB session, remember to include this module:

```
>> include IterationLab
=> Object
```

T11. Make a small array of integers to test the `each` iterator:
```
>> numbers = [3, 5, 2, 12, 7, 1, 14]
=> [3, 5, 2, 12, 7, 1, 14]
```

T12. As a reminder of what `each` does, use it to print the value of every item in `numbers`:
```
>> numbers.each { |x| puts x }
3
5
. . .
```

T13. Add a modifier to the expression in the block so it prints a number only if it is greater than 10:
```
>> numbers.each { |x| puts x if x > 10 }
12
14
```
Do you see how this expression works? Do you see why the first few numbers were not printed, but 12 and 14 were?

T14. Here is another call to the `each` method; in this version, the expression in the body prints the string "found it" if the number 7 is found anywhere in the array:
```
>> numbers.each { |x| puts "found it" if x == 7 }
```

T15. What do you suppose will happen if we change the 7 to a number that is not in the array? What will happen if we ask Ruby to evaluate this expression?
```
>> numbers.each { |x| puts "found it" if x == 13 }
```
Type the expression above. What did you see?

T16. Use `contains?` to do some searches, including the ones from the previous two problems:
```
>> contains?(numbers, 5)
=> true
>> contains?(numbers, 7)
=> true
>> contains?(numbers, 13)
=> false
```

T17. Try a few more searches of your own in this array to make sure you understand what a call to `contains?` does. Do you understand how this method works?

T18. What do you suppose will happen if you ask `contains?` to search an array of strings instead of an array of numbers? To find out, make an array of strings, and then call `contains?` a couple of times:

```
>> fruits = ["apple", "orange", "kiwi", "mango", "banana"]
=> ["apple", "orange", "kiwi", "mango", "banana"]
>> contains?(fruits, "kiwi")
=> true
>> contains?(fruits, "ugli")
=> false
```

So this method works with arrays of strings as well as arrays of numbers (Figure 4.5). In fact we can have an array of any type of object and use this method to search for objects of that same type.

T19. Use the `search` method (the RubyLabs implementation of the search that tells us where an item was found) to look for strings in the fruits array:

```
>> search(fruits, "kiwi")
=> 2
>> search(fruits, "ugli")
=> nil
```

Do you see why the result of the first expression is 2?

T20. An important part of our version of the `search` method is the increment operator. To see how this operator works, initialize a variable and use += in a few expressions:

```
>> n = 5
=> 5
>> n += 1
=> 6
>> n
=> 6
>> n += 3
=> 9
```

Can you see how the increment operator adds the value on the right side to the variable named on the left side?

Figure 4.5: *The* `search` *method will search through arrays of numbers or arrays of strings.*

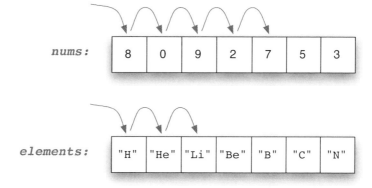

The next exercises refer to the source code for the `search` method, which is shown in Figure 4.4. You can also print a listing for yourself by typing `Source.listing("search")`.

T21. The key step in the `search` method is at line 4. When this line is executed, the variable `i` is the index of the item Ruby will check next. Let's attach a probe to this line to ask Ruby to print `i` before it executes this statement:

```
>> Source.probe("search", 4, "puts i")
=> true
```

T22. Now let's trace the execution of `search` again after setting the probe:

```
>> trace { search(fruits, "kiwi") }
0
1
2
=> 2
```

Can you see how `i` starts at 0, which is the leftmost location in `a`, and increases by 1 each time through the loop? And can you see how the iteration stops when `a[i]`, the value at the current location, matches the value we're looking for?

T23. Trace the execution of an unsuccessful search:

```
>> trace { search(fruits, "ugli") }
0
1
2
3
4
=> nil
```

Can you see why the iteration terminated in this case?

T24. Type this expression to print the `fruits` array with all the items from location 1 to the end enclosed in brackets:

```
>> puts brackets(fruits, 1)
 apple [orange  kiwi  mango  banana]
=> nil
```

T25. This expression does the same thing, but asks `brackets` to put the opening bracket before the item at location 4:

```
>> puts brackets(fruits, 4)
 apple  orange  kiwi  mango [banana]
=> nil
```

T26. To use `brackets` to monitor the execution of `search`, clear the current probes, and then make a new one that calls `brackets` at line 4:

```
>> Source.clear
=> true
>> Source.probe( "search", 4, "puts brackets(a,i)" )
=> true
```

T27. Now when we trace `search` again the new probe will show us two parts of `a`:

```
>> trace { search(fruits, "kiwi") }
[apple  orange  kiwi  mango  banana]
 apple [orange  kiwi  mango  banana]
 apple  orange [kiwi  mango  banana]
=> 2
```

Do you see what the probe is doing? The items before the opening bracket are the items that have been checked so far. The first item after the opening bracket is the current item, and `search` returns as soon as it notices this item is the one it is looking for.

T28. Try an unsuccessful search with the probe in place:

```
>> trace { search(fruits, "ugli") }
[apple  orange  kiwi  mango  banana]
 apple [orange  kiwi  mango  banana]
 apple  orange [kiwi  mango  banana]
 apple  orange  kiwi [mango  banana]
 apple  orange  kiwi  mango [banana]
=> nil
```

Can you see how every item is compared before `search` returns `nil`?

This concludes the experiments with the linear search algorithm, and you should be ready to move on to the next section. But if you are still unsure how the algorithm works, or if you want to play around with the program, here are some suggestions.

T29. Check out a copy of the `search` method so you have a copy you can modify:

```
>> Source.checkout("search")
Saved a copy of source in search.rb
=> true
```

T30. Use your text editor to add some lines to the program. One idea is to add statements that print the current value of `i` and `a[i]`, the current location in the iteration and the item in the array at this location. Add two new lines after the line with the word `while`, so your method looks like this:

```
while i < a.length
  puts i
  puts a[i]
  return i if a[i] == k
  i += 1
end
```

T31. Save your changes and tell Ruby to load your version of the program:

```
>> load "search.rb"
=> true
```

Note that after you load your file, the modified copy of `search` overwrites the one that came with RubyLabs.

T32. Now when you call `search` Ruby will use your modified version. Try it by doing a search in the `fruits` array:

```
>> search(fruits, "kiwi")
0
apple
1
orange
...
```

T33. Try a few more searches, including an unsuccessful search. Does this help you understand the finer details of how the algorithm works?

4.3 The Insertion Sort Algorithm

Sorting is one of the most studied problems in computer science. There are hundreds of different algorithms for sorting information. Insertion sort, the algorithm we will look at in this chapter, is a simple technique that is easy to understand and easy to implement. The main reason it was chosen is that the computation carried out by this algorithm is a simple extension of the iterative solution to the search problem, and it is another example of how a seemingly complex problem can be solved by repeatedly executing very simple steps.

To get an idea of how insertion sort works, imagine you are playing a card game. When a hand is dealt you want to sort the cards in your hand, putting them in order defined by their rank (we'll ignore suits for now). One way to do this systematically is to first make sure the two cards on the left are in the right order. Then pick up the third card and place it in the proper location among the first two. Then pick up the fourth card and place it among the first three cards. Repeat this operation of picking out a card and finding a location for it until all cards have been placed.

If we step back from this problem and think about what is going on each time we pick up a card, an important feature of this algorithm becomes apparent. At any point during this process there are two parts to the hand. The part on the left is already sorted, and the part on the right is still unordered. A succinct statement of the main step is that we remove the first card from the unsorted part and insert it into its proper location in the sorted part (thus the name of the algorithm, "insertion sort"). Note also that after each round of insertions the sorted part grows longer, and eventually the algorithm terminates when the last card has been placed where it belongs and the entire hand is sorted.

To implement this algorithm in Ruby, we just have to modify the description a little bit, replacing "hand" with "array." The overall structure of a method that sorts an array is shown in Figure 4.6. The main loop of the algorithm uses an index variable named `i`. At any point during the execution of the algorithm `i` is the dividing line between the sorted and unsorted portions of the array. On each iteration of the main loop the algorithm removes `a[i]`, which is the first item in the unsorted region, and searches back through the sorted region to find the right place to insert this item back into the array.

```
# Search for item k in array a

def search(a, k)
  i = 0
  while i < a.length

    # compare k to a[i]

    i += 1
  end
end
```

```
# Sort items in array a

def isort(a)
  i = 0
  while i < a.length

    # move a[i] to a location
    # between 0 and i

    i += 1
  end
end
```

Figure 4.6: *Outline of two iterative algorithms. Both algorithms use an index* i *as a place holder that marks a current location in an array.*

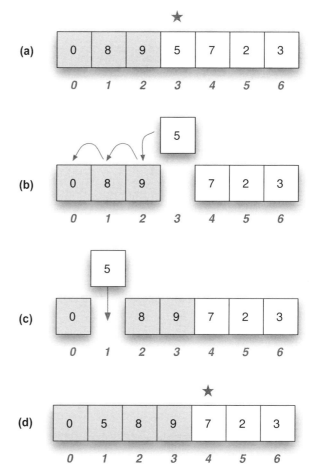

Figure 4.7: *A single iteration of the* `isort` *method.* ***(a)*** *At the start of the iteration,* `i` *is 3, and items to the left are already sorted.* ***(b)*** *The* `move_left` *method removes* `a[3]` *and scans left to find a place for it.* ***(c)*** *The item is inserted back into* `a`. ***(d)*** *At the start of the next iteration the sorted region is one item longer.*

The operation of finding the right place for the current item is itself an iterative algorithm. The idea is to do a search, but in the opposite direction. If *i* is the current location, we want to scan left from *i* − 1 and work back toward the beginning of the array.

The Ruby code that removes an item from a specified location in an array and puts it back somewhere to the left of that location has been implemented in a method called `move_left`. The important things to know about `move_left` are (a) a call of the form `move_left(a,i)` tells Ruby to remove `a[i]` and find a place for it somewhere to the left of location `i`, and (b) that this operation itself is a simple iteration very much like a linear search.

An example that gives a general idea of what the `move_left` method does is shown in Figure 4.7. The figure shows the state of the array when `i` is 3, meaning the sorted part of the array goes from `a[0]` up to `a[2]`. The call to `move_left` removes `a[3]`, the first item in the unsorted region, from the array, and scans left until it finds the correct location for it. After the item has been put back in the array, the sorted region is one item longer, and the next iteration of the main loop is ready to begin with `i` set to 4.

Given a working version of `move_left`, writing a method to implement the insertion sort algorithm is straightforward. We simply have to put a call to `move_left(a,i)` in the body of the main `while` loop, so that on each iteration the current item is removed and

```
# Sort the items in array a

1:      def isort(array)
2:          a = array.clone
3:          i = 1
4:          while i < a.length
5:              move_left(a, i)
6:              i += 1
7:          end
8:          return a
9:      end
```

Figure 4.8: *Ruby implementation of the insertion sort algorithm. The* `clone` *method, called on the first line in the body of the method, makes a copy of the array passed as an argument so* `isort` *does not modify the array.*

reinserted somewhere to the left. The code for this method, which we call `isort`, is shown in Figure 4.8. For the experiments in this section, we will attach a probe in the main loop, and watch how `move_left` picks up each successive item and finds a place for it in the sorted region.

Tutorial

T34. Make a small array of strings to use in the first experiments with `isort`:

```
>> cars = ["mazda", "ford", "bmw", "saab", "chrysler"]
=> ["mazda", "ford", "bmw", "saab", "chrysler"]
```

T35. The first test is just to make sure the method works as advertised:

```
>> isort(cars)
=> ["bmw", "chrysler", "ford", "mazda", "saab"]
```

T36. An important detail is that a call to `isort` does not modify the original array; instead, it returns a new array with all the items from the original array in sorted order. Verify this claim by asking for the value of the `cars` array again, to see that it hasn't changed:

```
>> cars
=> ["mazda", "ford", "bmw", "saab", "chrysler"]
```

A listing of the `isort` method is shown in Figure 4.8. If you want, you can print your own listing by calling `Source.listing("isort")`.

Recall from the previous section that a method named `brackets` will print brackets around part of an array. We can use `brackets` to show us the region of the input array that has not been sorted yet.

T37. As a reminder of how `brackets` works, call it to put brackets around the part of the test array that starts at location 2 and goes to the end of the array:

```
>> brackets(cars, 2)
=> " mazda   ford [bmw   saab   chrysler]"
```

T38. The first statement in the main loop is at line 5. Attach a probe to this line to ask Ruby to call `brackets` to show the already-sorted part (to the left of `i`) and the to-be-sorted part (from `i` to the end of `a`):

```
>> Source.probe( "isort", 5, "puts brackets(a,i)" )
=> true
```

T39. Use `trace` to monitor the progress of `isort` when it sorts the list of car names:

```
>> trace { isort(cars) }
   mazda [ford   bmw    saab   chrysler]
   ford   mazda [bmw    saab   chrysler]
   bmw    ford   mazda [saab   chrysler]
   bmw    ford   mazda  saab [chrysler]
=> ⌊"bmw", "chrysler", "ford", "mazda", "saab"]
```

Note how items to the left of the bracketed region are all sorted, how the unsorted region between brackets shrinks by one word each time through the loop, and how each iteration takes the first item from the unsorted region and finds the correct place for it in the sorted region to the left.

T40. If you'd like to do more tests on your own, try sorting some arrays of numbers. You don't need to save the array, just pass it directly to `isort`:

```
>> isort( [3, 5, 1, 7, 8, 0, 6] )
=> [0, 1, 3, 5, 6, 7, 8]
```

4.4 Scalability

The projects in the previous two sections are similar to the tests programmers use when they are developing software. By using "test data," in the form of small arrays, programmers can see how the algorithms work and verify the implementations are correct. But these experiments don't give any insight into how the methods will perform when they are used on real data, where the arrays might be much longer than the test arrays. Unfortunately, it is often the case that software will work well for short test data, but not for longer inputs. It would be a disaster for a company if the software developers use a dozen songs to test an application that manages music libraries, only to find the program crashes when users try to import thousands of tunes, or that it takes the program several minutes to sort lists with more than a few hundred songs.

The ability of an algorithm to efficiently solve increasingly larger problems is an attribute known as **scalability**. The term refers to the fact that we want algorithms to continue to work well as the size of the input "scales up."

The experiments in this section will give us some insight into how well the algorithms used in the `search` and `isort` methods work on longer arrays. Since the key step in each method is the one that compares two array elements, we will measure performance by counting the number of comparisons made when a method is called.

Counting comparisons in a linear search is straightforward. The method makes one comparison on each iteration, so the number of lines printed by `trace` corresponds to the number of comparisons. In the worst case, when `search` does not find what it is looking for, the method makes n comparisons, where n is the number of items in the array.

Counting the number of comparisons required to sort an array is not as simple. The output from the `trace` experiments in the previous section show one line for each iteration of the `while` loop, but the problem is the algorithm does not always do the same number of comparisons on each iteration. Each line printed by `trace` is the result of a call to `move_left`, the "helper method" that moves each item to its correct location, and this method might make a different number of comparisons each time it is called. If we want an accurate assessment of the total number of steps, we need to take a closer look at what happens inside `move_left`.

```
i = 1
while i < n
    move_left(a, i)
    i += 1
end
```

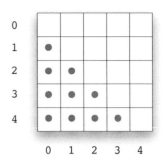

Figure 4.9: *The dots in the matrix represent potential comparisons made by* isort *as it sorts an array of five items. Each row represents comparisons made by a single call to* move_left, *the helper method that does the comparisons. The label next to a row is the value of* i *passed to* move_left. *For example, when* i *is 2,* a[2] *might have to be compared with* a[1] *and* a[0].

A call to move_left will remove an item from the array and then scan to the left to find a location to reinsert it. Each time move_left is called we pass it i, the index of the item that will be moved. The key observation is that move_left will make as many as i comparisons to find a home for item a[i]. For example, when i is 3, move_left might have to compare a[3] with every item to its left: a[2], a[1], and a[0].

If we add up the number of comparisons that isort would make to sort an array with n items, assuming that each item is compared the maximum number of times, the total will come to $n \times (n-1)/2$. This equation can be simplified to $n^2/2$ if we just want a rough estimate of the number of comparisons instead of an exact count.

The picture in Figure 4.9 shows why $n^2/2$ is an accurate estimate. The matrix on the right side of the figure has a series of rows, labeled 0 to $n-1$. The number in front of a row refers to an iteration, *e.g.*, the row labeled 3 shows what could happen when i is 3. The blue dots indicate the three potential comparisons made by move_left. There are dots in columns 0, 1, and 2 because move_left might have to compare a[3] with a[2], a[1], and a[0].

The isort method initializes i to 1, which means there will never be a call to move_left when i is 0. But there is a row labeled 0 in the diagram to emphasize the fact that the table has n rows and n columns, where n is the number of items in the array being sorted. If we were to draw a line from the upper left to the lower right, it would divide the matrix in half. As you can see, the dots fill the entire region below the diagonal, and since there are n^2 cells in the matrix the total number of dots is roughly $n^2/2$.

We now have formulas that estimate the number of comparisons made by our two algorithms as they operate on arrays of size n. Linear search will make up to n comparisons, and insertion sort up to $n^2/2$ comparisons. The equations are simply estimates, because the actual number depends on the contents of the arrays. Linear search will do fewer comparisons when the item it is looking for is found near the front of the array. The helper method used in insertion sort might only have to do one comparison, if an item is already in the right location.

When computer scientists write formulas to describe roughly how many comparison steps are made by these algorithms, they use the notation $\mathcal{O}(n)$ or $\mathcal{O}(n^2)$. The $\mathcal{O}$ comes from the phrase "on the order of." When we read these equations out loud we say "oh of n" or "oh of n squared." These equations refer to the *computational complexity* of an algorithm, and there is a very precise and formal definition for what it means to describe an algorithm as $\mathcal{O}(n)$ or $\mathcal{O}(n^2)$. In this book, however, we will just use the notation informally to give a rough sense of how the number of steps executed by an algorithm grows when it is applied to problems of a specified size.

The "big O" notation can be used to estimate scalability since it gives us an idea of what to expect as n, the size of the input, grows larger. For example, if we double the size of the input array passed to `isort`, we can expect it to make roughly four times as many comparisons to sort the longer list. When we increase the size of the input list by a factor of 10, the execution time will increase by a factor of 100. The reason the number of steps increases so dramatically is because the equation that predicts the number of comparisons includes the term n^2.

In the experiments in this section, we will use a probe to count the number of comparisons made during an insertion sort. The RubyLabs implementation calls the helper method `move_left` to look for the place to insert an item back into the list. The helper method calls another method, named `less`, to do the comparisons. If we want to count the number of comparisons made by `isort` all we have to do is attach a probe to the `less` method and use `count` to monitor the execution of a call to `isort`.

The arrays used to test the scalability of our two algorithms are a special type of array called a TestArray. To make a new test array, we just need to call `TestArray.new`, passing it a number that specifies how big to make the array. For example, to get an array with 100 numbers:

```
>> a = TestArray.new(100)
=> [778, 686, 840, ... 232, 408, 413]
```

Arrays for Testing Scalability

In experiments in this chapter and later in the book we are going to test searching and sorting methods using arrays with thousands of numbers. Since it would be inconvenient, to say the least, to type in an array with more than a few dozen values, RubyLabs has methods that make these arrays for us. The easiest way to make a TestArray is to call `TestArray.new`, passing it the size of the array you want:

```
>> a = TestArray.new(1000)
=> [8914, 7515, ... 2256, 4787]
```

You will get back an array with 1000 different random numbers. You can also get an array of car names, or fruits, or other types of strings:

```
>> a = TestArray.new(5, :colors)
=> ["sky blue", "firebrick", "forest green", "beige", "lime green"]
```

Tutorial

T41. If you need to, retype the expression from Problem T34 that makes a list of car names:
```
>> cars = ["mazda", "ford", "bmw", "saab", "chrysler"]
=> ["mazda", "ford", "bmw", "saab", "chrysler"]
```

T42. Recall from the last section what the trace method showed for the state of the array at the start of each iteration when isort was sorting the list of cars:
```
>> trace { isort(cars) }
 mazda [ford   bmw    saab    chrysler]
 ford   mazda [bmw    saab    chrysler]
 bmw    ford   mazda [saab    chrysler]
 bmw    ford   mazda  saab   [chrysler]
=> ["bmw", "chrysler", "ford", "mazda", "saab"]
```
isort made a total of eight comparisons. On the first iteration, "ford" was compared to "mazda", on the second "bmw" was compared to "mazda" and then "ford." But on the third iteration only one comparison was made ("saab" with "mazda"). Can you see why four comparisons were made on the last iteration, when looking for a place for "chrysler"?

T43. Type this expression to see the code for the less method:
```
>> Source.listing("less")
   1:    def less(x, y)
   2:       return x < y
   3:    end
=> true
```

T44. The listing shows the comparison is made on the second line, so attach the counting probe there:
```
>> Source.probe("less", 2, :count)
=> true
```

T45. Now we can ask Ruby to count the number of comparisons for us:
```
>> count { isort(cars) }
=> 8
```
So the count method confirms the analysis above, that 8 comparisons are required to sort this array of car names.

T46. Make an array of 10 numbers for testing isort by calling TestArray.new:
```
>> a = TestArray.new(10)
=> [36, 88, 0, 61, 35, 12, 54, 53, 92, 33]
```
Since the contents of the array are random the result you get will be different, but you should see an array of 10 different numbers.

T47. Pass your new test array to isort:
```
>> isort(a)
=> [0, 12, 33, 35, 36, 53, 54, 61, 88, 92]
```
Do you see a sorted copy of the array you created when you called TestArray.new?

T48. We can combine the two previous operations into a single step. We don't have to save the test array, we can just pass it directly to isort:
```
>> isort( TestArray.new(10) )
=> [6, 17, 20, 23, 27, 34, 42, 55, 61, 65]
```

Make sure you understand what Ruby did for that last expression: it called TestArray.new to make an array, and then that array was passed right to isort.

Repeat the previous expression a few times. You won't see the original test array, before it's sorted, but each time you enter that expression you should see a new random test array, printed in order because it was sorted by `isort`.

Next we'll use the counting probe attached earlier to count the number of comparisons made by `isort` when it sorts a test array.

T49. Enclose the previous command in braces and pass it in a call to `count` (this would be a good place to practice command line editing—all you need to do is go back one line, insert the braces and the word "count," and hit return):

```
>> count { isort( TestArray.new(10) ) }
=> 27
```

T50. Repeat the previous expression a couple more times. Do you see how each time Ruby executes this command it makes a random array of 10 items, sorts them, and prints the number of comparisons required to sort the array?

T51. We're now ready to do the experiments. Repeat the expression, but change the 10 to 100 to sort a test array of 100 numbers:

```
>> count { isort( TestArray.new(100) ) }
=> 2857
```

T52. Repeat the experiment, sorting lists of 200, 400, and 1000 items. Methods run noticeably more slowly when a probe is attached, so be prepared to wait a while for the longer sorts to complete.

T53. The section on scalability made the claim that doubling the size of the input list should quadruple the number of comparisons. Is the count for $n = 200$ approximately four times the count for $n = 100$? Is the count for $n = 800$ approximately four times the count for $n = 200$?

T54. When we increase the size of the array by 10, the number of comparisons should increase by a factor of 100. Is the count for $n = 1000$ around 100 times greater than the count for $n = 100$?

♦ If you have a spreadsheet application, make a table with one column for array size and another column for the number of comparisons used to sort an array of that size. Make a graph that shows how the number of comparisons grows as a function of array size. Do you see a curve that grows roughly like n^2 (Figure 4.10)?

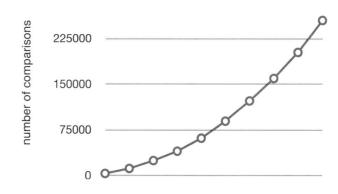

Figure 4.10: *Use a spreadsheet application to plot the number of comparisons needed to sort an array as a function of array size.*

4.5 ♦ Best Case, Worst Case

The discussion in the previous section, on the number of comparisons made by the insertion sort algorithm, used the worst case scenario to derive a formula for the number of comparisons: sorting an array with n items can need as many as $n \times (n-1)/2$ comparisons.

It's easy to see when this situation will arise. If the items in the array passed as an argument to isort are in reverse order, isort will have to move each item the maximum distance when it goes to find a place for that item.

This observation naturally leads to another question. What will happen if the array passed to isort is already sorted? In this case, the algorithm will make the fewest number of comparisons: the program only needs to make one comparison on each iteration, and it will take only $n-1$ comparisons to sort an array of n items.

These situations are both illustrated in Figure 4.11. The diagram on the left shows which items are compared when isort is given an array that is already sorted. Each row shows the state of the array at the start of an iteration. The current item, the one that will be moved to its proper place in the array, is highlighted in a dark gray square. Each item it is compared to is highlighted in a lighter gray. The picture clearly shows that isort only needs to compare each item with its neighbor to the left, doing one comparison per iteration, so in all only $n-1$ comparisons are made.

The diagram on the right uses the same highlight colors to illustrate the situation when the array is in reverse order. Each item is compared to every item to its left, until it is finally put back in the array at the leftmost location. As explained in the previous section, the total number of comparisons is $1 + 2 + 3 + \ldots + n - 1 = n \times (n-1)/2$.

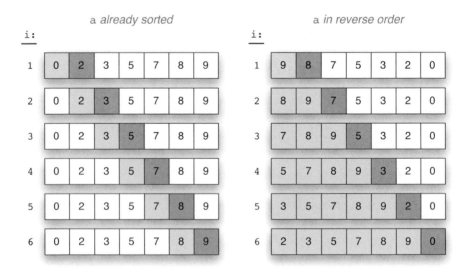

Figure 4.11: *The best case and worst case scenarios for insertion sort. A row labeled* i *shows comparisons made when the algorithm finds a place for* a[i]. *The dark gray square corresponds to* a[i], *the light gray squares are the items it is compared with.*

Tutorial Project

◆ Attach a counting probe to the helper method that compares items:
```
>> Source.probe("less", 2, :count)
=> true
```

◆ Make a test array with seven strings:
```
>> notes = ["do", "re", "mi", "fa", "sol", "la", "ti"]
=> ["do", "re", "mi", "fa", "sol", "la", "ti"]
```

◆ Count the number of comparisons required to sort this array:
```
>> count { isort( notes ) }
=> 12
```

◆ We can use Ruby's built-in `sort` method to make a sorted copy of the test array:
```
>> notes.sort
=> ["do", "fa", "la", "mi", "re", "sol", "ti"]
```

◆ To count how many comparisons are required to sort an array that is already in order, just pass `notes.sort` as the argument to `isort`:
```
>> count { isort( notes.sort ) }
=> 6
```

Do you see what Ruby did here? First, it called `notes.sort` to make the sorted copy of `notes`, and then it passed this copy as the argument to `isort`. As expected, there are 7 notes, and the sort was done with 6 comparisons.

It's easy to make an array with the items sorted in reverse order: first call `sort` to make a sorted copy of the test array, then call `reverse` on this sorted copy. Ruby will us do both operations in one expression.

◆ Type this statement to verify the fact that we can make a reverse-ordered copy of the test array with a single Ruby expression:
```
>> notes.sort.reverse
=> ["ti", "sol", "re", "mi", "la", "fa", "do"]
```

◆ Enter that expression as the argument to `isort` in the statement that counts comparisons:
```
>> count { isort( notes.sort.reverse ) }
=> 21
```

◆ Does the result printed in the previous problem agree with the formula for the worst case, *i.e.*, is it $n \times (n-1)/2$?

For the next set of experiments we are going to make several test arrays by calling `TestArray.new` to make a random array of numbers, and then passing that array to `isort`. These experiments will be easier to run if we have a method that combines these operations into a single step.

◆ Type this statement to make a new method named `istest` (which stands for "insertion sort test"):
```
>> def istest(n) isort( TestArray.new(n) ) end
=> nil
```

◆ Test your new method. A call to `istest(7)` should make an array of 7 random numbers and sort it:
```
>> istest(7)
=> [11, 13, 14, 66, 68, 70, 92]
```
You will see a different set of 7 numbers, but they should be printed in order.

♦ Now count the number of comparisons for a few different arrays of 7 numbers:

```
>> count { istest(7) }
=> 18
>> count { istest(7) }
=> 12
>> count { istest(7) }
=> 17
```

You will probably see a different result each time, since each call to `istest` makes a new test array.

♦ Are the counts you see from for these tests with $n = 7$ items between the minimum of $n - 1$ and the maximum of $n \times (n - 1)/2$?

♦ Here is an expression that tells Ruby to run the counting experiment 10 times:

```
>> 10.times { puts count { istest(7) } }
15
16
...
```

♦ What is the average value of the counts? Is it half way between the best case of $n - 1$ and the worst case of $n \times (n - 1)/2$?

♦ Repeat the previous exercise, but have Ruby count comparisons for test arrays with 100 or more items. What is the average count?

4.6 Summary

The main goals for this chapter were to introduce two important problems—searching and sorting—and to show how these problems can be solved by iteration, the same basic strategy used by the Sieve of Eratosthenes to make lists of prime numbers. For searching and sorting, the individual steps of an algorithm compare items in arrays or move items around in an array. What we learned is that repeated execution of these simple steps will eventually find an item in an array or rearrange the order of the items.

The two algorithms we looked at were linear search and insertion sort. An easy way to implement linear search in Ruby is to use an iterator method, like `each`, and let it "walk through" the array to compare items one after another. An alternative is to use a `while` statement to control the iteration, which is the approach taken in our implementation of a method called `search`.

The insertion sort algorithm has the same basic structure, with a `while` loop that iterates over each position in an array, but on each iteration it removes an item and reinserts it someplace closer to the beginning of the array. The main thing to understand about insertion sort is that at the start of each iteration an index variable points to the location of the element that will be moved. The region to the left of this location is already sorted, and the algorithm just has to find the correct location for the item somewhere in this sorted region.

An important new idea in computing introduced in this chapter is the concept of scalability. Linear search and insertion sort are easy to understand and easy to implement, so they are widely used for small data sets, but as the lengths of the input arrays grow longer the algorithms become less and less efficient. "Big-oh" notation, used by computer scientists to describe the computational complexity of an algorithm, is a quick way to estimate how many steps an algorithm will perform for a given problem size. Linear search is $\mathcal{O}(n)$, and

---

Concepts and Terminology Introduced in This Chapter

linear search	An algorithm that searches an array one item at a time, starting at the first location and working toward the end
insertion sort	An algorithm that sorts an array by repeatedly removing items and reinserting them in a continually growing sorted region at the front of the array
index	An expression that defines a location in an array; if `a` is an array in Ruby, the expression `a[i]` refers to the item at index location `i`
scalability	An attribute of an algorithm that determines how well it will perform on larger data sets
$\mathcal{O}$ **notation**	A formula that summarizes the number of steps an algorithm will execute for a particular problem size

---

the number of comparisons is proportional to the size of the array. Insertion sort is $\mathcal{O}(n^2)$, so the number of comparisons grows quadratically with the size of the array.

Scalability is an attribute of an algorithm, not a program. We did our experiments with Ruby methods that are based on the algorithms, but the same general results would have been seen no matter which language was used to write the methods that do a linear search or an insertion sort. Whether they are implemented in Java, C++, or any other language, these two algorithms will not give the best performance for large inputs. If an application is going to be searching or sorting very large arrays, we need to consider other algorithms, like the ones presented in the next chapter.

Exercises

1. Look around in the menus of some of the computer applications you use, and see if they have a search (or "find") function. What do you think the search function will do in a web browser? spreadsheet application? word processor? drawing program?

2. Do any of the menus in your applications have sort functions? What do they do?

3. Suppose an array is defined with this assignment statement:
   ```
   >> names = ["pete", "john", "paul", "george"]
   ```
   What are the values of the following Ruby expressions?
   a) `names.first`
   b) `names[0]`
   c) `names[2]`
   d) `names[4]`
   e) `names[2] == "george"`
   f) `names.last == "george"`

4. Do you think Ruby will let you change an item in an array using an assignment statement? What do you think will happen if you type this expression (assuming `names` has been defined as above)?

   ```
   >> names[0] = "ringo"
   ```

   Use IRB to check your answer.

5. Suppose an array is defined with this statement:

   ```
   >> gases = ["He", "Ne", "Ar", "Kr", "Xe", "Rn"]
   ```

   What are the values of the following Ruby expressions?

   a) `gases.include?("Ne")`

   b) `gases.include?("Fe")`

   c) `gases.index("Ne")`

   d) `gases.index("Xe")`

   e) `gases.index("Rb")`

6. Suppose an array is defined with this statement:

   ```
   >> languages = ["perl", "python", "ruby", "java", "c++"]
   ```

   Explain the meaning of each of the following expressions. First give a one-sentence description of what Ruby would do to evaluate the expression, then describe what would be printed in your terminal window if you were to type the expression in IRB.

   a) `languages.each { |x| puts x }`

   b) `languages.each { |x| puts x if x.length < 5 }`

   c) `languages.each { |x| puts "found it" if x == "fortran" }`

7. Briefly summarize, in one or two sentences, how Ruby would execute the following `while` loop:

   ```
   i = 0
   while i < 10
     puts i * i
     i += 1
   end
   ```

8. Suppose an array `a` is defined with this statement:

   ```
   >> a = [11, 0, 6, 12, 7, 8, 3, 15, 4, 10]
   ```

   How many comparisons will be made by the following searches using the linear search method?

   a) `search(a, 0)`

   b) `search(a, 3)`

   c) `search(a, 9)`

   d) `search(a, 12)`

   e) `search(a, 10)`

9. What would happen if the `search` method is asked to search through an array that contains duplicate entries? Explain what the method would do if it searches for 7 in this array:

   ```
   [6, 2, 7, 8, 2, 9, 7, 4, 5]
   ```

10. Use the insertion sort algorithm to sort a set of playing cards. Choose 7 cards from the same suit, and lay them out face up in a row on a table in front of you. As you work your way through the array, slide the cards around on the table, using the process illustrated in Figure 4.7.

11. Assume an array is defined with this statement:

    ```
    >> halogens = ["F", "Cl", "Br", "I", "At"]
    ```

    Explain how the array would be sorted by a call to `isort`. The easiest way to do this is to show the lines that would be printed when a probe is attached. Here are the first two lines, to get you started:

    ```
    >> trace { isort(halogens) }
     F [Cl  Br  I  At]
     Cl  F [Br  I  At]
    ```

    Show the remaining lines in the trace. How many lines will be printed before the array is sorted?

12. Repeat the previous exercise, using this array:

    ```
    >> heavy = ["U", "Np", "Pu", "Am", "Cm", "Bk", "Cf"]
    ```

13. We already know Ruby makes a distinction between upper and lower case letters in variable names and method names. What do you think it will do with string objects? If Ruby doesn't care about upper and lower case, the following comparison would result in `true`:

    ```
    >> "MPEG" == "mpeg"
    ```

    How do you think Ruby will evaluate this expression? Use IRB to check your answer.

14. Use IRB to ask Ruby to compare a string that starts with an upper case letter with a string that starts with a lower case letter:

    ```
    >> "Fred" < "fred"
    ```

15. Using what you learned from the previous exercise, how do you think Ruby would sort this array?

    ```
    >> names = ["Mendeleev", "Pasteur", "Pascal", "da Vinci", "Darwin",
                "von Neumann", "Galileo"]
    ```

16. ♦ What is the *fewest* number of comparisons that could be made by `isort` if it is called to sort an array with 20 strings? with 50 strings?

17. ♦ What is the *most* number of comparisons that could be made by `isort` if it is called to sort an array with 20 strings? with 50 strings?

18. ♦ Suppose you decide to run an experiment in which you make five different random arrays, each with 1000 elements, and then use `count` to determine the number of comparisons `isort` will make. What do you expect the average number of comparisons to be?

19. ♦ Write a Ruby method that computes the sum of the numbers in an array. Test your method with small arrays you make yourself, then try it on some larger arrays made with `TestArray`.

20. ♦ Write a method to compute the average of a group of numbers in an array, and test the method as described in the previous problem.

21. ♦ The `sieve` method from Chapter 3 fails with a strange-looking error message if we pass it a negative number:

    ```
    >> sieve(-1)
    Errno::EDOM: Numerical argument out of domain - sqrt
    ```

    A better way to deal with this situation would be to simply return an empty list. Check out a copy of the method and add a conditional statement that returns an empty list if the integer passed as the argument is less than 2. Load the new version into IRB and test it by calling `sieve(1)` and `sieve(-1)`.

Chapter 5

Divide and Conquer

A new strategy: breaking large problems into smaller subproblems

The previous two chapters introduced important ideas about algorithms and how they can be implemented in a programming language like Ruby. The algorithms presented in these chapters—the Sieve of Eratosthenes, linear search, and insertion sort—all used a very straightforward approach to solving their respective problems. At the heart of the algorithm is a simple operation, such as dividing one number by another or comparing two objects, and the problem was solved by repeating these operations until the the final result was produced.

The problem solving strategy used by these algorithms is often characterized as "brute force." There is nothing very sophisticated or clever going on. The algorithms are effective because a computer can perform divisions and comparisons very quickly, and it was a simple matter of repeating these operations over and over until the problem was solved.

Brute force works well for searching through small collections or sorting short lists, but as the problem size grows it is going to be necessary to use a better approach. The important new concept introduced in this chapter is a problem solving strategy called **divide and conquer**, which breaks a large problem into smaller subproblems and then works on each subproblem independently.

A search algorithm known as **binary search** uses divide and conquer to look for an item in a list. The word "binary" in the name of the algorithm comes from the fact that when the list is divided it is cut into two equal pieces. The strategy is similar to what you would use to look for a word in a dictionary or for a book on a library shelf. As an example of how this type of search works, suppose the goal is to find a recipe that starts with the word "eggplant" in a box of recipe cards sorted by title. First pick a card near the middle of the box, and compare the title on that card with "eggplant." Suppose the title on the card is "minestrone." Since the cards are sorted, any recipes starting with "eggplant" must be somewhere closer

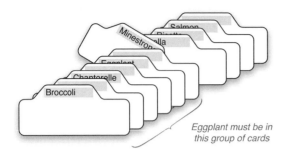

Figure 5.1: *If a set of cards is in alphabetical order we can use a binary search to find a card. Start in the middle of the collection, and then search the cards before or after the middle card.*

Eggplant must be in this group of cards

to the front of the box. Pick another card somewhere between the "minestrone" card and the front of the box and check again (Figure 5.1). Repeating this process of picking a card in the middle of the current group of cards, and then focusing on cards before or after the chosen card, will eventually zero in on the eggplant recipe. This is an example of divide and conquer because at each step the process splits the recipes into two groups, one before the card it just selected and one after, and in the next round the search can be limited to one group and the other group can be ignored.

Several different sorting algorithms also use a divide and conquer strategy. One famous algorithm, known as **quicksort**, applies the divide and conquer strategy by breaking a list to be sorted into smaller lists and then sorting each sublist. Suppose the recipe box is dropped on the floor and the cards need to be sorted again so they can be placed back in the box. To use the quicksort algorithm, divide the cards into two piles, the first containing all the recipes with titles that start *A* through *M*, and the second containing titles *N* through *Z* (Figure 5.2). There is no need to organize these piles when the cards are divided. The goal at this point is to just look at the title on a card, and then place it on top of one of the piles. Next, push the *N* through *Z* pile to the side of the table, saving it for later, and repeat this process with the *A* to *M* pile. Divide the *A* to *M* group into smaller piles, one for *A* through *F* and the other for *G* through *M*. Then push *G* through *M* to the side, and break *A* through *F* into two more smaller groups, *A* to *C* and *D* to *F*. Keep repeating this process of making two piles, saving one, and continuing with the other. Eventually there will only be three or four cards in a pile, and it can be sorted quickly and put back in the box. At this point, go pick up the most recently saved pile and start the dividing process again with this pile.

The sorting algorithm we are going to look at in this chapter is known as **merge sort**. Quicksort can be characterized as a "top down" approach, since it starts with a big problem and breaks it into smaller and smaller pieces. Merge sort takes the opposite approach. It works "bottom up" by starting with small groups and repeatedly merging them into bigger and bigger lists until it finally has just one complete sorted list. Both approaches have advantages, and depending on the situation either one might be the best algorithm to solve a particular problem. We are going to look at merge sort simply because it is easier to follow the state of the sort, watching how smaller pieces are combined into bigger pieces, and easier to count the number of comparisons that are made.

The obvious question at this point is whether these algorithms are really that much more efficient than the simple iterative techniques presented in the last chapter. Does the divide and conquer strategy lead to fewer steps, or are these new algorithms going to end up doing the same number of comparisons and take the same amount of time? The answer is that for large arrays, binary search is far more efficient than linear search, and merge sort is far more efficient than insertion sort. In this chapter we will do a set of experiments that will count the number of comparisons made by the algorithms, and the results of the experiments will supply empirical evidence that the divide and conquer strategy pays off. We will also develop equations that characterize the scalability of the new algorithms and show that divide and conquer is more effective from a theoretical point of view, as well.

"Divide and conquer" is a general term that can be applied to many different situations. Binary search and quicksort are examples of a special type of divide and conquer, where each subproblem is a smaller version of the main problem. When a problem can be broken into two or more smaller problems, where each subproblem is exactly the same type of problem as the original, we say the problem is **recursive** or **self-similar**.

When a programmer uses the term "recursive" it refers to a particular way of writing a method to solve a recursive problem. If a method is recursive, it means that somewhere in the body there is a statement where the method calls itself. For example, one way to write

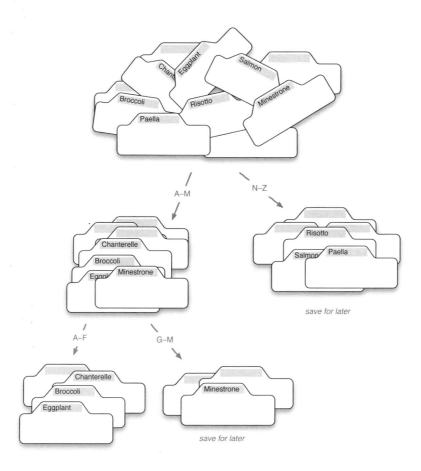

Figure 5.2: *To use the quicksort strategy to sort a set of cards, first divide them into two piles. Save the N–Z pile for later and use quicksort to sort the A–M pile. Then save G–M and use quicksort on the A–F pile. When a pile has only a few cards transfer them in order to the box, then go back and use quicksort to sort the most recently saved pile.*

a program to do a binary search would be to write a method named `rbsearch`. When the problem is reduced to searching one of two sublists, the new search can be done simply by passing the sublist in a call `rbsearch`, *i.e.*, the method calls itself.

The use of recursion can lead to some very elegant pieces of code that are easy to analyze and easy to understand. But learning to understand recursive methods takes a bit of practice, so the Ruby code we will use for the experiments in this chapter has been written without recursive calls. The important concept, of using divide and conquer to break a problem into smaller, more manageable pieces, will still be there, and it's accurate to say that the problems themselves have a recursive nature, it's just that to keep the programs simple we will use familiar iterative statements in our Ruby programs. For students who would like to see how recursive methods work, an optional section at the end of the chapter will show how binary search and quicksort can be written as recursive methods in Ruby.

5.1 Binary Search

The basic idea for a binary search is shown in Figure 5.3. The first comparison is made in the middle of the list. In this example, the number we're looking for is less than the one in the middle, so the second comparison is to the item halfway back toward the front of the list. The number there is less than the one we're looking for, so the algorithm moves forward, to a point halfway between the current location and the middle of the list. On each round, the sublist to check is half as big as the previous sublist.

There is an important caveat, however: for this type of search to work the collection must be arranged in order. If the items in the array are not sorted, the most reliable and efficient search is a linear search that checks each item, one after the other, starting at the front of the list. When we do experiments that apply binary search to arrays, we have to make sure the arrays are sorted or the method won't work.

To turn this general idea into an algorithm we need to figure out how to represent the idea of a "sublist" and how to break the current list into smaller sublists. We also need to address a detail that has been ignored in the discussion so far: how do we know to stop searching when the item is not in the list?

One way to implement sublists is to think of them as "regions" in the array passed as the argument to the search algorithm. The initial region will be the entire list, and as the

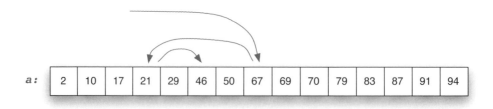

Figure 5.3: *The arrows show how a binary search looks for the number 46 in a sorted list of numbers. The first comparison is in the middle of the list; after that the search continually moves backward or forward, looking at sublists that are cut in half at each step.*

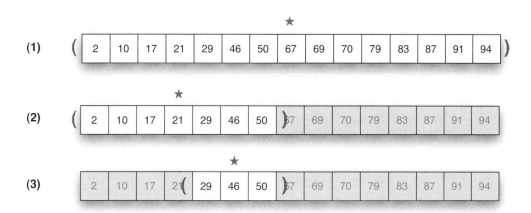

Figure 5.4: *The binary search algorithm uses three values to represent the current state of the search: the lower boundary of the current region (shown as an open parenthesis), the upper boundary (a closing parenthesis), and the index of the item in the middle of the region (indicated by the star). The figure shows three "snapshots" taken during the first three rounds of the search.*

algorithm progresses the region will get smaller and smaller. We will define two variables, named `lower` and `upper`, to mark the boundaries of the region that needs to be searched. At any point in the algorithm, `lower` will be one less than the index of the first item in the current region, and `upper` will be one more than the index of the last item in the region. Figure 5.4a shows the initial region for a search of an array with 15 items. Since the array indexes run from 0 to 14, the initial value of `lower` is −1 and `upper` is initialized to 15.

Finding the middle of the current region is simple: it's the index value that is halfway between `lower` and `upper`. A statement to compute this index is

$$\mathtt{mid = (lower + upper) / 2}$$

If the array item at this location, `a[mid]`, is the one we're looking for the search terminates with success.

The key step in the algorithm reduces the current region when the item we're looking for (let's call it `k`) is not at the middle of the region. The way to do this is to move one of the boundaries. If `k` is less than the value at `a[mid]`, we can restrict the region to search the lower region in the next round simply by setting `upper` to `mid`. On the next round of the search, `lower` will have the same value, but `upper` is now at a location we know is one past the last place where `k` can be found (Figure 5.4b). To search the upper half of the current region when `a[mid]` is less than `k` simply set `lower` to `mid` (Figure 5.4c).

The important thing to keep in mind about this algorithm is that `lower` and `upper` mark locations that are outside the region: `lower` is just to the left of the first item in the current region, and `upper` is just to the right. Each time the region shrinks, either `lower` or `upper` will move, and its new location is the place that was the middle of the current region. Since `mid` is half way between `lower` and `upper`, the region shrinks to half its size on each iteration of the algorithm.

```
# Search an array a to find an item k.  Return the location of k if it is found, otherwise nil.

 1:  def bsearch(a, k)
 2:    lower = -1
 3:    upper = a.length
 4:    while true
 5:      mid = (lower + upper) / 2          # find middle of region
 6:      return nil if upper == lower + 1   # fail if the region is empty
 7:      return mid if k == a[mid]          # succeed if k is at the mid
 8:      if k < a[mid]
 9:        upper = mid                      # next search: lower region
10:      else
11:        lower = mid                      # next search: upper region
12:      end
13:    end
14:  end
```

Figure 5.5: *Ruby implementation of the* bsearch *method.*

To answer the question about how to terminate the algorithm for an unsuccessful search, when the item we're looking for is not in the list the region will shrink all the way down to a sublist of size zero. When this happens, lower will be at some location i and upper will be at the adjacent location i + 1, and there will be nothing in the region between them. Thus the method will return nil when lower is one less than upper.

A listing of bsearch, our implementation of binary search in Ruby, is shown in Figure 5.5. After initializing lower and upper so they are next to the first and last elements in the array, the method goes into a while loop. At the start of each iteration, the method checks to see whether the current range has shrunk all the way down to zero items. If the item at the midpoint of the range is the one being searched for, the method returns the midpoint location. Otherwise the range will be cut in half, by moving the lower or upper boundary to the midpoint of the current range, and the execution continues on to the next iteration.

The Ruby code on lines 8 through 12 of Figure 5.5 are part of an **if statement**, a form of conditional execution we have not seen before. The section on conditional execution in the Ruby Reference (page 362) has more information about how these statements are executed.

Tutorial Project

The module that contains the methods used for the projects in this chapter is named RecursionLab. When you start an IRB session remember to include this module:

```
>> include RecursionLab
=> Object
```

Like the search method from the previous chapter, a call of the form bsearch(a, x) will look for x in an array a. The return value will be the location of x, if it is found, or else nil if x is not in a.

T1. Make a small array of strings to use for the initial tests to make sure the method works:

```
>> consonants = ["b", "c", "d", "f", "g", "h", "j"]
=> ["b", "c", "d", "f", "g", "h", "j"]
```

T2. Use `bsearch` to find the locations of some of these letters:
```
>> bsearch(consonants, "d")
=> 2
>> bsearch(consonants, "h")
=> 5
```

T3. Look for a letter that is not in the array:
```
>> bsearch(consonants, "e")
=> nil
```

In the experiments in Chapter 4 we used a method named `brackets` to print an array with brackets around some if the items. When we call `brackets` in this lab project we will pass it an additional location, which will indicate the position of the last item in a region (previously `brackets` just assumed the region went all the way to the end of the array).

T4. To see what the `brackets` method does, type this expression to print the test array with an opening bracket next to item 2 and a closing bracket next to item 4 (don't forget the first item in an array is item 0):
```
>> puts brackets(consonants, 2, 4)
 b  c [d  f  g] h  j
=> nil
```

T5. If you pass a third argument to `brackets`, it will be used as the value of the midpoint of a region, and `brackets` will put a star next to that item:
```
>> puts brackets(consonants, 0, 6, 3)
[b  c  d *f  g  h  j]
=> nil
```

We're now ready to trace the execution of a call to `bsearch`. What we want to do is attach a probe and have it print the state of the array (using Ruby's `puts` method) right after the method calculates the location of the midpoint of the current region.

According to the listing (Figure 5.5) the midpoint is computed on line 5, so we will attach the probe on line 6 (if you want, call `Source.listing("bsearch")` to print the code in your terminal window).

T6. Type the expression that attaches the probe (you can type the entire expression on a single line; it's broken into two lines here to fit within the margins of the book):
```
>> Source.probe( "bsearch", 6,
      "puts brackets(a, lower+1, upper-1, mid)" )
=> true
```

Note that `lower` and `upper` are locations next to a region, so the values passed to `brackets` are adjusted so it attaches the brackets to the proper items.

T7. With the probe in place, trace the execution of a successful search:
```
>> trace { bsearch(consonants, "d") }
[b  c  d *f  g  h  j]
[b *c  d] f  g  h  j
 b  c [d] f  g  h  j
=> 2
```

This output shows the search took three rounds. Notice how the region being searched shrinks in half on each round. The return value of 2 is what we expect, because the string "d" is at location 2 in the array.

T8. Now trace an unsuccessful search:
```
>> trace { bsearch(consonants, "e") }
[b   c   d *f   g   h   j]
[b *c   d]  f   g   h   j
 b   c  [d]  f   g   h   j
 b   c   d []  f   g   h   j
=> nil
```
Did this work the way you expected? Do you see why there is an empty region between the "d" and the "f" on the last line?

T9. Try several more searches on your own, doing searches for letters that are in the list and others for letters not in the list, until you are sure you know how the algorithm works.

T10. The comparison step in this method is on line 7. Let's attach a counter to this statement:
```
>> Source.probe( "bsearch", 7, :count )
=> true
```

T11. Count the number of comparisons made in the successful search:
```
>> count { bsearch(consonants, "d") }
=> 3
```
Look back at the output from the trace. Does this output look correct? Does bsearch make three comparisons when searching for the letter "d"?

T12. Count the number of comparisons in the unsuccessful search:
```
>> count { bsearch(consonants, "e") }
=> 3
```
Can you see why this is also 3? The trace showed four lines were printed for this search. But since the region has shrunk down to 0 items there is nothing to compare on the last iteration, and the method returns by executing the statement on line 6 before it gets to the probe.

T13. The probe you created in exercise T6 should still be active. Use a trace to see what happens if you try to do a search in an unsorted array:
```
>> ua = ["m", "i", "d", "t", "o", "w", "n"]
=> ["m", "i", "d", "t", "o", "w", "n"]
>> trace { bsearch(ua, "m") }
[m   i   d *t   o   w   n]
[m *i   d]  t   o   w   n
 m   i  [d]  t   o   w   n
 m   i   d []  t   o   w   n
=> nil
```
The result of this search should have been 0, not nil. Do you see why the algorithm went astray?

T14. Try a few more searches in this unsorted array. Some may succeed, just by luck, but it should be clear why this algorithm requires the input array to be sorted.

♦ If you would like to look at the execution in more detail, change the tracing probe so it prints the values of lower, upper, and mid during each iteration:
```
>> Source.probe( "bsearch", 6, "p [lower, upper, mid]" )
=> true
```

♦ Trace the successful bsearch again with this new probe in place:
```
>> trace { bsearch(consonants, "d") }
[-1, 7, 3]
[-1, 3, 1]
[1, 3, 2]
```
Does this help explain how the regions are implemented, and how they become shorter and shorter?

5.2 Binary Search Experiments

It is not uncommon to see a newspaper or magazine article describe something as "growing exponentially." In everyday usage the phrase simply means something is increasing very rapidly. To a mathematician or scientist, however, the term has a very specific meaning: if something has exponential growth, the equation that explains the rate of growth has a term with one of the variables in an exponent. One of the simplest exponential equations is $y = 2^x$. The y in this equation grows very quickly indeed. Each time x increases by 1 the value of y doubles, because by definition $2^{x+1} = 2 \times 2^x$.

Exponential growth is relevant to this chapter because it gives us a way to appreciate how truly efficient a binary search can be. To see why, let's first turn the problem around, and ask how big a list we can search if we are given a "budget" of a certain number of comparisons. Suppose we know already, perhaps by running a binary search and counting the number of comparisons, that it takes c comparisons to search for something in a list of length n. With one more comparison, the algorithm would be able to search a list of twice the size, *i.e.*, with $c + 1$ comparisons we can search a list of $2n$ items. That's because the comparison in the first iteration would cut the list down from $2n$ to n items, and we already know that finding the item in a list of size n takes only c comparisons. Since the list size n doubles each time c increases by one, the equation that defines the size of the list we can search with c comparisons is $n = 2^c$.

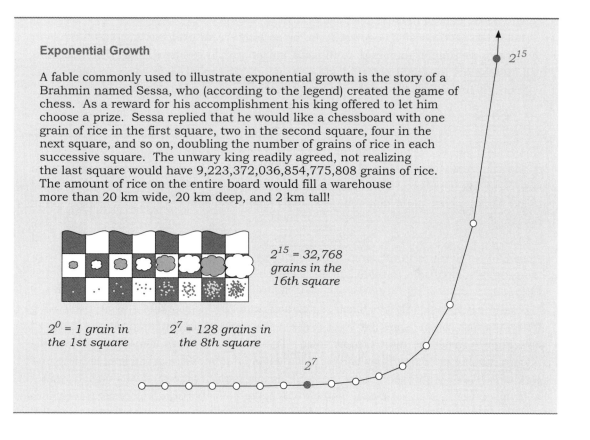

Exponential Growth

A fable commonly used to illustrate exponential growth is the story of a Brahmin named Sessa, who (according to the legend) created the game of chess. As a reward for his accomplishment his king offered to let him choose a prize. Sessa replied that he would like a chessboard with one grain of rice in the first square, two in the second square, four in the next square, and so on, doubling the number of grains of rice in each successive square. The unwary king readily agreed, not realizing the last square would have 9,223,372,036,854,775,808 grains of rice. The amount of rice on the entire board would fill a warehouse more than 20 km wide, 20 km deep, and 2 km tall!

$2^{15} = 32,768$ grains in the 16th square

$2^0 = 1$ grain in the 1st square

$2^7 = 128$ grains in the 8th square

2^{15}

2^7

n	$\lceil \log_2 n \rceil$
2	1
4	2
8	3
16	4
1,000	10
2,000	11
4,000	12
1,000,000	20
1,000,000,000	30
1,000,000,000,000	40

Table 5.1: *To estimate the number of comparisons required to search an array with n items, calculate* $\log_2 n$ *and round up to the nearest integer. A binary search through an array of 1,000,000 items will need at most 20 comparisons.*

Now let's go back to the original question, which is how many comparisons it will take to search a list of a given size. To answer this question, simply invert the equation we just derived. If $n = 2^c$, that means $c = \log_2 n$, using the definition of a logarithm. In other words, if we have a sorted list of n items, we can determine in $\log_2 n$ steps whether or not a particular item is in the list.

Unless n is a power of 2, $\log_2 n$ will not be an integer; for example, $\log_2 100$ is 6.64. In these cases we simply "round up" to the next integer, so our estimate is that 7 comparisons will be required to search an array with 100 items. The more formal way to express the relationship between c, the number of comparisons, and n, the length of the list, is

$$c = \lceil \log_2 n \rceil$$

where the notation $\lceil n \rceil$ means "the smallest integer greater than n," and is pronounced "the *ceiling of n*." This function grows very, very slowly. The list size n can be quite large—much larger than the number of items we can fit in a computer's memory—and a method based on the binary search algorithm will take only a few dozen comparisons to find any item in the list (Table 5.1).

In Chapter 4 the notation $\mathcal{O}(n)$ was used to describe algorithms where the number of steps grows linearly with the length of an input list. Because the binary search requires at most $\log_2 n$ comparisons to search an array with n items, this algorithm is characterized as $\mathcal{O}(\log_2 n)$.

Binary search could, like the linear search algorithm, be lucky and find what it is looking for on the first comparison. But what is the worst case? What situations would cause an algorithm to make the most comparisons? The answer, for both algorithms, is that the highest number of comparisons will be made when an item is not in the list. In Chapter 4 we didn't bother with experiments involving unsuccessful searches: it was obvious that if a list has n items, a linear search method will make n comparisons before returning `nil`. What is so impressive about binary search is that, even for an unsuccessful search, there will be $\log_2 n$ comparisons before the algorithm determines the item it is looking for is not in the list.

The experiments in this section will help explain why binary search makes at most $\log_2 n$ comparisons. We will attach a probe to count comparisons, and then purposely choose an item we know is not in the list when we call `bsearch`. The large arrays we will use in these tests will be a special type of array, known as a TestArray (introduced on page 93 in Chapter 4). To make an array with n different randomly chosen numbers, we simply have to call `TestArray.new`. If we want to use a TestArray object for experiments with binary search, we have to make sure the array is sorted. The easiest way to do this is to use Ruby's built-in `sort` method. This expression makes a test array with 100 numbers and sorts it before saving it with the name `a`:

```
>> a = TestArray.new(100).sort
=> [4, 20, 41, 44, ... 953, 977]
```

Arrays made by calling `TestArray.new` have a method named `random` that generates a random value to search for. When we call `random`, we pass it an option to specify whether we want a number that is in the array or a number that is not in the array. Here is an example from an IRB session that makes a test array and then asks it for random numbers:

```
>> a = TestArray.new(5)
=> [60, 31, 46, 83, 43]

>> a.random(:success)
=> 31

>> a.random(:fail)
=> 57
```

The symbol `:success` tells `random` to return a value that will lead to a successful search, and the symbol `:fail` tells `random` to pick a value that is guaranteed not to be in the array.

It may seem strange to ask a test array to find a random value and then turn around and call a second method to search for that same value. Here's one way to think of the situation. When a magician does a card trick, he asks a member of the audience to pick a card at random, look at it, and insert it back in the deck. He then does his "magic" to find the chosen card. We're doing something similar here. A call to `random` selects a number at random, and we can control whether or not the return value is actually in the array. After we have the value we will ask Ruby to do its magic, in the form of a binary search method, to look for that number. The reason we're doing this, of course, is that we want to know how many comparisons the binary search method makes, since we already know whether or not the item is in the array.

Tutorial Project

T15. The RubyLabs module has a method named `log2` that will compute $\log_2$ of a number:
```
>> log2(8)
=> 3.0
```
The result is 3 because $8 = 2^3$ and thus, by definition, $\log_2 8 = 3$.

T16. Call `log2` to compute the logarithms of some numbers that are not powers of 2:
```
>> log2(10)
=> 3.32192809488736

>> log2(50)
=> 5.64385618977472
```

T17. The Ruby math library has a method named `ceil` that computes the ceiling of a number. This is one of the methods that is called by writing the method name after an object. Modify the expressions typed in the previous experiment to get the nearest integer greater than the logarithms of those numbers:

```
>> log2(10).ceil
=> 4

>> log2(50).ceil
=> 6
```

T18. Use `log2` and `ceil` to compute some of the values shown in Table 5.1.

T19. Make a small array of numbers for the first set of tests. Don't forget to include the `.sort` at the end of the line:

```
>> a = TestArray.new(15).sort
=> [2, 6, 10, 11, 16, 17, 22, 33, 47, 57, 62, 64, 70, 85, 96]
```

As usual the array you get will be different, but you should see a sorted list of 15 numbers between 0 and 99.

T20. Ask the array for a random number that occurs somewhere in the array:

```
>> a.random(:success)
=> 17
```

T21. Repeat the call to `random` a few times. You should see a few different values from random locations throughout the array.

T22. Ask for some numbers that are not in the array. Repeat this expression a few times, making sure you get back a random number that is not in `a`:

```
>> a.random(:fail)
=> 41
```

When we use `trace` to monitor the progress of an unsuccessful search we want to do two things: select a random item and then call `bsearch` with this item. Ruby lets us put more than one statement between braces if we separate the statements with a semicolon. Because it might be helpful to know what `bsearch` is looking for, we are going to put three Ruby statements between the braces: one to get a random number, one to print the number, and one to call `bsearch`.

T23. Attach a probe to the first line in the main loop (this is the same statement that was used in exercise T6):

```
>> Source.probe( "bsearch", 6,
        "puts brackets(a, lower+1, upper-1, mid)" )
=> true
```

T24. Use `trace` to monitor an unsuccessful search:

```
>> trace { x = a.random(:fail); puts x; bsearch(a, x) }
67
[2   6   10   11   16   17   22  *33   47   57   62   64   70   85   96]
 2   6   10   11   16   17   22   33  [47   57   62  *64   70   85   96]
 2   6   10   11   16   17   22   33   47   57   62   64  [70  *85   96]
 2   6   10   11   16   17   22   33   47   57   62   64  [70]  85   96
 2   6   10   11   16   17   22   33   47   57   62   64  []  70   85   96
=> nil
```

This output shows us x (the item to look for) was set to 67. The next five lines were produced by the probe attached to `bsearch`. The last line shows the search region has shrunk down to zero elements in the place where 67 would be if it were in the list. Note also that four comparisons were made. No comparison was done in the last iteration (since there is nothing to compare).

T25. This same pattern should occur no matter where the missing item would be found. `bsearch` will always start with brackets around the entire array, and then each iteration will cut the region in half, until finally, on the fifth line, the region is empty. Repeat the previous expression several more times until you are convinced an unsuccessful search in this array of 15 items will always make four comparisons.

T26. If you want to try more experiments, make some new test arrays. Array sizes that are one less than a power of two (7, 15, 31, 63, *etc.*) are best for illustrating the divide and conquer strategy, but `bsearch` will work on any size array.

T27. Attach a counting probe to the line that does comparisons (this is the same statement that was used in exercise T10):

```
>> Source.probe( "bsearch", 7, :count )
=> true
```

T28. Count the number of comparisons required for an unsuccessful search:

```
>> count { x = a.random(:fail); bsearch(a, x) }
=> 4
```

So the probe agrees with the statement above—an unsuccessful search in an array of 15 items requires four comparisons.

T29. Is the number of comparisons for this array of 15 items the same as $\lceil \log_2 15 \rceil$?

```
>> log2(a.length).ceil
=> 4
```

T30. Let's repeat the experiment with a bigger array. Make a new test array, and then repeat the call to `count` from exercise T28:

```
>> a = TestArray.new(1000).sort
=> [1, 5, 11, 16, ... ]
>> count { x = a.random(:fail); bsearch(a, x) }
=> 10
```

T31. Ask IRB to compute $\lceil \log_2 1000 \rceil$? Is 10 comparisons right, *i.e.*, will `bsearch` make a maximum of 10 comparisons to search an array of 1000 items?

T32. How many comparisons will it take to search an array of 1,000,000 items? In other words, what is `log2(1000000).ceil`?

♦ Try some more experiments on your own, using even bigger arrays if you want. Is the number of comparisons for an array with n items equal to $\lceil \log_2 n \rceil$?

♦ Use the `time` method to measure the execution time for some of your test cases. Here is an example that shows how to compare the execution time of `search` and `bsearch`, assuming a test array `a` has already been defined (in this case the tests were done with an array of 1,000,000 numbers). The first line calls `search`, and the second line calls `bsearch`:

```
>> time { x = a.random(:fail); search(a, x) }
=> 0.467882
>> time { x = a.random(:fail); bsearch(a, x) }
=> 7.2e-05
```

The output of the second expression is Ruby's way of printing 7.2×10^{-5} seconds; see the sidebar on scientific notation on page 290.

♦ Do a systematic test of the number of comparisons and execution time of the two algorithms, using arrays of size 1000, 2000, *etc.* up to 10,000, and plot the results using a spreadsheet or some other graphing application.

5.3 Merge Sort

The merge sort algorithm uses the same general divide and conquer strategy that makes binary search so efficient. As explained in the introduction, the process works "bottom up" by first dividing the input array into several small chunks, and then combining the chunks into bigger and bigger groups until the final merged group includes the full array. In this section we will take a look at how to turn this general description into an algorithm, and see how the algorithm is implemented in Ruby. In the next section we will analyze the algorithm to come up with an equation that predicts the number of comparisons, and we will see that this strategy actually does lead to a much more efficient sort.

The key step in the algorithm is an operation that merges two small lists that are already in order into a single bigger list. Figure 5.6 shows how a merge is done, using stacks of cards with numbers on them to illustrate the process. In this example, there are two stacks of four cards, and each stack is in order, with the smallest card on top. The result of the merge will be a row of eight cards, in order from smallest to largest. The idea is to repeatedly take a card from one of the stacks and move it to the output area, stopping when all the cards have been moved. Since the two stacks are sorted, we only have to look at the top card in each stack when choosing the next item to place in the output. At each step, the merge process just moves the smaller of the two cards to the end of the output list—it doesn't have to look at any of the other cards.

The easiest way to understand how merge sort uses this list-merging process to rearrange the items in an array is to watch it in action. The RubyLabs implementation of merge sort is a method named `msort`. When we run `msort`, we can attach a probe that tells Ruby to print the array before each round of merges. This probe, like the ones in the previous experiments, will print brackets to show the progress of the algorithm. In the case of `msort`, that means printing brackets around each of the groups.

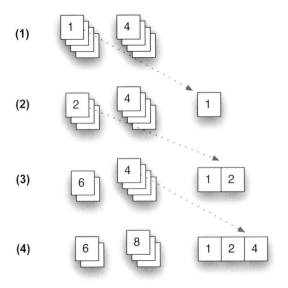

Figure 5.6: *Example of a merge operation that combines two stacks of cards into a single output list. If each stack is sorted, it is only necessary to compare the two cards on top of each stack to decide which card to move to the output list. In steps 1 and 2, the card on top of the left stack is smaller, so it is moved to the output. In step 3, the card on top of the right stack is copied.*

Figure 5.7: *This drawing shows how groups of size two are merged into groups of size four during a merge sort. To make it easier to see the groups, all the items in a group have the same color. The two-item groups start at locations 0, 2, 4, and 6. After the merges have been done, there are two groups of four items, at locations 0 and 4.*

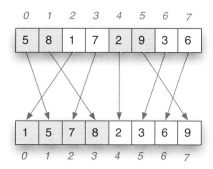

Here is an example that shows how an array of eight items is sorted:

```
>> a = [8, 5, 1, 7, 9, 2, 6, 3]
=> [8, 5, 1, 7, 9, 2, 6, 3]

>> trace { msort(a) }
[8] [5] [1] [7] [9] [2] [6] [3]
[5 8] [1 7] [2 9] [3 6]
[1 5 7 8] [2 3 6 9]
=> [1, 2, 3, 5, 6, 7, 8, 9]
```

The first line of the trace shows how there is initially one item in each group. The next line shows what happened when these groups were merged to form groups of size two. Notice how `[8]` and `[5]` were merged to form `[5 8]`, and how `[1]` and `[7]` were merged to make `[1 7]`. The following line shows how `[5 8]` and `[1 7]` were merged into a group of four items, namely `[1 5 7 8]`. The last step merged two four-item groups into the final sorted array with all eight numbers in order.

Figure 5.7 shows the second round of merges from this example in more detail. At the start of the round, the groups of two are adjacent pairs of numbers, where the first group starts at location 0, the second group at location 2, and so on. After the merge has been done, the array has groups of size four, starting at locations 0 and 4. The most important thing to notice about the rows in the figure is that the items within a group are always in order, so that on the next iteration they are ready to be merged into larger groups.

The Ruby code for `msort` is shown in Figure 5.8. A variable named `size` holds the current number of items in a group. It is initialized to 1, because the first round of merges should combine groups of size 1 into groups of size 2. The main loop of the algorithm, controlled by the `while` statement on line 4, keeps iterating until the group size has grown to include the entire array. The statement on line 6 multiplies `size` by 2, *i.e.*, it doubles the group size to get ready for the next round.

The merging operations are performed by a "helper method" named `merge_groups`. This method will find the starting location of each group, based on the current group size, and do all the movements required to merge adjacent groups. If you would like to learn more about how `merge_groups` works, you can read the description in the following optional section, otherwise skip ahead to the tutorial project to run `msort` and watch how it sorts a few test arrays.

```
# Return a sorted copy of array a

1:   def msort(array)
2:     a = array.clone
3:     size = 1
4:     while size < a.length
5:       merge_groups(a, size)
6:       size = size * 2
7:     end
8:     return a
9:   end
```

Figure 5.8: *Ruby implementation of the merge sort algorithm. The variable* size *represents the group size, which doubles on each iteration. The "helper method" named* merge_groups *combines groups of the specified size.*

♦ Implementation Details

The definition of `merge_groups` is shown in Figure 5.9. The method uses a `while` loop to iterate over all the groups in the array. On each iteration, the variable named `i` is the index of the first item in the first group to merge, and `j` is the index of the last item in the second group. For example, when the group size is two, groups start at locations 0, 2, 4, 6, *etc.* The first iteration merges the groups that start at location 0 and 2, so `i` is 0 and `j` is 3. On the next iteration, `i` will be 4 and `j` will be 7.

Another helper method named `merge` does the actual merge operation. Each call combines two groups, returning a new array with all the items in the merged group. The key step is the assignment statement on line 5:

```
a[i..j] = merge(a, i, gs)
```

This tells Ruby to replace the current items in locations `i` through `j` of the array with the items in the array returned by the call to `merge`.

Tutorial Project

T33. Make a small test array:

```
>> a = TestArray.new(8)
=> [7, 58, 89, 73, 87, 16, 54, 23]
```

As usual, your array will have a different set of random values, but you should see an array of 8 numbers.

```
# Merge all adjacent groups of size gs to form groups of size 2*gs

1:   def merge_groups(a, gs)
2:     i = 0                          # start of first group
3:     while i < a.length
4:       j = i + 2*gs - 1             # end of second group
5:       a[i..j] = merge(a, i, gs)    # merge groups at a[i] and a[i+g]
6:       i += 2*gs                    # start of next group
7:     end
8:   end
```

Figure 5.9: *The "helper method" that merges adjacent groups during merge sort.*

T34. To see how merge sort is called, use `msort` to sort the test array:
```
>> msort(a)
=> [7, 16, 23, 54, 58, 73, 87, 89]
```

T35. A statement of the form `puts msort_brackets(a, n)` asks Ruby to print a with brackets around each group of n items. Type this statement to print your test array with groups of size 2:
```
>> puts msort_brackets(a, 2)
[7 58] [89 73] [87 16] [54 23]
=> nil
```

T36. Repeat the previous statement, but use a group size of 4. Can you see how `msort_brackets` displays the array with brackets around groups of the specified size?

T37. Attach a probe to the first line inside the `while` loop in `msort`:
```
>> Source.probe( "msort", 5, "puts msort_brackets(a,size)" )
=> true
```

T38. With the probe in place, trace a call to `msort`:
```
>> trace { msort(a) }
[7] [58] [89] [73] [87] [16] [54] [23]
[7 58] [73 89] [16 87] [23 54]
[7 58 73 89] [16 23 54 87]
=> [7, 16, 23, 54, 58, 73, 87, 89]
```
Do you see how items within a group are always in order, and that groups in each line are formed by merging two adjacent groups from the line above it?

Notice how the trace produced three lines in the example above. Since the group size doubles on each iteration, it only took three rounds to have the group size go from 1, to 2, and then to 4. After merging the two groups of four, the entire array of eight items was sorted.

T39. This statement will create a test array of size 16 and sort it:
```
>> trace { msort( TestArray.new(16) ) }
[3] [66] [33] [60] [67] [25] [24] [10] ...
[3 66] [33 60] [25 67] [10 24] ...
...
=> [3, 5, 10, 17, 24, 25, 26, 33, 36, 49, 60, 64, 66, 67, 73, 79]
```
How many iterations did it take to sort this array?

T40. Repeat the previous statement, but make an array with 32 elements. How many lines are printed in each case? Be careful when you count lines. The lines printed by the trace will be long, and may wrap around on your terminal, but you should count 5 lines in all.

T41. It's easier to follow the "divide and conquer" strategy if the size of the array is a power of 2, but `msort` will work for any size array. Watch what happens when we ask it to sort an array of size 10:
```
>> trace { msort( TestArray.new(10) ) }
[60] [33] [1] [23] [56] [57] [47] [26] [48] [43]
[33 60] [1 23] [56 57] [26 47] [43 48]
[1 23 33 60] [26 47 56 57] [43 48]
[1 23 26 33 47 56 57 60] [43 48]
=> [1, 23, 26, 33, 43, 47, 48, 56, 57, 60]
```
The "extra" items form a small group of 2 that is put in order on the first round. The small group hangs around until all the items to the left are merged, and finally on the last step this small group is merged with the main group to make the final result.

5.4 Merge Sort Experiments

The trace printed by calls to `msort` in the previous section followed a pattern that should be familiar by now: when the array size is some value n, the number of iterations of the outer loop is $\log_2 n$. This should not be surprising. Each time through the loop the group size doubles, a hallmark of exponential growth. Since the `while` loop is controlled by the group size, the loop makes only $\log_2 n$ iterations before the group size reaches n.

As was the case for the other searching and sorting algorithms, we would like to develop an equation that predicts the amount of work required to sort an array of n items. The most straightforward estimate is to simply assume the inner loop does n steps, since each item in the current array is part of a group, and every group in the array is merged with one other group. If there are $\log_2 n$ iterations, and each iteration does n steps, the equation for the number of operations is $n \times \log_2 n$.

If we want to be more precise in the way we compare merge sort to insertion sort (the algorithm described in Chapter 4), we should count the number of comparisons made by merge sort, not simply the number of items moved around by the merge step. Here the situation is similar to insertion sort. Depending on how the input array is ordered, the merge operation will make anywhere between g and $2g$ comparisons to merge two groups of size g (see the sidebar on finishing a merge). But since the worst case is equivalent to making one comparison for each item copied to the output list, we know the number of comparisons is at most $n \times \log_2 n$.

Using the "big O" notation introduced in Chapter 4, we can characterize the scalability of merge sort as $\mathcal{O}(n \times \log_2 n)$. This is a significant improvement over the $\mathcal{O}(n^2)$ for insertion sort. Table 5.2 shows the maximum number of comparisons that will be made by both

---

Finishing a Merge

When the merge operation at the heart of merge sort gets to the end of one of the lists, it can simply move the remainder of the other list to the output, without doing any comparisons.

In the example shown here, the cards in the stack on the right are all less than 7, so they are all moved to the output list before the 7 is moved.

After the 5 is moved, the stack on the right is empty. Now all the cards from the stack on the left can be moved to the output list without any more comparisons.

In the best case, a merge of two groups of size g requires only g comparisons.

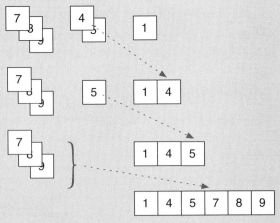

n	$n^2/2$	$n \times \log_2 n$
8	32	24
16	128	64
32	512	160
1,000	500,000	10,000
5,000	12,500,000	65,000
10,000	50,000,000	140,000

Table 5.2: *Estimates of the number of comparisons required to sort an array of* n *items using insertion sort (roughly* $n^2/2$ *comparisons) and merge sort (roughly* $n \times \log_2 n$ *comparisons).*

algorithms for some different array sizes. In each case, merge sort should make fewer comparisons. For arrays of 1000 or more items the difference is dramatic.

In the tutorial for this section we are going to do two sets of experiments. One will count the number of operations that move items to the output list as part of a merge, and the other will count the number of comparisons made during the sort. The results of these experiments will be similar to what we saw in the experiments with insertion sort in the last chapter: the equation $n \times \log_2 n$ is an upper bound that exactly describes the total number of steps, but for random arrays of data, the actual number of comparisons is lower than this upper limit.

To count the number of times an item is copied to the output array being built by the `merge` helper method we can use a new technique for attaching a probe:

```
Source.probe("merge", '<<', :count)
```

Instead of supplying a line number, as we have in previous experiments, we pass a string. This command tells RubyLabs to attach a probe to any line in the `merge` method that has a << operator. It turns out there are two of these lines, since there is one statement that copies from the first group used in a merge and a second statement that copies from the other group.

To count the number of comparisons, we will do the same thing we did in the last chapter, and attach a counting probe to the `less` method, which is the method that compares two items.

Tutorial Project

T42. If you are still in the same IRB session used for the previous exercises type this expression to remove all the existing probes:

```
>> Source.clear
=> true
```

T43. Attach a counting probe to every line in `merge` that uses the << operator:

```
>> Source.probe("merge", '<<', :count)
=> true
```

The `merge` method is fairly complicated, but if you're curious you can see the Ruby code by typing `Source.listing("merge")`. You'll see two statements that have the << operator, one for each group being merged.

T44. To make sure the probes were attached get a list of all current probes:

```
>> Source.probes
merge  7: count
merge 10: count
=> true
```

T45. Make a test array with eight items and sort it:

```
>> count { msort( TestArray.new(8) ) }
=> 24
```

T46. The number you see should be exactly $n \times \log_2 n$, because `msort` makes $\log_2 n$ iterations and each iteration copies n items to the merged groups. What is $\log_2 8$? Is 24 the correct count?

T47. Repeat the experiment, using arrays of size 16, 32, and 64. Is the number of operations consistently equal to $n \times \log_2 n$?

T48. Clear the counting probes:

```
>> Source.clear
=> true
```

T49. In Chapter 4 we saw that items from the input array are compared by a method named `less`. This statement will attach a counting probe to that method:

```
>> Source.probe("less", 2, :count)
=> true
```

T50. Repeat the statement that counts operations in a call to `msort`:

```
>> count { msort( TestArray.new(8) ) }
=> 17
```

You will probably get a different count, since the number of comparisons depends on whether or not a call to `merge` can simply copy the items at the end of one group.

T51. Repeat the previous expression several more times. The output should always be less than 24, since the number of calls to `less` should always be less than the number of times an item is appended to the output array. Is that what you saw?

T52. Try this new experiment several times each with arrays of size 16, 32, and 64. Is the number of comparisons always less than $n \times \log_2 n$?

If you did the optional exercises at the end of Section 4.4 and recorded the number of comparisons made by the insertion sort method (`isort`) in a spreadsheet, you can repeat the experiments here, except call `msort` instead of `isort`. Add a new column to the spreadsheet, and make a chart that displays the number of comparisons made by the two methods.

♦ Repeat this command once for each row in your spreadsheet ($n = 100, 200, \ldots, 1000$):

```
>> count { msort( TestArray.new(100) ) }
=> 551

>> count { msort( TestArray.new(200) ) }
=> 1318

>> count { msort( TestArray.new(300) ) }
=> 2218

. . .
```

♦ Plot your results. Do you get a general trend like the one shown in Figure 5.10?

Figure 5.10: *For arrays with fewer than 100 items there might not be much difference between insertion sort ($\mathcal{O}(n^2)$) and merge sort ($\mathcal{O}(n \times \log_2 n)$). For $n > 100$ there is a dramatic difference in the number of comparisons, as shown in this plot, which compares results from experiments with* `isort` *and* `msort`*.*

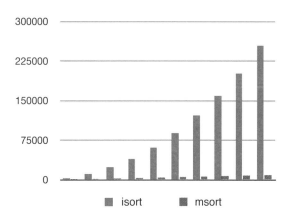

5.5 ◆ Recursive Methods

In programming, a recursive method is one that calls itself. At first this might seem like a recipe for disaster. Not only is it hard to imagine what such a thing might mean, it can also lead to an infinite loop. Here is the general outline of a recursive implementation of binary search; we'll call it `rbsearch`:

```
def rbsearch(a)
    ...                    # start here
       rbsearch( ... )
    ...
    end
```

When `rbsearch` is first called, it executes the statement on the line with the comment "start here," and then it makes the recursive call. On this call, Ruby again executes the line with the comment, and then makes another recursive call. Each time Ruby reaches the recursive call it goes back to the beginning, and it appears the computer has been caught in an endless cycle.

Infinite recursion is a problem to watch out for, but there are two simple rules to make sure it does not happen. First, the parameter value passed in the recursive call must be different than the one passed in the original call. In most cases recursion is used to implement a divide and conquer strategy, so the recursive calls are made with one of the smaller pieces of the input problem. Second, the recursive call needs to be part of an "if statement" that tests whether or not to make the call. When the recursive call gets smaller pieces of the input problem, the conditional checks to see if the problem has shrunk all the way down to the smallest possible problem. If so, the problem is solved without a recursive call, otherwise it is further divided and the pieces handed off to the recursive calls.

In the case of the two recursive methods we'll look at in this section, binary search obeys the two rules because it stops making recursive calls either when it finds what it is looking for or the region to search shrinks all the way down to the empty list, and the recursive call always looks at a region that is one half as big. Quicksort follows the same rules: it first checks to make sure the array has more than one item, and if so the array is divided into two smaller pieces for the recursive calls.

The Ruby code for these two methods is shown in Figure 5.11. Both of the methods illustrate a new construct that hasn't been used yet in this book. When a method is defined, we can specify default values for some of the parameters. For example, the statement on line 1 of `qsort` says the first parameter is always a reference to the array to sort, but the second two are optional. If they are not supplied, `p` will be initialized to 0 and `r` to one less than the length of the array. Since these two parameters mark the boundaries of the region to sort, this definition basically says that if a call to `qsort` contains only the name of an array the region should include the entire array. The body of `rbsearch` is very similar to `bsearch`. The main difference is that instead of using a `while` loop to make the region to search smaller and smaller the method uses a recursive call to search one of the regions.

All the hard work in `qsort` is implemented by a "helper method" named `partition`. In the brief description of quicksort given in the introduction to this chapter, the partitioning step was described as putting names that started with *A* to *M* in one pile and names starting *N* to *Z* in a second pile. That might work if we always know the method will sort strings, but what should it do for an array of numbers? The solution is to use the first item in the region as what is known as a "pivot" value. The partition step will move all the values in the array that are less than the pivot to the left side of the region and will move the pivot and all the values greater than the pivot to the right side. The dividing line between these two subregions is returned as the result of the call to `partition`, so now `qsort` just needs to make recursive calls to sort these two subregions.

```
# Search array a to find an item k in the region between lower and upper

 1:    def rsearch(a, k, lower = -1, upper = a.length)
 2:       mid = (lower + upper) / 2
 3:       return nil if upper == lower + 1        # fail if region empty
 4:       return mid if k == a[mid]               # succeed if k found
 5:       if k < a[mid]
 6:          return rsearch(a, k, lower, mid)
 7:       else
 8:          return rsearch(a, k, mid, upper)
 9:       end
10:    end
```

```
# Sort the region of array a bounded by p and r

 1:    def qsort(a, p = 0, r = a.length-1)        # sort region p..r
 2:       a = a.dup if p == 0 && r == a.length-1  # copy input array
 3:       if p < r
 4:          q = partition(a, p, r)     # q is boundary regions
 5:          qsort(a, p, q)             # sort small items in range p..q
 6:          qsort(a, q+1, r)           # sort large items in range q+1..r
 7:       end
 8:       return a
 9:    end
```

Figure 5.11: *Ruby implementation of two recursive methods. Note that the body of each method contains a call to the same method.*

Tutorial Project

♦ Make a small test array for `rbsearch`. Note that it must be sorted, since `rbsearch` implements the binary search algorithm:

```
>> a = TestArray.new(8).sort
=> [6, 21, 24, 28, 37, 39, 45, 47]
```

♦ Call `rbsearch` to try it out:

```
>> x = a.random(:success)
=> 21
>> rbsearch(a, x)
=> 1
```

♦ Try attaching some of the probes used to test `bsearch`, and run some more tests on your own to trace the execution and count the number of comparisons. You should see that `rbsearch` does exactly the same thing—it's just coded differently.

♦ Make an array for testing `qsort`:

```
>> a = TestArray.new(10)
=> [38, 0, 33, 39, 45, 2, 63, 36, 20, 15]
```

♦ Call `qsort` to sort the test array:

```
>> qsort(a)
=> [0, 2, 15, 20, 33, 36, 38, 39, 45, 63]
```

♦ Attach a probe to print brackets around the region to sort at the start of each call:

```
>> Source.probe("qsort", 2, "puts brackets(a, p, r)")
=> true
```

♦ Trace a call to `qsort`. Your output will be different, but you will see something like this:

```
>> trace { qsort(TestArray.new(10)) }
[16   55   17   2   13   58   25   9   48   43]
[9    13   2]  17  55   58   25   16  48   43
[2]   13   9   17  55   58   25   16  48   43
 2   [13   9]  17  55   58   25   16  48   43
 2   [9]  13   17  55   58   25   16  48   43
 2    9  [13]  17  55   58   25   16  48   43
 2    9   13  [17  55   58   25   16  48   43]
 2    9   13  [16]  55   58   25   17  48   43
 . . .
```

The first line shows that for the top level call the region is the entire array. The pivot is the first item in the array, in this case 16. The call to `partition` rearranges the array into two parts, putting numbers less than 16 on the left side and numbers greater than or equal to 16 on the right side. It then makes a recursive call to sort the left part of the array. The second line shows that all the numbers less than 16 have been moved to the front of the array and this region is now going to be sorted by the recursive call.

The third line follows the same pattern: the number 9 is the pivot, and the only value less than 9 has been moved to the front, and the recursive call is sorting this small region.

The most difficult thing to understand about recursion is that when a recursive call finishes, the method "pops up" to return to the most recent pending call. In this case, after sorting the region that has only the number 2, the program still has some unfinished business. It has to sort the values from the right side of the second call. That's shown on the fourth line above, where the current region has the two numbers 13 and 9. But don't forget that there is still another pending sort: the entire second half of the original array. The program eventually gets around to this item on its "to do list," as shown on line 7 of the output.

♦ Try tracing a few more examples, experimenting on slightly larger arrays.

♦ It may be hard to believe given this complicated looking output, but the total amount of work done by `qsort` is about the same as `msort`. Both algorithms make $\mathcal{O}(n \times \log_2 n)$ comparisons. It turns out that in practice quicksort is often even more efficient than merge sort, mainly because it uses the space allocated for arrays more efficiently. Try some timing experiments that sort an array with `msort` and then again with `qsort` to see the difference.

5.6 Summary

This chapter introduced two new algorithms, binary search and merge sort. These algorithms solve the searching and sorting problems that were introduced in the last chapter. The difference between the new algorithms and the searching and sorting algorithms from the last chapter is the problem-solving strategy: linear search and insertion sort use "brute force" iteration to repeatedly compare items and move them around in an array, but binary search and merge sort are based on a more sophisticated divide and conquer strategy.

Binary search is very similar to what people do when looking for an item in a sorted list, such as a dictionary or phone book. The search starts in the middle of the array. If the item being searched for happens to be at that location, the search is done. Otherwise, since the collection is sorted, the problem has now been divided in two. The search will continue in one half of the array, and the other half can be ignored. By repeatedly checking the middle of a region and continuing with a new region that is half the size the problem will be solved with far fewer comparisons. In the worst case, only $\lceil \log_2 n \rceil$ comparisons are needed to search a sorted array of n items.

Merge sort starts by repeatedly merging small groups, each of which is already sorted, into larger groups. The fact that the small groups are sorted makes the merge an efficient operation, since the comparisons only need to be made to the items at the start of each group. By the time merge sort has finished, it will have made at most $n \times \log_2 n$ comparisons to sort an array with n items.

Scalability, a concept introduced in the last chapter, is the motivation for studying binary search and merge sort. For small arrays containing only a few dozen items, linear search and insertion sort are perfectly adequate. In fact, since they are simpler to implement, they may even run faster. But as arrays become longer, the more sophisticated algorithms will be much more efficient and do far fewer comparisons.

Binary search is a natural way to solve the problem of finding an item in an ordered collection, and it is something people have been doing intuitively for many years, probably since the first dictionaries were published. But merge sort and quicksort were invented by computer scientists who analyzed the problem of sorting arrays of numbers and were motivated to find more efficient algorithms that would work on large arrays. Merge sort was first described in 1945, by John von Neumann, and quicksort in 1960, by C.A.R. Hoare.

This raises an interesting question: if merge sort is such an effective algorithm for computers, wouldn't it also be an efficient way for people to sort real objects? If you drop your box of recipes, and you need to sort several hundred cards, would a merge sort be more effective than a more intuitive insertion sort as a way to reorganize the cards so you can put them back in the box? If you would like to pursue this question further, Exercises 15 and 16 below have some suggestions for how to organize a merge sort using real cards.

Concepts and Terminology Introduced in This Chapter

divide and conquer	A problem-solving strategy that breaks a problem into smaller pieces and addresses each subproblem separately
binary search	A divide and conquer search algorithm; it divides an array into smaller regions so it can search one region and ignore the other
merge sort	An algorithm that sorts an array by combining small groups into larger groups, using a bottom-up application of the divide and conquer strategy
quicksort	An algorithm that sorts an array through a top-down application of the divide an conquer strategy
recursive problem	A problem that can be broken into one or more subproblems that are each smaller instances of the main problem
recursive method	A method, written in a programming language like Ruby, where a statement in the body of the method is a call to the same method

---

Exercises

1. Do a web search for newspaper or magazine articles with the phrase "exponential growth." Is the term being used in a mathematical sense, *i.e.*, can the situation be described by an exponential equation, or is it being used in the colloquial sense, of "growing very rapidly"?

2. Here is an array with 15 numbers:

   ```
   1  3  9  25  26  27  29  32  48  53  64  82  88  94  95
   ```

   What sequence of values is compared by a binary search algorithm when it searches for the following values? You can either print the array with brackets and a star, the way Ruby would if `bsearch` is called with a probe attached, or draw a picture with arrows, as in Figure 5.3, or just list the sequence of numbers used in each comparison.

   a) 53
   b) 25
   c) 26
   d) 95
   e) 42

3. Show how a binary search would look for values in the following array of element names (see the note at the end of the previous problem):

   ```
   Ce  Dy  Er  Eu  Gd  Ho  La  Lu  Nd  Pm  Pr  Sm  Tb  Tm  Yb
   ```

   a) Pm
   b) Nd
   c) Dy
   d) Rb
   e) Ce

4. Here is an unordered array of numbers:

    23 53 39 71 11 92 88 65 16 56 79 95 18 68 86

    The following numbers are in the array, but binary search won't find them. Show which items are compared to explain why bsearch gets lost:

    a) 88

    b) 18

    c) 39

5. Are there any items in the array for the previous problem that would be found, by luck, even though the array is not sorted?

6. Does bsearch always make fewer comparisons than search (the linear search method)? Given the array shown in Problem 2, is there any value that search would find with fewer comparisons than bsearch?

7. The equation given in Section 5.2 for the number of comparisons made by bsearch is correct in most cases, but there is one special situation where it is not quite right. Create a test array with 16 numbers, set a probe so bsearch prints the array with brackets on each iteration (as shown in Exercise T6), and do several searches for items that are not in the array, using this expression:

    >> trace { x = a.random(:fail); puts x; bsearch(a, x) }

    According to the equation, the searches should all use $\lceil \log_2 16 \rceil = 4$ comparisons. If you repeat the expression above often enough, you will encounter a search that takes 5 comparisons. Repeat the searches several more times, until you have three or four examples of searches that require 5 comparisons. Can you spot a pattern? Can you explain why these searches take 5 comparisons?

8. ♦ The issue described in the previous problem only occurs when n, the size of the array, is a power of 2. Can you devise a formula for the actual number of comparisons made by an unsuccessful search that will always be correct, even when n is a power of 2?

9. Below are several test arrays with 16 numbers. Show how the arrays would be sorted by a call to msort. The easiest way to do this is to show the groups before each round of merges, the way Ruby prints them when a probe is attached.

    a) 1 99 3 47 50 37 79 71 15 51 87 28 19 93 91 70

    b) 25 69 64 92 10 7 27 51 54 12 71 65 59 74 79 46

    c) 66 38 79 70 45 20 16 69 52 67 72 13 5 28 39 82

    d) 37 61 40 53 89 10 72 68 99 17 67 74 47 36 3 23

    e) 14 27 32 57 34 9 56 79 44 89 35 90 84 43 59 41

10. How many comparison or copy operations will msort make when it sorts an array of 128 numbers? Check your answer by attaching a counting probe to lines that match << in merge (see Exercise T43) and counting the number of operations in a test array with 128 numbers.

11. Repeat the previous problem, but with an array of 200 numbers, then with an array of 500 numbers.

12. ♦ What will happen if merge sort is asked to sort an array that is already sorted? Will it make fewer comparisons than for a random test array?

13. ♦ Which method will make fewer comparisons when it is passed an array that is already sorted, isort or msort? Explain why.

14. ♦ Can you develop a formula for the number of comparisons msort will make if it is passed an array that is already sorted?

15. If you want to try your hand at using merge sort to organize a real-world collection of objects, here is one way to do the sort. You can try this method with any group of objects that can be ordered: a deck of playing cards, recipe cards, homework papers that have been graded and need to be entered into a gradebook, *etc.*

    a) In the first round, pick up two items, put the smaller one on top, and set them face down on a table. Pick up two more items, put the smaller one on top, and put this group face down on top of the first group, but turn them so they are at a right angle to the first group. Keep picking up pairs of items, sorting them, and placing them at right angles on top of the pile until all items are in a group of two (unless there is an odd number, in which case just put the last item on the pile in its own group of one).

    b) Turn the pile over, so it is face up.

    c) Take the top two groups off the pile and place them side by side on the table. Remove the smaller of the two items showing on the top of a group and place it face down on the table. Repeat this step until the two groups have been merged into a new group, which will be face down.

    d) Repeat the previous step, but this time, the new group should be at a right angle to the first one. Keep removing two groups, merging them, and building up the output pile with each new group at right angles to the previous one.

    e) When the last groups have been merged, begin a new round of merges by repeating this process from step (b). Eventually the output pile will be a single group with all the objects in order, at which point you are done.

16. ♦ If a person uses insertion sort to sort a group of real word objects (cards, papers, *etc.*) do you think they would do a linear search to find the place to insert each new object? Or could they use a binary search?

17. ♦ In Chapter 4, the process used by the helper method named `insert_left` was described as an iteration that required up to i steps, where i marks the current location in the sort. Since all the items in the array from 0 up to $i - 1$ are already sorted, would it be possible to use a binary search for this operation? Would the new version of insertion sort then be $\mathcal{O}(n \times \log_2 n)$?

Chapter 6

When Words Collide

Organizing data for more efficient problem solving

The binary search algorithm introduced in Chapter 5 is an efficient method for finding an item in an ordered collection. A real-world example of where this type of search is effective is looking for a book in a bookstore, where books in each section are arranged by the author's last name. To find a book on algorithms by an author named Knuth, we could pick a starting location near the middle of the computer science section, and then move forward or backward in the section, at each step narrowing the range of places where the book might be found.

Sorting the books by author name allows us to quickly find a book when we know who wrote it, but how do we find a book if we don't know the author? If all we know is that the book is called *Fundamental Algorithms*, we would have to scan the books one by one, comparing each title to the one we're looking for. No matter how the books are arranged, we can make an efficient search using the same criterion as the one used to order the books on the shelf, but searches based on other criteria will be much more time-consuming.

Libraries solve this problem by using a book numbering system to assign a unique number (the "call number") to each book and by arranging books on shelves according to their call number. In a modern library, a patron can type what they know about a book into a computer, and the system will return the call number so the person can search the stacks using the number.

Before computers, libraries kept descriptions of books in card catalogs, which were collections of cards that had the essential information about a book, including its call number. A typical library had three catalogs, one for titles, one for authors, and one for subjects. To find a book by Knuth, a patron would go to the author catalog, find the drawer for authors whose names started "Knu," and then search within that drawer—using a divide and conquer strategy, since cards were in alphabetical order—for cards labeled "Knuth." To find *Fundamental Algorithms*, the person would go to the title catalog (Figure 6.1), find the drawer labeled "Fun," and search within that drawer.

The point of this story is that how information is organized plays an important role in the solution of a problem. Putting descriptions of books on cards and storing the cards in several different catalogs allowed library patrons to use a new algorithm to search for a book. Without the catalog, a patron could do a binary search by author (if the books were sorted by author name) but could only do a linear search when looking for titles. With the additional information stored in the catalogs, a catalog-based algorithm made searches based on author, title, or keyword equally efficient.

While card catalogs may have gone the way of the dinosaurs, the need for organization has not. Computer programs that manage information about what is in a library use the computational equivalent of a card catalog. More generally, database management applications are designed to store millions of records. The people who use the information in a database often need to access it using a variety of different criteria. The applications cannot afford to spend a lot of time running search algorithms, so database management systems use sophisticated techniques to organize the data.

The important concept in computing introduced in this chapter is the idea of a **data structure**. The term refers to the fact that we often have a choice in how to organize the data used in a computation. The algorithms we've seen so far have all been based on one

Figure 6.1: *Libraries used to have a set of card catalogs to help patrons locate books. If a person wanted to find a book with a particular title, they would search for a card in the title catalog, get the book's call number from the card, and then search for the book in the stacks.*

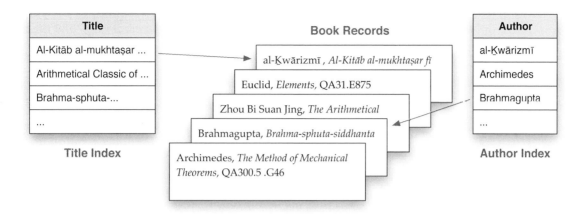

Figure 6.2: *An organization for information about books in a library. The main records have complete descriptions of each book, including author, title, publication date, call number, and other data. An index is a separate data structure with only author name or title, along with a link to one of the main records. A search for a book by a particular author finds an entry in the author index, and then retrieves the complete description in the record linked from the index.*

of the simplest data structures, in which items are arranged in a linear array. But often the most effective way to solve a problem requires more than simply coming up with a clever set of operations to move data around in an array. We also need to think of new ways of organizing the data. By arranging data differently, or adding additional information that helps make a more sophisticated organization, we can develop more effective algorithms that can be used to solve a wider variety of problems.

Algorithms and data structures both play an important role in developing computational solutions to problems. Just as having descriptions of books in card catalogs allowed library patrons to use a new algorithm for finding a book, creating new ways to organize information in a computer's memory will allow us to explore new algorithms based on alternative approaches to structuring the data. Understanding how the data for a problem can be organized, and knowing the properties of different data structures, is a crucial part of the process of solving the problem.

6.1 Word Lists

One way a library might organize the descriptions of its books is shown in Figure 6.2. The centerpiece of this organization is a set of records, each containing the essential information about a single book. For this example, we can think of a record as a piece of text that includes the author's name, title, call number, and any other information the library thinks is important. If we were to implement this structure in Ruby, we could use a single Ruby string object for each book, and make an array object to hold all the strings.

The computational equivalent of a card catalog is a secondary data structure, often called an **index**. The name comes from the fact that the data structure is similar to the index at the back of a book. The organization in Figure 6.2 has two indexes, one for book titles and one for author names. If patrons type a name in the "author" box in a web form, the system will search the author index for a matching name, and if it finds one, follow the link to the records for books by that author. Users can also type strings in a box labeled "title," and the system will search the title index.

The tutorial project in this chapter will explore a data structure that is often used to create an index. To keep things simple, we won't try to manage the entire organization shown in Figure 6.2, which would have records with complete book descriptions and arrows that link from the index to the main record. Instead, we'll just focus on the structure of the index itself. We will make a data structure, called a **hash table**, and use it to store a set of strings like the author names and book titles, but we won't try to associate those strings with any other information.

The fact that our data structure does not contain links to other data means it won't be an effective index, but it will still be a useful way of organizing information. One example of how to use such a structure would be as a **word list**. A word list is simply a collection of words, without reference to their meanings. Common examples in the real world are lists of words maintained by a spell checker or a crossword puzzle dictionary. People (or computer programs) who use these lists just want to know if a string is in the list or not. If it is, they can assume it is a correctly spelled word or a word that can be used in a puzzle.

The exercises in the first few sections will show how a hash table is organized, and how to add words to a table or search a table to see if it contains a word. At the end of the chapter we will be ready to see how well our hash tables work on large lists. The experiments will test the effectiveness of the structure when we try to add over 200,000 words from an English dictionary.

The words we will use for these experiments are accessible through a type of array called a TestArray, which was used in Chapters 4 and 5 to build arrays of random numbers. For the projects in this chapter, we will want arrays of random words. The way to make such a list is to pass a second argument to `TestArray.new`, the method that builds the array. The argument is a Ruby symbol that specifies what kind of words to include in the list.

Web2

A data file included with most versions of the Unix operating system since the 1980s is named web2. It contains a list of all the words from the 1934 edition of *Webster's New International Dictionary*, commonly referred to as "Webster's 2nd." The dictionary is now in the public domain, since the copyright has expired.

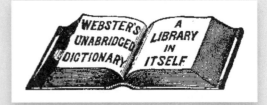

The RubyLabs software package includes a modified copy of web2. Single-letter words and capitalized words have been removed. The result is a file with around 210,000 English words with American spellings. Many of the words seem contrived by today's standards, and of course the list is missing new words ("downland" is in the list, "download" is not). In spite of these flaws, the word list is a useful source of data for experiments based on random words.

Here is an example. If we want a list of 5 car names, we pass the symbol `:cars` in the call to `TestArray.new`:

```
>> a = TestArray.new(5, :cars)
=> ["jaguar", "chevrolet", "porsche", "maserati", "audi"]
```

To make a list of 10 random words from the dictionary, pass the symbol `:words` as the second argument:

```
>> a = TestArray.new(10, :words)
=> ["shrewly", "palliobranchiate", ... "calligraphically"]
```

We can get the entire list of words of a specified type by using the symbol `:all` instead of a number. This is how we would get the complete list of car names:

```
>> a = TestArray.new(:all, :cars)
=> ["acura", "alfa romeo", ... "rolls-royce", ... "volkswagen"]
```

As the output from this last example shows, our definition of a "word" is somewhat flexible, since we are including hyphenated names and compound names like "Alfa Romeo" or "Aston Martin."

Tutorial Project

The module with the methods used for the projects in this chapter is named HashLab. When you start an IRB session remember to include the module:

```
>> include HashLab
=> Object
```

T1. Type this expression to make a list of names of 5 different cars:
```
>> TestArray.new(5, :cars)
=> ["mazda", "fiat", "lexus", "mg", "audi"]
```

T2. Repeat the previous command a few times. You should see a different list each time.

T3. This command will show a list of the symbols you can pass to `TestArray.new`:
```
>> TestArray.sources
=> [:cars, :colors, :fruits, :words]
```
More options may be added when the RubyLabs software is updated.

T4. Make a list of 10 random words from the English dictionary:
```
>> TestArray.new(10, :words)
=> ["measly", "aphidolysin", ... "amissness"]
```

T5. Repeat the previous command a few times.

You will see some odd-looking strings when you ask for a list of random words from the dictionary ("amissness" is a word?), but these strings will make a nice test set for the hash table experiments at the end of the chapter.

0	apple
1	
2	mango
3	
4	elderberry
5	
6	
7	
8	strawberry
9	

$h(\text{apple}) = 0$

$h(\text{mango}) = 2$

$h(\text{elderberry}) = 4$

$h(\text{strawberry}) = 8$

Figure 6.3: *A hash table with 10 rows. A row is either empty (indicated by the gray boxes in this figure) or it contains a single word. The location for each word is defined by a hash function.*

6.2 Hash Tables

Applications that use a hash table typically start with an empty table and then add entries over time. A spell checker might have an empty table for each user's specialized vocabulary, and as the user adds new words to their personal list the words would be inserted into the table. If the table is used for an index structure in a product database, new entries would be created as new products become available, but there could also be situations where data is removed from the table as products are discontinued.

A small hash table is shown in Figure 6.3. The table consists of a set of **rows**, each of which can contain a single word. Note that most of the rows in this table are empty; the table has 10 rows, but only four have been filled.

From the example it may look like words have been added to random locations in the table, but in fact each word has a designated location. The row where a word is stored is defined by a function, called a **hash function**. The input to the function is a word to store in the table, and the output is a row number. For example, the hash function used to build the table in Figure 6.3 determined that the word "mango" would go in row 2:

$$h(\text{mango}) = 2$$

The same hash function is used to look up words. To look for a word s, simply compute $i = h(s)$ and then check row i of the table. If s was added to the table earlier, we will find it in row i, but if the row is empty, it means s is not in the table. For example, to look up the word "strawberry," compute

$$h(\text{strawberry}) = 8$$

Since "strawberry" is in row 8 of the table, we have found the word we are looking for. To see whether "guava" is in the table, compute

$$h(\text{guava}) = 6$$

Since row 6 is empty we can conclude "guava" is not in the table.

The big question, of course, is how to define the function that does the mapping from words to row numbers. There are several other questions, including the issue of what to do if the hash function wants to store two different words in the same row, but we'll put these questions aside for the moment. The tutorial exercises in this section will explore the basic table structure, and we'll take up the question of how to define a hash function in the next section.

If we want to build a hash table in Ruby, a natural choice is to use an array. Ruby arrays are collections of objects, and we can store any type of data in an array. The words we store in our hash tables will be represented by Ruby string objects. Hash functions will be implemented by Ruby methods that take a string as a parameter and return a number between 0 and $n - 1$, where n is the size of the array.

In previous chapters, we saw how to make array objects by writing a set of values between square brackets or by passing a range to a method that made a list of all numbers in a certain range. A third way to make an array is to call a method named `Array.new`, passing it an integer that specifies how big to make the array. This expression creates an array of size 10 and saves it in a variable named `t`:

```
>> t = Array.new(10)
=> [nil, nil, nil, nil, nil, nil, nil, nil, nil, nil]
```

Each cell in this array contains the special object `nil`, which stands for "nothing" or "no object," so calling `Array.new` is exactly what we need to make an initially empty hash table. In previous chapters `nil` was used to indicate a search was unsuccessful, but here we are using it as a placeholder to use in a row that does not yet contain a string.

The HashLab module has a method named `h` that implements the hash function used to add words to the table shown in Figure 6.3. To build the example table, we just need to pass a string to `h` to find a location, and then use an assignment statement to put the string in the table at that location:

```
>> i = h("mango")
=> 2
>> t[i] = "mango"
=> "mango"
```

To see if a table row is empty, we can use a method named `nil?` (the question mark is part of the method name). For example, after storing a string in location 2 of the array `t`, `nil?` shows us that row is no longer empty, but others are still empty:

```
>> t[2].nil?
=> false
>> t[3].nil?
=> true
```

```
# Add string s to hash table t

1:  def insert(s, t)
2:    i = h(s)
3:    if t[i].nil?
4:      t[i] = s
5:      return i
6:    else
7:      return nil
8:    end
9:  end                    initial version
```

```
# Return the location of s in hash table t

1:  def lookup(s, t)
2:    i = h(s)
3:    if t[i] == s
4:      return i
5:    else
6:      return nil
7:    end
8:  end                    initial version
```

Figure 6.4: *Preliminary versions of hash table methods. The* `insert` *method calls the hash function to find where to put a string, and the* `lookup` *method calls the hash function to see if a string is stored in that location. Use* `Source.checkout` *to obtain copies of these methods if you want to do further experiments on your own.*

A method named `insert`, shown in Figure 6.4, will add a string to a hash table. This initial version checks to see whether there is already a string in the table at the location computed by the hash function. If so, `insert` returns `nil` to indicate it could not insert the string. If the method was able to store the string in the table it returns the row number where s was saved. This expression adds a second string to our table `t`:

```
>> insert("strawberry", t)
=> 8
```

The result of this expression means the string was saved in row 8 of the table.

A method named `lookup`, also shown in Figure 6.4, will check to see if a word is in table. This expression asks Ruby to look for a string that was inserted previously:

```
>> lookup("mango", t)
=> 2
```

If the string is not in the table, the method returns `nil`:

```
>> lookup("kiwi", t)
=> nil
```

The easiest way to see the contents of a table is to just ask Ruby to print the array. This is what our table `t` looks like after adding the two strings:

```
>> t
=> [nil, nil, "mango", nil, nil, nil, nil, nil, "strawberry", nil]
```

(don't forget the first cell is `t[0]`, so location 2 is actually the third item in the array).

When we start experimenting with larger tables it's going to be hard to determine which cells are full simply by printing the entire table. The HashLab module defines a method named `print_table` that will display a table in the terminal window, printing a row number at the front of a line and skipping any empty rows:

```
>> print_table(t)
   2: "mango"
   8: "strawberry"
=> nil
```

Tutorial Project

T6. Type the following expression to define an array to represent a hash table with 10 rows:

```
>> t = Array.new(10)
=> [nil, nil, nil, nil, nil, nil, nil, nil, nil, nil]
```

Notice how the rows are all initially empty.

T7. Use the hash function to find where the following strings will be placed in the table:

```
>> h("grape")
=> 6
>> h("lime")
=> 1
>> h("plum")
=> 5
```

T8. Call the `insert` method to add the strings from the previous problem to `t`:

```
>> insert("grape", t)
=> 6
>> insert("lime", t)
=> 1
>> insert("plum", t)
=> 5
```

Can you see, by looking at the return value for these calls, how `insert` placed each string at the location defined by the hash function?

T9. The first location in the table should be empty:

```
>> t[0].nil?
=> true
```

T10. The second row, however, has the string `"lime"`, so it should not be `nil`:

```
>> t[1].nil?
=> false
```

T11. Print the complete array:

```
>> t
=> [nil, "lime", nil, nil, nil, "plum", "grape", nil, nil, nil]
```

Do you see how the strings are at the locations specified by the hash function?

T12. Print the array again, this time using the `print_table` method:

```
>> print_table(t)
   1: "lime"
   5: "plum"
   6: "grape"
=> nil
```

T13. Use the `lookup` method to see if one of the strings inserted in a previous exercise is in the array:

```
>> lookup("grape", t)
=> 6
```

Note the return value is 6, which means the method found the string in location 6 in the array (which agrees with the output above).

T14. Look for a string that is not in the table:

```
>> lookup("lemon", t)
=> nil
```

T15. The hash function wants to put the string "blueberry" in the same row as "lime":

```
>> h("blueberry")
=> 1
```

T16. If we try to insert "blueberry" the return value will be `nil`, because this row is already occupied:

```
>> insert("blueberry", t)
=> nil
```

If a hash table has more than 10 rows, we need to pass the table size as a second argument in a call to h. For example, to find the place for `"lime"` in a table with 30 rows the call is h(`"lime"`,30).

T17. Find out where `"mango"` would be stored in a hash table with 30 rows:

```
>> h("mango",30)
=> 12
```

From the examples in the text we saw that `"mango"` is placed in row 2 of a table with 10 rows, but this output shows the same string would be put in row 12 if the table has 30 rows.

♦ Make a test array with 15 random strings:

```
>> a = TestArray.new(15, :fruits)
=> ["gooseberry", "guava", "watermelon", ... "pomegranate"]
```

Your array will be different of course, since the strings are random, but you should see 15 different names.

♦ This expression tells Ruby to sort the array and print each string, followed by its row number in a 30-row hash table:

```
>> a.sort.each { |x| puts x; puts h(x,30) }
apple
0
blackberry
1
...
```

♦ Do the results of the previous expression give you a hint about how the hash function is defined?

6.3 The mod Function Again

To recap what we've seen so far:

- the goal is to create a data structure, called a hash table, to hold a word list;

- a hash function figures out where to place a word in the table;

- in the Ruby experiments, we are using an array to represent the table, strings to represent words, and calling a method that implements a hash function to compute the location for a string.

Whether or not this is a viable approach to implementing a word list clearly depends on the hash function. In the previous section we used a trivial function that was defined for tables with only 10 rows, but now it's time to look at functions that work for tables with thousands of rows that are capable of storing all the words in a dictionary.

A key observation is that the function does not need to put similar words close to each other. There is no reason why two words that are next to each other in a dictionary, like "abacus" and "abate," have to be in adjacent rows in the table. All that matters is that we can determine quickly whether a word is in the table by computing h(abacus) or h(abate); it doesn't matter if the two rows are widely separated.

What does matter is that different words be placed in different locations in the table. Ideally, if we have a collection of n words, we could put them all in a table that had exactly n rows, and the hash function would assign each word a different row number between 0 and $n-1$. It is possible to define what is known as a "perfect hash function," but it is a very difficult procedure that is effective only for small groups of words. If the goal is to create a word list to hold all the words in a dictionary, a more practical technique is required.

In a typical application, programmers make a table with lots of extra space, and then design a function that spreads the words around in the table. But even with the extra space, there will be situations, known as **collisions**, where the hash function puts two words in the same row. We'll put off until the next section figuring out what to do when words collide. For now, we'll focus on the basic technique for choosing a row based on the letters in a string.

A single letter can be converted to a number between 0 and 25, using the convention that a = 0, b = 1, and so on, up to z = 25. A common name for this function that converts a letter to a number is *ord*, which stands for "ordinal." It just means we want the relative position of the letter in the English alphabet.

A trivial hash function uses the first letter of the input word to figure out which row to place the word in. We'll call this function h_0:

$$h_0(s) = ord(s_0)$$

where the notation s_0 means "the first character in string s."

The function h_0 puts words in different rows as long as they all start with different letters, but as soon as we try to add a word that starts with the same letter as one already in the table there will be a collision. A slightly more complex function uses the same basic idea, but it works with the first two letters in a word. If a word starts with "aa" it will go in row 0; if it starts "ab" it will go in row 1, and so on, up to words starting "az," which will go in row 25. Continue by using row 26 for a word that starts with "ba," row 27 for words starting "bb," and so on.

Hash Functions

The "hash" in the term "hash function" comes from cooking, where hash is a dish made by chopping up and mixing together different ingredients.

In a program, the first step in implementing a hash function is to chop up a string by considering each letter individually. A selected combination of pieces are then reassembled mathematically into a row number.

"mango" ⟶ m / a / n / g / o ⟶ 12

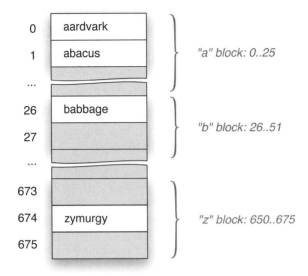

Figure 6.5: *The hash function based on the first two letters in a word computes row numbers between 0 and 675. The first 26 rows are for words that begin with "a," the next 26 for words starting with "b," and so on.*

To use this function a table needs $26 \times 26 = 676$ rows, numbered from 0 to 675. One way to look at the new table, shown in Figure 6.5, is that it is a set of 26-word blocks, where each block has words that start with the same letter. The first block of 26 rows is reserved for words that start with "a," the next block for words that start with "b," and the last block for words starting with "z." From this point of view, the function uses the first letter to find the start of the block, and the second letter to find a row within the block.

We can express this idea of using the first letter as a block number and the second letter as a row number with an equation:

$$h_1(s) = ord(s_0) \times 26 + ord(s_1)$$

where the notation s_0 again means "the first letter in s" and s_1 stands for "the second letter in s." Some examples of how this function would be applied to words in a dictionary are

$$
\begin{aligned}
h_1(\text{bed}) &= 1 \times 26 + 4 = 30 \\
h_1(\text{cnidarian}) &= 2 \times 26 + 13 = 65 \\
h_1(\text{zymurgy}) &= 25 \times 26 + 24 = 674
\end{aligned}
$$

It turns out this method of assigning a numeric value to a set of two letters is the same method used to determine the value of two-digit numbers in a positional number system. For example, in the octal (base 8) number system the digits "47" represent the number 39, because the digit 4 in the "eights column" has a weight of $4 \times 8 = 32$ and the digit 7 in the "ones column" has a weight of 7. If we are using hexadecimal (base 16) the same digits represents the number 71 because the 4 has a weight of 16, and $4 \times 16 + 7 = 71$.

The new two-letter hash function is equivalent to finding the value of a "number" in a base-26 number system, where the "digits" are the letters from "a" to "z." In this base-26 number system, the columns have weights that are powers of 26. The number corresponding to the first letter has a weight of 26^1, and the number for the second letter has a weight of 26^0.

Radix-26

To store a string in a hash table, we need to compute a location for it by making an integer from the letters in the string.

The technique used in this book is to treat the letters A to Z as "digits" in a number system, where A is 0, B is 1, and so on. Then a string is simply the value of the sequence of digits in base 26. Mathematicians refer to the base as a *radix*, so we call this technique radix-26.

Here is an example of how to compute the value of a string of digits in octal (base 8), by multiplying each digit by a power of 8. To figure out the value of 314_8:

8^2	8^1	8^0
3	1	4

$$3 \times 8^2 + 1 \times 8^1 + 4 \times 8^0 = 204_{10}$$

To make an integer from a sequence of letters we do the same thing, just using powers of 26. To make an integer from the string "BED" just compute the value of BED_{26}:

26^2	26^1	26^0
B	E	D

$$B \times 26^2 + E \times 26^1 + D \times 26^0 =$$

$$1 \times 26^2 + 4 \times 26^1 + 3 \times 26^0 = 783_{10}$$

The new function h_1 does a better job of spreading words around than the original function h_0, but it clearly has the same basic problem. All words that start with "be" will go in row 30 of the table, all words starting "cn" will go in row 65, *etc.* But now that we have the general idea that letters in a word correspond to digits in a base-26 number system, it's easy to extend the function to work with words of any length. In a three-letter word, the number for the first letter is multiplied by 26^2, the second by 26^1, and the third by 26^0. In the general case, the function assigns a weight of 26^i to the letter that is i places from the right end of the string. A common name for this function is **radix-26** (abbreviated r_{26}), to reflect the fact that it is the same as assigning a value to a number in a base 26 number system.

Here are some examples of the radix-26 values of a few words from an English dictionary:

$$
\begin{aligned}
r_{26}(\text{do}) &= 3 \times 26 + 14 = 92 \\
r_{26}(\text{duck}) &= 3 \times 26^3 + 20 \times 26^2 + 2 \times 26^1 + 10 = 66,310 \\
r_{26}(\text{zymurgy}) &= 25 \times 26^6 + 24 \times 26^5 + \ldots + 24 = 8,013,894,328 \\
r_{26}(\text{cnidarian}) &= 2 \times 26^8 + 13 \times 26^7 + \ldots + 13 = 524,574,935,989
\end{aligned}
$$

The r_{26} function assigns a unique integer to every word, or, more precisely, to any sequence of the 26 letters of the English alphabet.

Now we have a new problem: the radix-26 values for long words are very large numbers. When a word has more than six or seven letters, its radix-26 value is over 10^9. These values are far too big to be used as row numbers for a table we want to fit inside a computer's memory.

There is a way to use r_{26} in a hash function, however: simply "cut the numbers down to size" using the mod operator. We first saw the mod function in the Sieve of Eratosthenes project. x mod y is defined to be the remainder after dividing x by y. In the Sieve project, the mod function was used to help determine when x is a multiple of y, but in this project we use it to restrict the range of values produced by r_{26}. The useful fact about division modulo n for this project is that the result is always a number between 0 and $n-1$.

The mod function is exactly what we need for the hash function, since we want the hash function to compute a number between 0 and $n-1$ for a table that has n rows. All we have to do is use the r_{26} function to turn the string into a number, and then use the mod function to trim the result to a value between 0 and $n-1$:

$$h_r(s) = r_{26}(s) \bmod n$$

The subscript r in the name refers to the fact that the hash function is based on the radix-26 value computed using all the letters in the string.

The r_{26} function and the three hash functions described in this section have all been implemented in the HashLab module. When we call a method that implements a hash function, we pass it a string and a table size. The result returned by the method will be the row number where the string should be stored. For example, to use the h_r function to find a place for the word "hello" in a table with 1000 rows the call would be

```
>> hr("hello", 1000)
=> 872
```

In order to implement a hash function in Ruby, we need to be able to access the individual letters in a string object. In previous projects we saw how it is possible to select individual items from an array using the index operator. We can do the same thing with strings. If s is a string, an expression of the form s[i] refers to the character at location i. For the projects in this chapter, we want to know the ordinal values of the letters, which we can find by calling a method named ord. This example shows how to find the ordinal value of the first letter in a string:

```
>> s = "duck"
=> "duck"
>> s[0].ord
=> 3
```

The ord Method

The method named ord we are using in this project to determine the ordinal value of a character is part of the RubyLabs gem you installed when you set up your computational workbench. It maps letters A to Z from the English alphabet to the corresponding numbers from 0 to 25.

Because some of the "words" in our word lists might by hyphenated or be compound names (e.g., "lime green" from the list of colors) the method simply ignores any character that is not a letter. If you're curious, try some experiments on your own, comparing the radix-26 values of strings like "lime green" with "lime-green" or "limegreen."

If you took the challenge at the end of the tutorial project in the previous section, of trying to figure out the simple hash function used to make the table in Figure 6.3, you probably noticed the method was only using the first letter in the input string. When there were 30 rows, it was pretty obvious that each word was being placed in a row that depended on the ordinal value of the first letter.

What may not have been quite as obvious was that for smaller tables, the hash function used the mod operator to limit the row numbers to a value between 0 and $n - 1$. For the table with 10 rows, the row numbers need to be between 0 and 9. The method that created the table in Figure 6.3 put words starting "a" through "j" in rows 0 to 9, but then it "wrapped around." Words starting with "k" went in the first row, "l" in the second row, and so on.

Counting up to a maximum value and then restarting at 0 is another property of the mod operator. A hash table with 10 rows is a like a clock or a dial with 10 numbers. After counting up to 9, the maximum value, the next count starts all over at 0 again. The method named h that was used to find a location for a string in an array with 10 locations simply used the remainder of the ordinal value of the first letter after dividing by 10:

```
def h(s)
    return s[0].ord % 10
end
```

Tutorial Project

T18. Make a string to test methods that implement hash functions:

```
>> s = "mango"
=> "mango"
```

T19. The first test is to look at the ordinal values of the letters:

```
>> s[0].ord
=> 12
>> s[1].ord
=> 0
```

Does these results look correct, remembering the ordinal value of "a" is 0?

T20. You can call a method named each_byte to evaluate an expression for every character in the string:

```
>> s.each_byte { |x| puts x.ord }
12
0
...
```

T21. The method named h0 implements the hash function that looks at only the first letter in a string. Call h0 to see where it would place the string "guava" in a table with 10 rows:

```
>> h0("guava", 10)
=> 6
```

That's what we expect: the ordinal value of "g" is 6, so any string starting with "g" goes in row 6.

T22. Repeat the previous expression, but this time with the string "mango":

```
>> h0("mango", 10)
=> 2
```

Do you see why this string will go in row 2? The earlier exercises show the ordinal value of "m" is 12, and 12 mod 10 = 2.

T23. Call h0 a few more times, using the words from Figure 6.3, to make sure you understand why these strings were placed where they were.

T24. The Ruby version of the function r_{26} is a method named radix26. Test the method on a few short strings:

```
>> radix26("i")
=> 8

>> radix26("in")
=> 221

>> radix26("ink")
=> 5756
```

T25. Print the ordinal values of the three letters in "ink":

```
>> "ink".each_byte { |x| puts x.ord }
8
13
10
=> "ink"
```

T26. Plug these values into the radix-26 equation: use your calculator (or IRB) to compute $8 \times 26^2 + 13 \times 26 + 10$. Is the result the same as the value returned by calling radix26("ink")?

T27. Get the radix-26 value of the string "bed":

```
>> radix26("bed")
=> 783
```

T28. Since you know radix26("bed") from the previous exercise, can you predict what Ruby will print as the value of radix26("bee") without asking IRB to evaluate the expression? Call radix26("bee") to check your answer.

T29. Make a test array with 15 random strings:

```
>> a = TestArray.new(15, :fruits)
=> ["gooseberry", "guava", "watermelon", ... "pomegranate"]
```

We can combine two separate Ruby expressions into a single statement if we separate them by a semicolon. In the following exercise, we are going to ask Ruby to iterate over the new TestArray. For each string in the array, we want Ruby to print the string and then a hash function value for that string. The hash function is the function that uses the r_{26} function to figure out where to place a string in a table with 1000 rows.

T30. Type the following command to ask Ruby to print each string in a, followed by the row number where the string would be placed in a hash table with 1000 rows:

```
>> a.each { |s| puts s; puts hr(s, 1000) }
gooseberry
94
guava
922
...
```

T31. Call radix26 for some of the strings in your test array. Can you see how hr always converts the radix-26 value to a number between 0 and 999? Does it appear that the strings are scattered randomly throughout the table?

T32. Try calling radix26 with the longest words you can find. How big a number can you make?

T33. Do you think the hash function will treat uppercase letters differently than lowercase? Type a few expressions in IRB to figure out the answer to this question.

6.4 Collisions

If we want to use a hash table to implement a word list, we need to make sure all the words find a place in the table. In the previous section, we learned that the first step is to make a large array, with more rows than there are words, and to define a hash function that scatters the words around in the array.

It might seem that to avoid collisions—situations where the hash function wants to place two words in the same row—we simply have to make a table large enough that the probability of a collision will be very small. One would think a table with 1000 rows would have more than enough room for a hash function to find locations for 100 words with a low probability of a collision.

Unfortunately, this strategy does not work. Even if a hash function uses techniques similar to those used to generate random numbers, so words are distributed randomly and uniformly throughout the table, there is a high probability that two words will be put in the same row. Using well-established results from probability theory, if we use the hash function to store 100 words in a table with 1000 rows, the probability two words will be placed in the same row is over 99%. When a table has 10,000 rows, so there are 100 times as many rows as there are words, there is a 39% chance that two words will go in the same location (Table 6.1).

Table Size	Probability of Collision
1,000	99.29%
10,000	39.04%
100,000	4.83%
1,000,000	0.49%

Table 6.1: *The probability of a collision when adding 100 words to a hash table.*

So if collisions are inevitable, what can we do when when the hash function chooses a row that already has a word in it? One approach, which we will explore in this section, is to alter the data structure, so there is a place to store two or more strings in each row. The idea is that, instead of storing a single string in each row, we make each row a reference to an array, called a "bucket." When we add a string to the table, we just append it to the end of the bucket for its row (Figure 6.6).

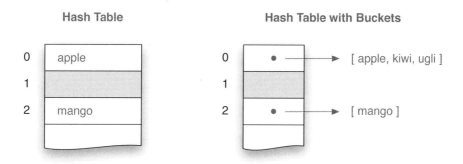

Figure 6.6: *In the hash tables we have been using so far, each row holds a single string. In the new organization, each row will have an array of strings, called a "bucket," so more than one string can map to the same row number.*

The Birthday Paradox

Imagine there is a lecture hall with 365 chairs, each labeled with a different day of the year. As students arrive for a class, they are directed to the chair that is labeled with their birthday.

The first student can take his seat when he arrives. The second student can usually take a seat, but there is a slight chance (1/365, or 0.27%) she has the same birthday as the first student and will find her seat taken already. Three students will probably all have different birthdays, but there is a slightly higher chance (0.82%) of a conflict between two of them.

The question is, at what point is there a 50/50 chance that two students will want to take the same seat? The answer is surprising: when there are only 23 students, the probability that two of them have the same birthday is 50.7%. With a group of 50 students, the probability that two will share a birthday is over 97%.

It's tempting to think that, with 22 students in the room, the probability of a conflict is the same as the probability of the 23rd student finding someone in her seat, which is only 22/365, or 6.02%. But this line of reasoning overlooks the fact that two of the students who are already there might have the same birthday. The question is about the entire group of students, not just the last to arrive.

The same probabilities apply to storing strings in a hash table. Even with a well-designed hash function that assigns entries to random rows, the probability that two of them will be entered in the same row are surprisingly high. If 100 items are stored in a table with 1000 rows, the probability of a collision is 99.3%. Even with 1,000,000 rows there is still a slight chance (0.49%) of two items being assigned to the same row.

```
# Add string s to hash table t

  def insert(s, t)
    i = h(s, t.length)
    t[i] = Array.new if t[i].nil?
    t[i] << s
    return i
  end
```

```
# Return the location of s in hash table t

  def lookup(s, t)
    i = h(s, t.length)
    if t[i] && t[i].include?(s)
      return i
    else
      return nil
    end
  end
```

Figure 6.7: *Final versions of hash table methods, using buckets to save more than one string in each row. These methods are used by a new type of object, called a HashTable, that is defined in the HashLab module.*

As before, a newly created table is is empty, so it is simply an array where every cell is nil. The first time a string is stored in a row, a new bucket is created for that row, and the string becomes the first item in the bucket. After that, other strings that hash to that row are appended to the end of the bucket.

The new versions of the insert and lookup methods are shown in Figure 6.7. Surprisingly, the Ruby code is actually simpler than the code for the preliminary versions. That's because every call to insert will succeed. The key step in this new version of insert is the one that creates a new bucket when a row is empty. Adding a string to the end of bucket is trivial, since we can just use Ruby's << operator.

The code for the lookup method introduces a new operator. In Ruby, && means "and." There is also an operator written with a single ampersand, but it has a different meaning, so be careful when you type expressions; in this chapter, we always want the version with a double ampersand. The key operation in the new version of lookup returns a row number if two conditions are both met: the row is not nil, and the bucket in that row contains the string we're looking for. If both of these are true, the method returns the row number, otherwise it returns nil, meaning the string is not in the table.

The experiments in this section use a new type of object, called a HashTable, that is defined as part of the HashLab module. Whenever HashLab is included at the start of a session with IRB it will automatically load the software that defines these objects.

Each time we want to make a hash table, we will call a method named HashTable.new to make a new object. Then, to add a string to the table, we call a method associated with the object. For example, to make a table with 1000 rows, and then add the word "apple" to the table, we type these expressions into IRB:

```
>> t = HashTable.new(1000)
=> #<RubyLabs::HashLab::HashTable: 1000 rows, :hr>
>> t.insert("apple")
=> 70
```

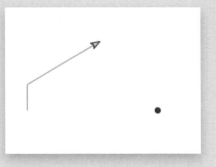
There are two reasons for using specially defined objects instead of using arrays like we
did in the previous sections. HashTable objects have additional methods that will be useful
in experiments, and they have methods that will draw a table on the RubyLabs canvas.
Being able to see drawings of tables and watching how buckets are created and extended
when strings are added to the table is an effective way of learning about the data structure.

The default hash function used by the HashTable objects is h_r, the function based on the
full radix-26 value of the string. It's also possible to make a table that uses the simple one-
letter hash function h_0 by passing an option in the call to `HashTable.new`. To make the
small table shown in Figure 6.7 the expression is

```
>> t = HashTable.new(10, :h0)
=> #<RubyLabs::HashLab::HashTable: 10 rows, :h0>
```

After we create a HashTable object, we can ask the HashLab module to draw a picture of
it by calling a method named `view_table`:

```
>> view_table(t)
=> true
```

When a table has been drawn on the canvas, each time we call `insert` to add a new string
to the table the drawing will be updated to show the string at the end of the bucket it was
added to.

The projects in this section will make some small tables and draw them on the canvas to
illustrate how buckets are used to resolve collisions. In the next section we will do some
further experiments to test the performance of these hash tables when we build a word list
for all the words in our English dictionary.

Tutorial Project

T34. Create a new HashTable object:

```
>> t = HashTable.new(10, :h0)
=> #<RubyLabs::HashLab::HashTable: 10 rows, :h0>
```

T35. Call the table's `insert` method to add two strings:

```
>> t.insert("apple")
=> 0
>> t.insert("mango")
=> 2
```

T36. The `print_table` method will print the contents of the table in the terminal window:

```
>> print_table(t)
   0: ["apple"]
   2: ["mango"]
=> nil
```

Note how each row is now an array. In this table, the two nonempty rows are each arrays containing one string.

T37. Call `view_table` to open a new graphics window and draw the table on the canvas:

```
>> view_table(t)
=> true
```

T38. Now add a string that collides with one already in the table:

```
>> t.insert("kiwi")
=> 0
```

With the previous version of `insert` the return value would have been `nil`, but the value 0 here means the string was added to the bucket in row 0.

T39. You should see "kiwi" at the end of the array in row 0 on the canvas. If you want, you can also call `print_table` to see the table in the terminal window:

```
>> print_table(t)
   0: ["apple", "kiwi"]
   2: ["mango"]
=> nil
```

T40. Make an array with some more strings to add to the table:

```
>> a = ["orange","strawberry","eggplant","ugli"]
=> ["orange", "strawberry", "eggplant", "ugli"]
```

T41. Add the strings to the table:

```
>> a.each { |x| t.insert(x) }
=> ["orange", "strawberry", "eggplant", "ugli"]
```

Does the table on the canvas look like the one in Figure 6.8?

T42. The `lookup` method should find a string no matter where it is in the bucket:

```
>> t.lookup("orange")
=> 4
>> t.lookup("eggplant")
=> 4
>> t.lookup("strawberry")
=> 8
```

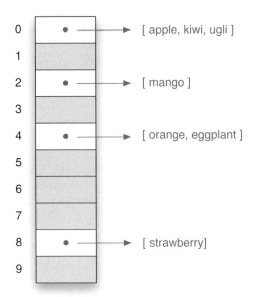

Figure 6.8: *A hash table that uses buckets to resolve collisions has a list of strings in each row.*

T43. As a final test, try looking up some strings that are not in the table. First verify `lookup` returns `nil` if a table row is empty:

```
>> t.lookup("tangerine")
=> nil
```

Do you see why this call returned `nil`? Strings starting with "t" will be in the row after the strings starting with "s," and this row is currently empty.

T44. We should also get back `nil` if the row has a bucket, but the string is not in the bucket:

```
>> t.lookup("apricot")
=> nil
```

6.5 Hash Table Experiments

The project for this section will do some experiments with the HashTable objects introduced in the previous section. We will make HashTable objects of varying sizes and then see how well they work when we try to add a large number of words.

Recall from previous projects that we get an array of strings from the RubyLabs word list by calling `TestArray.new`. For example, to get a list of 10 random words we call

```
>> a = TestArray.new(10, :words)
=> ["geognosis", "bounding", ... "displant"]
```

We can get the entire list of words by using the symbol `:all` instead of a number. But before you try to make a list of all the words in the dictionary, a word of warning: if you call `TestArray.new` to make the full list, IRB will print the entire array on your terminal.

Here's a trick programmers use to prevent IRB from printing a huge array like this in the terminal window. Since we can type more than one command on a line, make the call to `TestArray.new` the first expression, then type a semicolon, and then type any other Ruby expression. IRB will do both operations, but only print the result of the second. This is the recommended way of reading all the words from the RubyLabs dictionary and saving them in a variable named `words`:

```
>> words = TestArray.new(:all, :words); puts "done"
done
=> nil
```

Note there are two Ruby commands on this line. The first is the assignment statement that defines a variable named `words`, and the second is a call to `puts`. We can tell IRB executed the call to `puts`, since Ruby printed "done." To see that the assignment statement was also executed, ask Ruby to print the length of `words`:

```
>> words.length
=> 210653
```

So we now have an array with over 210,000 strings in it.

The first experiment will be based on a subset of the words, just to make sure we know how to make a hash table from words in the list. After that we will make some larger tables, using all of the words, in order to understand what effect the number of rows has on the efficiency of the methods that insert and look up words.

One way to assess the performance of the hash table methods is to simply look at the structure of the table after all the words have been added. If the hash function has done a good job, the words will be spread evenly through the table. There will be few empty rows, and the buckets will have roughly the same number of strings. We will not do any formal comparisons of the `lookup` method with the search algorithms from the previous two chapters. The number of steps made by `lookup` is simply the number of comparisons it needs to search through one of the buckets, so the worst case number of comparisons is determined by the number of strings in the longest bucket.

Tutorial Project

T45. Define a variable named `words`, assigning it an array that has all of the words in the Ruby-Labs dictionary:
```
>> words = TestArray.new(:all, :words); puts "done"
done
```

T46. To make sure the array was created, ask Ruby to print the number of words in the list:
```
>> words.length
=> 210653
```

T47. Look at the first 10 words in the list:
```
>> words[0..9]
=> ["aa", "aal", "aalii", "aam", "aardvark", ... "abacate"]
```

T48. Look at the last 10 words (the expression `a[-1]` refers to the last item in `a`, so the expression `a[-10..-1]` refers to the last 10 items in `a`):
```
>> words[-10..-1]
=> ["zymotechnical", ... "zymurgy", "zythem", "zythum"]
```

T49. Make a small hash table with only 25 rows:
```
>> t = HashTable.new(25)
=> #<RubyLabs::HashLab::HashTable: 25 rows, :hr>
```
The output shows this table will use h_r, the hash function based on radix-26.

T50. Initialize the canvas with a drawing of the table:
```
>> view_table(t)
=> true
```
You should see a table with 25 empty rows.

T51. Add the first 10 words in the word list to the table:
```
>> words[0..9].each { |s| t.insert(s) }; puts "done"
done
=> nil
```

Look at the table in the graphics window (or print the table on your terminal by calling `print_table`). Does it look like the words are distributed fairly evenly? Some rows might have two words, but there shouldn't be any row with three or more words.

Is there a pattern to the words, or are they more or less random? Are the words in alphabetical order in the table, or do some of the words at the end appear earlier in the dictionary than some words at the beginning of the table?

T52. Make a table with 100,000 rows:
```
>> t = HashTable.new(100000)
=> #<RubyLabs::HashLab::HashTable: 100000 rows, :hr>
```

T53. You can pass this table to `view_table`, but it will only show the first few rows:
```
>> view_table(t)
=> true
```
Even though you won't see the whole table, watching what happens in the first 20 or so rows is useful.

T54. Add all the words in the word list to the table:
```
>> words.each { |s| t.insert(s) }; puts "done"
done
=> nil
```
This will take a few seconds, so you won't see any output from IRB for a bit.

T55. Call a method named `print_stats` to get some information about the table:
```
>> t.print_stats
shortest bucket: 0
longest bucket:   15
empty buckets:    18298
mean bucket length:  2.58
=> nil
```

With 210,000 words and 100,000 rows, we would like the hash function to scatter the words well enough that there would be an average of two words in each bucket. The output from `print_stats` shows the average is 2.58, so that's not too bad. But it also shows one bucket has 15 strings in it.

T56. Make a hash table with 500,000 rows, and pass it to `view_table` to view the first few rows:
```
>> t = HashTable.new(500000)
=> #<RubyLabs::HashLab::HashTable: 500000 rows, :hr>

>> view_table(t)
=> true
```

T57. Add all the words to this table:
```
>> words.each { |s| t.insert(s) }; puts "done"
done
=> nil
```

T58. Print the table statistics:

```
>> t.print_stats
shortest bucket: 0
longest bucket:  7
empty buckets:   336216
mean bucket length:  1.29
=> nil
```

As you might have expected, using more rows allows the hash function to spread the words around more. The average bucket length is now just over 1.0, which means the majority of words are the only word in their row. The longest bucket is now only seven strings long. But this improvement has come at a cost, since about 2/3 of the rows are empty.

The remaining exercises, which are all optional, explore this relationship between table size and efficiency.

The next experiment shows what happens if the table size is a prime number. First, we'll make a small table with 10,000 rows, and then compare it to a table that has almost that many rows, except the number of rows will be a prime number. According to the Sieve of Eratosthenes, the largest prime less than 10,000 is 9973.

♦ Make a table with 10,000 rows:
```
>> t = HashTable.new(10000)
=> #<RubyLabs::HashLab::HashTable: 10000 rows, :hr>
```

♦ Add the words to this table, and print the table statistics:
```
>> words.each { |s| t.insert(s) }; t.print_stats
shortest bucket: 1
longest bucket:  75
empty buckets:   0
mean bucket length:  21.07
=> nil
```

♦ Now make the table with a prime number of rows:
```
>> t = HashTable.new(9973)
=> #<RubyLabs::HashLab::HashTable: 9973 rows, :hr>
```

♦ Repeat the line with the expressions that add words and print table statistics:
```
>> words.each { |s| t.insert(s) }; t.print_stats
shortest bucket: 5
longest bucket:  42
empty buckets:   0
mean bucket length:  21.12
=> nil
```

So it looks like having a prime number of rows makes a difference. The shortest bucket has 5 strings, instead of just 1, and the longest bucket has 42 strings, instead of 75. Overall, the hash function, which finds the remainder after dividing by the table size, does a much better job of scattering words evenly when the table size is a prime number. The reason is that when the table size is a prime number, the equation that defines the hash function is very similar to an equation used by a random number generator. We'll come back to this issue in an optional exercise in Chapter 9.

The next experiment shows what happens if the table size is a power of 26.

♦ Ask Ruby to compute 26^4:

```
>> 26**4
=> 456976
```

♦ Make a table with that many rows:

```
>> t = HashTable.new( 26**4 )
=> #<RubyLabs::HashLab::HashTable: 456976 rows, :hr>
```

As you can see, the new table has just over 450,000 rows, nearly as many as the table from Exercise T56.

♦ Add the words to this table, and print the statistics:

```
>> words.each { |s| t.insert(s) }; t.print_stats
shortest bucket: 0
longest bucket:   6932
empty buckets:    440375
mean bucket length:  12.69
=> nil
```

That's a significant difference! Even though the table is not much smaller—it is about 90% as big as the table from Exercise T58—it has far worse performance. Over 95% of its rows are empty, and one of the buckets has almost 7000 strings.

♦ A method named `long_rows` will return an array of row numbers, where each row has more than a specified bucket size. This expression prints the rows that have more than 1000 words:

```
>> t.long_rows(1000)
=> [966, 7746, 12951, ... 339902]
```

♦ Row 966 is one of the very long rows. This expression will print all the strings in the bucket in row 966:

```
>> t.table[966]
=> ["abandonable", "abatable", ...  "yokeable"]
```

What do these words have in common? Try printing some of the other long rows, to see if the words in those rows also have anything in common.

Can you explain why a table size that is a power of 26 gives such terrible performance? Hint: the hash function finds the radix-26 value of a string, and then computes the remainder mod n, where n is the table size. Using the fact that $26^i \div 26 = 26^{i-1}$, explain what happens when the radix-26 value of a string is divided by 26. What is the remainder of that division? What happens when the table has $26^2 = 676$ rows? $26^4 = 456,976$ rows?

6.6 Summary

The important new idea in computing introduced in this chapter is that data structures play an important role in how algorithms solve problems. Our ability to solve a problem computationally depends not only on the sequence of steps defined by an algorithm, but also by the way in which the data is organized. Alternative ways of structuring data open up possibilities for new approaches for addressing a problem, and an important part of computational problem solving is considering several different data structures and the algorithms that work with them.

A real-world example of a data structure is a card catalog in a library. The "algorithm" for finding a book in the library was a two-step process that made use of this data structure: first do a search through a catalog to look up a call number, and then go to the shelf labeled with the call number. In a modern library the first phase is done by a computer-based search, where we type a title, or author name, or keyword into a form on a web page, and the system looks up the call number for us.

Using a card catalog or its computational equivalent to help manage a collection is an example of indexing. An index is a data structure that contains brief descriptions of the items in the collection; for a set of books, an index would have author names, or keywords, or other descriptive information. The data structure we looked at in this chapter, the hash table, is one way to create an index, but there are several other widely used techniques.

In our experiments with hash tables, we used a Ruby array to represent a table. A hash table was implemented as an array with a predetermined number of rows. Each row is itself an array, called a bucket. To store a word in the table, we used a hash function to figure out which row to store it in, and then added the string representing the word to the end of the bucket.

The efficiency of this scheme for creating an index depends on the hash function. If the hash function spreads words around evenly, the buckets will have more or less the same number of strings, but a poorly designed function will cluster too many strings in just a few buckets. Long buckets reduce the efficiency of the method that looks for words in the table. The method needs to do a linear search of a bucket, so it is important to keep buckets as short as possible.

Hash tables and other index structures play a vital role in information management. The Online Computer Library Center (OCLC) is a cooperative effort that maintains a catalog of over 1 billion items maintained by thousands of libraries around the world. A local library that participates in OCLC will often connect to the worldwide catalog, so that when we search for an item we can get information on whether it is in the local library or another OCLC library close by. Scientific datasets are also managed by large index structures. The National Center for Biotechnology Information (NCBI) maintains dozens of different

---

Concepts and Terminology Introduced in This Chapter

data structure	A technique for organizing data so it can be accessed efficiently by an algorithm
hash function	A function that maps a string (sequence of characters) to an integer
hash table	A data structure, similar to an array, where strings are stored at a location determined by a hash function
collision	A situation where two strings are mapped to the same location in a hash table
bucket	A data structure for resolving collisions, in which all the strings that map to the same location in a hash table are saved in an array

databases of gene sequence information, and complex index structures make it possible for researchers to connect information found in one database to data contained in other databases.

Web search engines are another type of information service where indexing is essential. In this case, the index consists of a collection of words, and there is a link from each word to the URLs of web pages that contain the word. By typing a few key words into a form in a web browser, we can get information from a "search engine" that is essentially a massive index of the entire Internet.

Exercises

1. What are the ordinal values of the following letters?

   q   w   e   r   t   y

   Note: you can use IRB to check your answer:

   ```
   >> "qwerty".each_byte { |x| puts x.ord }
   ```

2. Draw a hash table with 10 rows, then add the following strings, using hash function h_0:
   a) "bmw"
   b) "fiat"
   c) "suzuki"
   d) "lotus"
   e) "peugeot"

3. Compute the radix-26 values of the following strings:
   a) "um"
   b) "ump"
   c) "lump"
   d) "lumpy"
   e) "clumpy"

   Does knowing the radix-26 value of one of the strings make it easier to compute the value for the next string in the sequence?

4. Compute the radix-26 values of the following strings:
   a) "ironic"
   b) "clown"
   c) "green"
   d) "bundle"
   e) "garden"

5. Draw a hash table with 10 rows, then add the strings from the previous problem, using the hash function h_r.

6. Can you come up with a formula for giving a rough estimate of the radix-26 value of an n-letter word?

7. What do you think will happen if you pass a string that has nonalphabetic characters to `radix26`? Try some experiments to figure out what the `ord` method returns for +, :, and some other symbols, then pass a string containing these symbols to `radix26`.

8. Can a hash table have only one row? What will Ruby do if you call `HashTable.new(1)` to make a new HashTable object? What would happen if you add the strings from Problem 4 to this table?

9. Is looking for a string in a one-row hash table the same as doing a linear search in an array? Explain.

10. Is using a hash table to organize a word list with 210,000 words as efficient as sorting the words and saving them in an ordinary array, and then using binary search to see if a word is in the list?

11. Suppose a friend is planning on writing a program that will play a word game, and as part of this project she needs to decide how to represent the word list. What would you advise her to use? What are some of the pros and cons of a hash table compared to a simple array of strings?

12. ♦ Explain how you might use a hash table in an algorithm that checks to see if an array of strings has any duplicates.

13. ♦ Check out a copy of the `radix26` method, and modify it so it uses digits, spaces, and punctuation characters as part of the calculation. The goal is to define a method that maps strings like `"flip-flop"` and `"flip flop"` and `"flipflop"` to different integers.

14. ♦ Implement a method that will delete an item from a hash table. Hint: first call `lookup` to see which row an item is in (if it's in the table at all), then remove it from the bucket in that row.

15. ♦ What would happen if you made a hash table with 26 rows, and then used the hash function h_1 to store words in the table? Would there be any sort of pattern to the row assignments for words added to the table?

Chapter 7

Bit by Bit

Binary codes and algorithms for text compression and error detection

Computer systems use a variety of technologies to store and transmit data. Processor and memory chips are based on semiconductor technology; disks are made from magnetic materials spread over the surface of a thin, circular plate; removable disks have a reflective surface with tiny pockmarks that are detected by lasers; digital cameras and cell phones store digital information in "flash" memories that retain their data even when the device is turned off; networks transfer information in the form of light waves over fiber-optic cables.

As diverse as these technologies are, they have one thing in common that allows them to be used to store or transmit information: they are physical devices that can be set to one of two different states. For example, a semiconductor memory cell will be in a state where we can measure one of two different voltage levels, and a section of a DVD will be smooth or have an indentation, which can be detected by a laser scanning the surface of the disc.

The role of computer science is to figure out how to represent information in a form that can be stored or transmitted using these technologies. Real-world objects, things like pieces of text from a book, images captured by a camera, or sounds recorded in a studio, can be described by a sequence of symbols. Information can be stored or transmitted using any of the technologies mentioned above by equating the two possible states with different symbols. Traditionally, the two symbols are 0 and 1, the digits of the binary number system.

Exactly how the two states are managed physically is in the realm of computer engineering, or, as programmers like to say, is "a hardware problem." Engineers designing a memory system can decide what voltage levels will be used inside the chips, and might assign the symbol 1 to a level of 3.5 volts and the symbol 0 to a level of 0 volts. Computer scientists figure out how to represent text and other information as a sequence of binary symbols and devise algorithms that manipulate these symbols, knowing that engineers are building systems that will store them and move them from place to place.

Codes (Secret and Otherwise)

In everyday usage the word "code" is commonly associated with secrecy or privacy, as in "diplomatic code" or "personal access code."

In computer science, however, a code is simply a sequence of 0s and 1s used to represent data inside a machine or in a message being sent over a network.

A **coding scheme** defines codes for a particular set of items. A good example is ASCII, which is a common scheme for representing text. To transmit a message, each letter is translated into a code, a process known as **encoding**. At the other end of the network, a **decoding** operation translates codes back into text. There is no attempt to keep the message secret, since anyone with a table of ASCII codes can recover the original string.

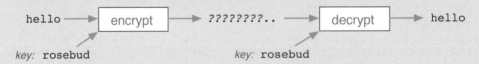

If someone wants to keep information private they would use **encryption**. An encryption algorithm uses a password (usually called a **key**) to scramble the message so it looks like a random string of 1s and 0s. The only way to recover the original message is to pass the same key to a **decryption** algorithm.

hello ⟶ encrypt ⟶ ????????.. ⟶ decrypt ⟶ hello

key: rosebud   *key:* rosebud

Programmers also use the word "code" to simply mean "program," as in "the Ruby code is in this file" or "the compiler generated the object code."

The process of transforming information into a sequence of 0s and 1s is known as **encoding**. An example of how data can be encoded is a system known as Unicode, an international standard for representing text in a computer. Unicode has a set of tables that define encodings for each letter of the alphabet. There are also codes for punctuation marks, digits, and other symbols.

Going in the opposite direction, from the coded form back to the original data, is a process known as **decoding**. For text, decoding is usually just a matter of reverse translation, of finding the letters or symbols that correspond to each binary code.

Encoding text is a fascinating and surprisingly complex topic. While it might seem like a very straightforward process—just look in the table that defines a code to find the sequence of 0s and 1s for each character and store the result in memory—there are many difficult and subtle issues. To take just one example, how should the encoding scheme deal with accents? Should we use two different codes for *é* and *e*? Or should we have one code for the letter *e*, and a second code for the accent, so that when *é* appears in a text it is encoded with two codes? How will this decision affect a sorting algorithm when it compares letters? In French, putting an accent on a letter does not have any effect on the ordering of words, so *école* comes before *elegant* in a dictionary and the scheme that encodes accents separately from letters would make sense. But in Swedish, *å* and *a* are considered to be two different letters. Words starting with *å* follow words starting with *z* in a dictionary, which is an argument in favor of using two different codes for the accented and unaccented letters.

The projects in this chapter explore algorithms that work with encoded text. We will begin by looking at basic methods for encoding characters as binary numbers, and see how strings in Ruby are basically sequences of encoded characters.

Errors occur with surprising frequency, both when data is transmitted over a network or is simply sitting in memory. No matter what sort of technology is used, data will be lost or changed in transmission, or the storage media can be corrupted. One way to deal with potential errors is to store extra information along with the data, and then to use this extra information to see if an error has occurred. In Section 7.3 we will look at algorithms that figure out the necessary extra information and use it to detect errors.

Another issue related to binary representation is data compression. In some cases, data can be compressed by coming up with an alternative representation that captures all the essential information. A good example is audio compression. MP3 files are approximately 1/10 the size of the original digital recordings, but in most cases (especially when played through earbuds or inexpensive headphones) the music sounds almost as good as the original. With text, however, we can't afford to lose any information. Changing a single letter or substituting a single punctuation mark can greatly modify the meaning of the text. Although we can't replace the original text with an approximate copy, there are ways to reduce the size of the encoding without losing any information. In Sections 7.4 and 7.5 we will look at a simple but effective algorithm that generates alternative codes for letters, resulting in an encoding that is shorter yet retains all the information from the original text.

7.1 Binary Codes

The 0s and 1s in a binary encoding are commonly referred to as **bits**. In this context the word "bit" is an abbreviation for "binary digit." Since the devices that store information or transmit it from one system to another are binary (two-state) devices, a bit is the smallest unit of information that can be stored or transmitted.

Unicode

An international standard known as **Unicode** is a scheme that defines codes for thousands of letters, symbols, and ideograms from all over the world. As of 2009, the Unicode standard includes codes for over 100,000 characters, covering most modern languages and several ancient languages, including Egyptian heiroglyphics.

å	é	あ
Latin small letter A with ring above Unicode 00E5	Latin small letter E with acute Unicode 00E9	Hiragana letter A Unicode 3042

One of the fundamental relationships in computer science is between the number of bits in a code and the number of items that can bc cncodcd. The relationship is defined by the formula $n = 2^k$, which says that n, the number of different codes we can construct, grows exponentially with k, the number of bits in a code.

To see why, start with the fact that there are two possible 1-bit codes, 0 and 1. With two bits, there are four codes: 00, 01, 10, and 11. When written in this order, one can see a pattern: the first two codes begin with 0, and the second two begin with 1.

Let's use this same pattern to write out all the 3-bit codes (Figure 7.1). Write the four 2-bit codes on a piece of paper, and then write them again, so there are two groups of codes next to each other. Put a 0 in front of each code in the first group, and a 1 in front of each code in the second group, and you will have all eight possible 3-bit codes. The general rule is that we can write twice as many codes using $n + 1$ binary digits as we can with n digits, because we write each n-bit code twice, and then write a 0 in front of half the codes and a 1 in front of the other half. There are 2^2 2-bit codes, $2 \times 2^2 = 2^3$ 3-bit codes, and, more generally, 2^n n-bit codes.

As a practical example of this relationship, the ASCII code described in the next section is a 7-bit code, which means we can encode $2^7 = 128$ different characters. This gives us enough codes for ordinary English text, since we can use 26 of the codes for lowercase letters, another 26 for uppercase letters, and still have several codes left over for punctuation marks.

The inverse equation is also an important formula. Often we are given the size of a set, and we want to know how many bits will be required to encode each item in the set with a unique sequence of bits. Since $n = 2^k$, it follows that $k = \log_2 n$. For example, if we want to design a code for 26 uppercase letters and 6 punctuation marks, for a total of 32 characters, we would need only $\log_2 32 = 5$ bits.

When the logarithm is not an integer, we just "round up" to the next higher integer. The more formal way to express the relationship between the size of a set, n, and k, the number of bits required to encode each item in the set, is

$$k = \lceil \log_2 n \rceil$$

(recall from Chapter 5 the notation $\lceil x \rceil$, pronounced the "ceiling of x," means "the smallest integer greater than x").

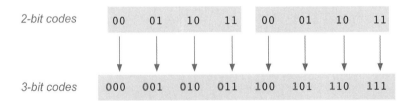

Figure 7.1: *To create a three-bit code, make two copies of the two-bit code and put a 0 bit in front of one copy and a 1 bit in front of the other copy.*

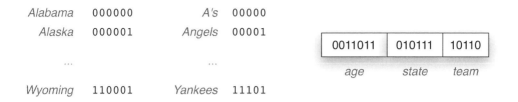

Alabama	000000		A's	00000
Alaska	000001		Angels	00001
...			...	
Wyoming	110001		Yankees	11101

0011011	010111	10110
age	state	team

Figure 7.2: *Responses from a survey can be encoded in binary by using $\lceil \log_2 n \rceil$ bits for a field that has n alternatives. The encoding needs 7 bits for numbers between 0 and 127, 6 bits to represent one of the 50 U.S. states, and 5 bits for one of 30 different baseball teams.*

Binary codes can be used to represent more than just numbers and characters. Any type of data based on a finite set of values can be encoded so it can be stored in a file and analyzed by a computer. As an example, suppose a research group is collecting data through a survey, and the questionnaire has places for respondents to fill out their age, the state they live in, and their favorite baseball team. Age is a number, so techniques for encoding integers can be used to represent a respondent's age. If the researchers assume everybody who fills out the survey will be less than 127 years old, they only have to use $\lceil \log_2 127 \rceil = 7$ bits to encode the age of the respondent. To encode the response for the state a person lives in, the researchers need to figure out how many bits to use for the encoding, and then assign a unique pattern to each state. There are 50 U.S. states, which means $\lceil \log_2 50 \rceil = 6$ bits are needed to encode a state name. One way to assign codes is to go in alphabetical order, so Alabama is 000000, Alaska is 000001, and so on. The same sort of reasoning would lead to unique 5-bit codes for each of the 30 major league baseball teams (Figure 7.2).

The projects for this chapter begin with experiments on binary codes. The BitLab module defines a method named `code` that will create a sequence of 1s and 0s corresponding to the pattern of bits used to encode a number. For example, to see how the number 23 is encoded in binary, we simply write

```
>> 23.code
=> 10111
```

There will be situations later in the chapter where we want a code that has a specific length, so we can pass a number of bits as a parameter. To see the same code, but as an 8-bit number, the expression is

```
>> 23.code(8)
=> 00010111
```

This is the same pattern as before, but it has extra 0s at the front (the equivalent of writing *02* instead of *2* when entering a month in a form).

Another method defined for BitLab is `make_codes`, which assigns a unique binary code to each item in a set. Suppose we want to design an encoding scheme for musical notes. Start by defining an array with one Ruby symbol for each note:

```
>> notes = [:do, :re, :mi, :fa, :sol, :la, :ti]
=> [:do, :re, :mi, :fa, :sol, :la, :ti]
```

Associative Array (*aka* Hash)

Ruby has a special type of array where items are accessed not by their location, but
instead by their name.

In a regular array, items are stored in order, and we can ask
Ruby to get the item at a specified location:

0	fee
1	fie
2	foe
3	fum

```
>> a = ["fee", "fie", "foe", "fum"]
=> ["fee", "fie", "foe", "fum"]
>> a[2]
=> "foe"
```

When we make a hash object, we tell Ruby the name of each object, and when we
want something from the hash we specify its name, not its location:

```
>> gp = { :a => 4, :b => 3, :c => 2, :d => 1 }
=> {:d=>1, :a=>4, :b=>3, :c=>2}
>> gp[:a]
=> 4
>> gp[:c]
=> 2
```

:d	1
:a	4
:b	3
:c	2

The name "hash" comes from the fact that these objects are sometimes stored in
memory as small hash tables (Chapter 6). Ruby can choose whatever order it wants
to save the items; all we care about is that we can get the value of an item associated
with a name.

Read more about hash objects in the Ruby Reference at the end of the book.

To define a code that assigns each note its own binary encoding call `make_codes`, passing
the array of symbols as an argument:

```
>> code = make_codes(notes)
=> {:re=>001, :do=>000, :fa=>011, ... :ti=>110}
```

Each item in this list shows a note on the left side of an arrow, and its encoding on the right
side of the arrow. The `make_codes` method figures out how may items are in the array, and
thus how long each code has to be, and it then just makes a new code for each item, starting
with the code that has all 0s. Since there are seven notes in this list, there are $\lceil \log_2 7 \rceil = 3$
bits in each code.

After we have the array of codes we can find the encoding of a syllable by looking it up
with the index operator:

```
>> code[:re]
=> 001
```

This example shows that `code` is an array, just like arrays we've seen in earlier chapters, but
the difference is that instead of accessing an item by its position, we access it by name. In
this case, since we want the code for a note, we put the name of the note in between the
square brackets in the index operator (see the sidebar on Associative Arrays).

Tutorial Project

Include the BitLab module when you start a new IRB session for the projects in this chapter:

```
>> include BitLab
=> Object
```

T1. BitLab defines a method that shows the binary representation of an integer. Type this expression to see the binary form of the number 12:

```
>> 12.code
=> 1100
```

T2. In previous chapters we used the iterator named `each` to print all the items in an array. We can also use it to print every number in a specified range:

```
>> (0..3).each { |n| puts n }
0
1
2
3
=> 0..3
```

T3. Repeat the previous exercise, but this time ask Ruby to print the binary representation of the number:

```
>> (0..3).each { |n| puts n.code }
0
1
10
11
=> 0..3
```

T4. By default the `code` method shows the minimum number of bits needed to encode an integer. We can designate a number of bits to use as a parameter. Type this expression to print a 2-digit representation of each code:

```
>> (0..3).each { |n| puts n.code(2) }
00
01
10
11
=> 0..3
```

T5. Repeat the exercise again, but this time ask Ruby to print eight numbers, and change the argument so you get a list of 3-bit codes:

```
>> (0..7).each { |n| puts n.code(3) }
000
001
...
110
111
=> 0..7
```

By looking at the bit patterns printed in the last example, can you see why there are twice as many 3-bit numbers as there are 2-bit numbers? The list of 3-bit numbers has two parts. The first four lines contain each of the 2-bit numbers, except they are preceded by a 0. The last four lines also have each of the 2-bit numbers, but these are all preceded by a 1.

This same pattern will occur if you ask Ruby to print all 16 4-bit patterns, and then all 32 5-bit patterns. Does this exercise help convince you that the number of codes grows exponentially with the number of bits?

T6. Make a list of four color names:

```
>> a = [:yellow, :green, :white, :black]
=> [:yellow, :green, :white, :black]
```

T7. Call make_codes to create a binary code for each color:

```
>> code = make_codes(a)
=> {:yellow=>00, :green=>01, :white=>10, :black=>11}
```

T8. Use print_codes to get a list that shows the code for each color:

```
>> print_codes(code)
00   yellow
01   green
10   white
11   black
=> true
```

Notice that each of the possible 2-bit binary numbers was used to make the codes.

T9. Add a fifth color to the end of the array:

```
>> a << :steel
=> [:yellow, :green, :white, :black, :steel]
```

T10. Make a new code for the extended array, and print the new code:

```
>> code2 = make_codes(a); print_codes(code2)
000   yellow
001   green
010   white
011   black
100   steel
=> true
```

Notice that since every 2-bit pattern was needed for a list of four colors, Ruby has to make 3-bit patterns for the list of five colors.

T11. The formula for figuring out how many bits are needed to give each item in a set a unique code is $\lceil \log_2 n \rceil$. Ask Ruby to compute the value of the formula for a set of four items:

```
>> log2(4).ceil
=> 2
```

T12. For a set of five items we have to "round up" to use one more bit:

```
>> log2(5)
=> 2.32192809488736

>> log2(5).ceil
=> 3
```

T13. How many bits do you think it will take to give each of 18 different colors a unique code? First try answering this question using a call to ceil, and then verify your answer by making an array of 18 color names and passing it to make_codes:

```
>> a = TestArray.new(18, :colors)
=> ["lavender", "azure", ... "sky blue"]

>> code = make_codes(a); print_codes(code)
=> ...
```

T14. Is the code for the first item in your set of 18 colors the binary representation of the number 0? Is the last code the binary representation of the number 17?

7.2 Codes for Characters

Computer processors generally work with several bits of data at once. When data is transferred from memory to the processor, or vice versa, it is in a group of 32 or 64 bits, called a **word**. A single word can represent a piece of data, in the form of encoded numbers, characters, or other information. A word can also hold the binary form of an instruction that tells the processor what to do for a single step in an algorithm. Words are often divided into smaller units, called **bytes**. There are eight bits in a byte, so a single word will hold four or eight bytes.

In the early days of commercial computing, transferring data between computers was rare, and networks were nonexistent, so each system was able to use its own technique for encoding data. The size of a word varied greatly from computer to computer, and it was not uncommon to find systems with 12-bit, 18-bit, or even 35-bit words. Not surprisingly, the number of bits in a byte also varied from system to system; common sizes were six, seven, or eight bits.

As people started sharing data across multiple computer systems the need for a standard encoding scheme became apparent. One of the first such standards was the American Standard Code for Information Interchange, or **ASCII**, which was formed by a committee representing several American computer manufacturers. ASCII used seven bits for each character, with codes for upper and lowercase letters and the symbols commonly found on a QWERTY keyboard.

Eventually computer designers settled on 32 bits as a typical word size. Later, as technology improved, the standard word size increased to 64 bits. An even number of 8-bit bytes will fit in either 32 or 64 bits, so eight bits became the standard size for a byte. Because it was convenient to store each code in its own byte when writing data to a file, text files were collections of 8-bit codes, where the first bit was always 0 and the remaining bits held an ASCII code. This naturally led to suggestions for extending ASCII to an 8-bit code. Eventually ISO, the International Standards Organization, defined a scheme called Latin-1 that includes codes for accented letters used in European languages, Greek letters, and a variety of math symbols.

ASCII

The American Standard Code for Information Interchange (ASCII) was developed in the 1960s, and formally adopted as a standard in 1968. It has codes for the symbols found on a QWERTY keyboard in the U.S.

	0100000	0	0110000	@	1000000	`	1100000
!	0100001	1	0110001	A	1000001	a	1100001
"	0100010	2	0110010	B	1000010	b	1100010
#	0100011	3	0110011	C	1000011	c	1100011
$	0100100	4	0110100	D	1000100	d	1100100

UTF-8

Unicode defines codes for thousands of letters, symbols, and ideograms from all over the world. As of October 2009, the Unicode standard includes codes for over 100,000 characters, so according to the formula there should be 17 bits in each code.

A clever scheme named UTF-8 allows the size of the code to vary, so the most common characters require only 8 bits. The code uses 8-bit groups called **octets**.

If the first bit in a code is 0 it means the entire code fits in a single octet, and the remaining 7 bits are used to hold the ASCII code. This is the encoding of the letter "a":

| 0 | 1100001 | *one octet*

If the first bit is a 1, the code continues with the next octet, and the remaining 15 bits encode the character. Here is the code for "å" (an "a" with a circle over it):

| 1 | 100001110100101 | *two octets*

There can be up to four octets in a code. The latest version of Unicode now includes codes for most of the languages in the world, and several ancient languages, including Egyptian heiroglyphics.

In the late 1980s researchers began to consider how to design an encoding scheme that would be able to represent a much wider collection of characters, including Arabic and Hebrew alphabets and the ideograms used in Chinese, Japanese, and Korean. Their effort led to the definition of Unicode, which is now the most widely used scheme for representing text. Most operating systems, including versions of Microsoft Windows and Mac OS X, use Unicode as the default format for encoding text files. E-mail programs and web browsers are also based on Unicode, which makes it possible to transmit files from one system to another while preserving all the letters and symbols that were written by the person who created the text.

Since Unicode supports such a wide variety of symbols—over 100,000 in the latest version of the standard—you might guess that it takes more bits to represent a single character than in ASCII. But people who work with Unicode came up with a scheme, called UTF-8, that allows just one byte to be used for all the symbols that are defined in ASCII. The idea is that if the first bit in a byte is 0, the remaining seven bits can simply hold an ASCII code. If the first bit is a 1, however, the code continues into the byte next to it. Accented letters and letters from Cyrillic, Chinese, Japanese, Arabic, Hebrew, and all the other alphabets require anywhere from two to four bytes.

In the experiments in this chapter the characters we work with will all be from the original ASCII code. Since they are represented in Ruby as 8-bit (UTF-8) codes, we will show them as 8-bit patterns of 0s and 1s. To check on the output from a Ruby expression we just need to look up a code in an ASCII table, which is far simpler than finding a code in Unicode, and we will continue to refer to letter codes as simply ASCII codes.

Ruby's index operator applies to strings as well as arrays. If a variable `s` refers to a string, an expression of the form `s[i]` returns the character at location `i` in the string. As with arrays, the first location is 0, so the expression `s[0]` refers to the first character in `s`. In Ruby, the index operator returns the numeric code for a character.[1] Since Ruby assumes we normally want to see decimal numbers that's what it prints:

```
>> s = "Angstrom"
=> "Angstrom"
>> s[0]
=> 65
>> s[1]
=> 110
```

To see the binary encoding of these numbers just call the `code` method:

```
>> s[0].code(8)
=> 01000001
>> s[1].code(8)
=> 01101110
```

The 8 passed as an argument to `code` tells it to make an 8-bit code.

Because long binary numbers are hard to read, many tables that show ASCII codes also give the hexadecimal (base 16) version of the codes. The `code` method will make the hexadecimal form if we pass it an option. These expressions ask Ruby to make the 2-digit hexadecimal code for a letter:

```
>> s[0].code(:hex, 2)
=> 41
>> s[1].code(:hex, 2)
=> 6E
```

Several of the codes in ASCII are referred to as **control characters**. The name comes from the fact that these characters were once used to control output devices. Control characters told line printers to start a new line of output, or told a teletype to ring a bell. Some of these control characters are still used. A character called **newline** is inserted into text to mark the end of a line. When you type text in a text editor application, newlines are included in the document when you hit the return key.

In Ruby, the `puts` method automatically attaches a newline character to the end of the string. When the output from a Ruby program is saved in a file, and the file is later opened with a text editor, the strings are shown as separate lines in the editor window. Without this newline character we wouldn't know where one string ended and the next began. When a string of n characters is written to a file, the number of bits in the encoding of the string is $8 \times (n + 1)$ to account for the fact that the newline character is part of the encoding.

---

[1]This behavior changed in version 1.9 of Ruby. If you are using Ruby 1.9 see the Lab Manual for an explanation of how you can access the encoding of individual characters in a string.

Tutorial Project

Before you start the projects in this section, use your web browser to find a reference page that has
a table of ASCII codes. The Wikipedia entry for ASCII is one source, or you can do a web search for
"ASCII" to find others.

T15. Make a string that has a combination of letters, digits, and punctuation marks:
```
>> s = "The 19th Century (1801-1900)."
=> "The 19th Century (1801-1900)."
```

T16. Notice that Ruby includes the spaces and parentheses as part of the string. To verify this,
count the number of characters between the quotes, and then ask Ruby to print the length of
the string:
```
>> s.length
=> 29
```
Does your count agree with Ruby's count?

T17. This expression will return the code for the first character in s:
```
>> s[0]
=> 84
```

T18. Use the code method to get the ASCII code for the character:
```
>> s[0].code(8)
=> 01010100
```

T19. The method that iterates over each item in a string is called each_byte. Type this expression
to see the ASCII code of each character in s:
```
>> s.each_byte { |x| puts x.code(8) }
01010100
01101000
01100101
...
```

T20. If you want to see the codes in hexadecimal, repeat the previous expression, but change the
arguments passed to code:
```
>> s.each_byte { |x| puts x.code(:hex, 2) }
54
68
65
...
```

Look up these codes in the ASCII table shown in your web browser. Most tables show the binary forms
as 7-bit numbers, so ignore the leading 0 printed by Ruby (or if you want, repeat the expression in
Problem T19 and tell Ruby to print 7-bit codes). Can you find each of the characters in s, including the
spaces and the punctuation marks? Did you notice there are different codes for upper and lowercase
letters, e.g., the letters T and t?

The following optional exercises refer to "Unix commands" that you can type in a terminal window.
These commands can be typed into any terminal emulator on Linux or Mac OS X. If you use Microsoft
Windows, you can type these commands in a command window that is running with cygwin.

♦ The paragraph at the end of this section said Ruby's puts method includes a newline when
it prints a string. Make a string to test this claim:
```
>> greeting = "hello"
=> "hello"
```

♦ Use `puts` in two separate statements, both entered on a single line typed into IRB:

```
>> puts greeting; puts greeting
hello
hello
=> nil
```

If `puts` did not attach a newline the output would be `hellohello` with all the characters in the string printed on a single line.

♦ Use your text editor to make a plain text file that has a single line containing the "19th Century" test string (without the quotes). Save the file as `century.txt`, and then find how big the file is. Open a new terminal window and type this Unix command:

```
% wc century.txt
       1       4      30 century.txt
```

`wc` stands for "word count." The output shows the file has 1 line, containing 4 words and a total of 30 bytes. The 30 bytes are for the 29 characters in the string plus an extra one for the newline character, which will be included if you hit the return key after typing the line in your text editor (on a Windows system there might be two characters to mark the end of the line).

♦ A Unix command named `file` looks at the contents of a file and tries to guess what is inside. Type this command to find out what is in `century.txt`:

```
% file century.txt
century.txt: ASCII English text
```

This output shows that all the characters are part of the ASCII encoding scheme.

♦ Figure out how to use your text editor to type accented letters, and add this sentence to your file:

> Anders Jonas Ångström was born in 1814.

Save the file and run the `file` command again:

```
% file century.txt
century.txt: UTF-8 Unicode English text
```

Since the accented letters are not part of the original 7-bit ASCII, the text editor had to use the UTF-8 scheme to encode the two accented letters. If you run `wc` you should see that the two accented letters take up two bytes each. You can verify this by replacing the $Å$ with A and the $ö$ with o and running the command again. The unaccented letters will require only one byte so the new file will be two bytes shorter.

7.3 Parity Bits

No matter what technology is used to store or transmit character codes, there is always a chance that errors will occur. Noise can interfere with network communications, or a bump might disrupt the motion of a disk. Magnetic tapes, and even CDs and DVDs, can lose information over time, and new data can be stored in a flash memory card only a certain number of times before the card begins to lose data.

There are a variety of ways of encoding text, music, and images in order to deal with potential errors. The simplest techniques are for **error detection**. The idea is to add extra descriptive information along with the code for the data itself. When the data is retrieved, an algorithm can compare the description with the data to determine whether anything has changed. If an error is detected, the data is retrieved again, *e.g.*, in a network the receiver might send a message back to the sender, to request that a message be sent again.

More complex schemes allow for **error correction**. When these codes are used, enough descriptive information is added to a message to allow the receiver to fix the error, without having to ask the sender to retransmit.

In this section we will look at the simplest form of error detection. It will work for any type of data, but we will use character codes in our experiments. The idea is to add a single bit, called a **parity bit**, to the end of a code before saving it or transmitting it. The word "parity" refers to the number of 1 bits in a binary pattern. If there are an even number of 1s we say the pattern has **even parity**. For example, the string *0101* has even parity, because two of the four bits are 1s, but *0111* has odd parity because it has three 1s.

If we start with an 8-bit character code, we will have a 9-bit extended code after adding a parity bit. When we add the extra bit we have a choice: we can either attach a 1 or a 0 to the end of the original code. The idea is to attach the bit that will make the total number of 1s in the new 9-bit code an even number, *i.e.*, we want the new code to have even parity. At the other end, the receiver will count the number of 1s in the 9-bit extended code, and if the count is an even number the receiver assumes no bits were changed during the transmission. The original 8-bit code can be extracted by removing the parity bit from the end of the 9-bit extended code (Figure 7.3).

This simple scheme works as long as there is only a single error within any 9-bit code. If an error occurs one of the bits will change: either a 1 will turn into a 0, or a 0 will turn into a 1. Either way, the total number of 1s in the code seen by the receiver will not be an even number, and the 9-bit packet can be marked as an error. But, as the last column in Figure 7.3 shows, the receiver can be misled if two errors occur. Again it doesn't matter if a 1 changes to a 0 or *vice versa*; the 9-bit code seen by the receiver will have an even number of 1s, and the receiver can't tell that two of the bits were not the same as the ones originally transmitted.

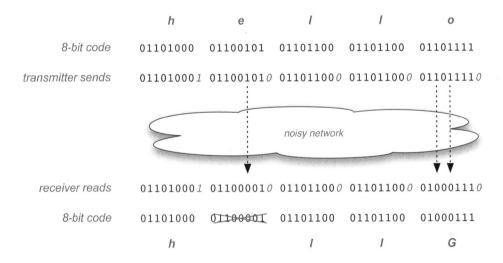

Figure 7.3: *When a message is transmitted the sender adds a parity bit to each 8-bit character to make a 9-bit code. The receiver counts the number of 1s in the entire code. If the code has an odd number of 1s an error is detected, otherwise the character is taken from the first 8 bits of the 9-bit code.*

In the experiments for this section we will create a message like the one shown in Figure 7.3 and then see what happens as errors are deliberately added. Methods defined in the BitLab module will count 1s in a code and attach a parity bit to the end of a code. A method named `flip` will let us change an individual bit (programmers who work with low-level machine languages refer to "flipping a bit," as in "flipping a switch" to turn a light switch on or off).

The BitLab module defines a new type of object to hold a list of codes. We will call a method named `encode` to create a message object, passing it a string to encode and an option that specifies how to encode each letter. For example, to make a message using ASCII characters, the call would be

```
>> msg = encode( "hello", :ascii )
=> 01101000 01100101 01101100 01101100 01101111
```

To ask that a parity bit be added to the end of each letter, pass the option `:parity` instead of `:ascii` in the call to `encode`:

```
>> msg = encode( "hello", :parity )
=> 011010001 011001010 011011000 011011000 011011110
```

Decoding a message, *i.e.*, creating a string of characters from a sequence of binary codes, is done by a method named `decode`. The parameters for this method are a coded message and the scheme originally used to use to encode the message:

```
>> decode( msg, :parity )
=> "hello"
```

A method named `garbled` simulates a noisy network. This method will make a copy of a message, add random errors, and return the modified message. For example, if `msg` is the message created in the example above, this statement will make a copy of it, add one random error, and save the result in a variable named `recvd`:

```
>> recvd = garbled( msg, 1 )
=> 011010001 011001010 111011000 011011000 011011110
```

To see what is encoded by this new message, simply pass it to `decode`:

```
>> decode( recvd, :parity )
=> "he•lo"
```

The bullet symbol in the third letter means `decode` found a parity error in the third code.

If we pass two objects in a single call to `puts` Ruby will print them on separate lines; this will allow us to compare the original encoding with the copy:

```
>> puts msg, recvd
011010001 011001010 011011000 011011000 011011110
011010001 011001010 111011000 011011000 011011110
```

If you compare the two codes in the third column, you can find where a 0 changed to a 1, and as a result the third code in the message now has odd parity.

Tutorial Project

T21. Create a test string:

```
>> s = "hello, world"
=> "hello, world"
```

T22. Save the codes of the first two letters in variables named c0 and c1:

```
>> c0 = s[0].code(8)
=> 01101000
>> c1 = s[1].code(8)
=> 01100101
```

How many 1 bits do you see in each of these codes?

T23. Use the `parity_bit` method to see which bits need to be added to make even parity codes:

```
>> c0.parity_bit
=> 1
>> c1.parity_bit
=> 0
```

Do you see why the parity bit for the first code is a 1, and the parity bit for the second code is 0?

T24. Call a method named `add_parity_bit` to attach parity bits to each code:

```
>> c0.add_parity_bit
=> 011010001
>> c1.add_parity_bit
=> 011001010
```

Can you see how each code is one bit longer? Do the extended codes each have even parity?

T25. A method named `even_parity?` will check to see if a code has even parity:

```
>> c0.even_parity?
=> true
>> c1.even_parity?
=> true
```

T26. The index operator will work on codes. To see the value of the first bit (bit 0) in code c0:

```
>> c0[0]
=> 0
```

T27. Call the `flip` method to change the first bit in c0:

```
>> c0.flip(0)
=> 111010001
```

Do you see how the first bit changed from a 0 into a 1? How many 1 bits do you see in the new code?

T28. There are now an odd number of 1s in this code:

```
>> c0.even_parity?
=> false
```

T29. We just saw how an error that converts a 0 to a 1 causes a change in the parity. Let's repeat the experiment, this time changing a 1 to a 0 in the second code word:

```
>> c1
=> 011001010
>> c1.flip(2)
=> 010001010
```

Do you see how bit 2 (the third bit from the left) changed?

T30. This change from a 1 to a 0 also changed the parity:
```
>> c1.even_parity?
=> false
```

If you want to try some more experiments on your own, make codes for the letters *e*, *l*, and a few other characters from the test string. First figure out what parity bit needs to be added to these codes, then check your answer by calling `add_parity_bit`.

T31. Make a code for the fifth letter in the test string:
```
>> c4 = s[4].code(8)
=> 01101111
```

T32. Verify that this binary number is the code for the letter *o* by calling a method named `chr`:
```
>> c4.chr
=> "o"
```

T33. Add a parity bit:
```
>> c4.add_parity_bit
=> 011011110
```

T34. Change two bits in this code, making the changes shown in Figure 7.3:
```
>> c4.flip(2)
=> 010011110

>> c4.flip(4)
=> 010001110
```

T35. After two bits change the code will have even parity:
```
>> c4.even_parity?
=> true
```

T36. Notice what happens when the receiver decodes this binary number:
```
>> c4.chr
=> "G"
```

Make sure you understand what happened in this last exercise. Because the code has even parity, a receiver can't tell errors occurred, and it will accept the code. After it removes the parity bit, it has the binary number 01000111, which is the 8-bit code for the letter *G*. If you're not sure how this happened try some more experiments on your own, where you make a code from a character in the test string, add two errors, and decode the result.

T37. Call `encode` to make a message with parity bits attached to each character:
```
>> msg = encode( s, :parity )
=> 011010001 011001010 011011000 ...
```

T38. Make a copy of the message with `garbled`, asking it to add a single random error:
```
>> recvd = garbled( msg, 1 )
=> 011010001 011001010 011011100...
```

T39. Decode the copy:
```
>> decode( recvd, :parity )
=> "hello, wor●d"
```

Since the error is introduced in a random location you will probably get a different output, but you should see one error somewhere in your new message.

T40. Print the two messages so you can compare the bit sequences:

```
>> puts msg, recvd
011010001 011001010 011011000 011011000 ...
011010001 011001010 011011100 011011000 ...
```

Find the code that corresponds to the bullet symbol printed in the decoded message, and compare this code to the original code. Can you find the one-bit error?

T41. Garble the message again, but this time add 10 errors:

```
>> recvd = garbled( msg, 10 )
=> 011010001 000001010 011011001 ...
```

T42. Decode this new message:

```
>> decode( recvd, :parity )
=> "he•l•• •o?ld"
```

Did the errors go in 10 different codes, *i.e.*, are there 10 bullet symbols?

T43. Print the two bit sequences:

```
>> puts msg, recvd
011010001 011001010 011011000 ...
011010001 000001010 011011001 ...
```

Each group of nine bits corresponds to a single character in the message. You should be able to account for all 10 errors. Places where there is a bullet in the decoded string identify characters where a single error was introduced, resulting in a parity error. Places where the second string does not have a bullet, but there is a difference between the original sequence and the copy, should show two errors.

T44. Try this experiment again, but introduce 30 errors instead of 10. Did you find character positions where three errors were introduced? Four? What happened at these positions? Was a parity error detected or was a character decoded?

7.4 Huffman Trees

As stated in the introduction to this chapter, the goal of text compression is to rewrite a piece of text in a new form that requires fewer bits. One simple way to do this, and the approach we will take in this chapter, is to simply change the encoding scheme to one that uses fewer bits. The technique we will look at is known as *Huffman encoding*, named after David A. Huffman (1925–1999), who originally described the method in 1952.

A Huffman code relies on the fact that some characters are more common than others. In ordinary English text the letters *e*, *t*, and *a* appear much more often than *q*, *j*, and *z*. An encoding scheme that uses fewer bits for the most common letters might be able to encode a piece of text with a smaller total number of bits than one that uses the same number of bits for every letter.

To test this idea we are going to use Huffman's algorithm to develop a special purpose code for Hawaiian words. The Hawaiian alphabet has a total of 13 letters: five vowels, seven consonants, and a symbol called the *okina* that sometimes appears between two vowels. An encoding scheme that uses the same number of bits to code each letter in this alphabet would require $\lceil \log_2 13 \rceil = 4$ bits per letter. Our goal is to devise a coding scheme that uses fewer bits for the common letters and more bits for the uncommon letters. After we build the code, we will test it with a few Hawaiian words to see if in fact the new code uses fewer bits. Of course, unless we also have a code for commas, periods, spaces, and other

The Hawaiian Alphabet

If you've ever traveled to Hawaii you may have noticed that geographical names, names of cities, and other Hawaiian words didn't have all the letters of the English alphabet.

It turns out Hawaiian words are spelled with the same five vowels -- A, E, I, O, and U -- but use only seven consonants: H, K, L, M, N, P, and W.

A 13th symbol, called the *okina*, is used between two vowels when they are to be pronounced as separate syllables. For example, *a'a* is pronounced "ah-ah" (it's one of the words for lava).

The table at right (from the American Cryptogram Association) shows the frequency of each letter. A is by far the most common, making up 26% of all the letters in Hawaiian words. W is least common, used just 1% of the time.

'	0.068
A	0.262
E	0.072
H	0.045
I	0.084
K	0.106
L	0.044
M	0.032
N	0.083
O	0.106
P	0.030
U	0.059
W	0.009

punctuation we won't be able to encode complete sentences, but this simple scheme will be enough to encode a word list like the one we used in Chapter 6.

Huffman's algorithm uses two data structures that we have not seen yet in this book. The first is known as a **priority queue**. Structurally, a priority queue is just like an array: it's a container that basically consists of a list of references to other objects. What distinguishes a priority queue from a regular array is that the objects in the list are always sorted. Each time we add a new item to a priority queue, the item is automatically saved in a location that preserves the order.

The priority queues we will be using for the project in this section are a type of object called a PriorityQueue. This statement will make a new, empty, queue:

```
>> pq = PriorityQueue.new
=> []
```

The method that adds an item to a queue has the same operator, <<, as the method that adds an item to the end of an array. It's important to be aware of the fact that there are actually two different methods—one for array objects, the other for priority queue objects—even though both methods are identified by the same operator. When Ruby evaluates an expression of the form obj << x, the first thing it does is figure out what sort of object is referred to by the variable named obj. If the object is an array, Ruby uses the method that attaches x to the end of the array. If the object is a priority queue, Ruby uses a method that compares x to the other objects already in the list and inserts it into the correct location. We've already seen this operation: it's part of the inner loop of the insertion sort algorithm.

Because priority queues decide for themselves where items go, there is no method named insert for these kinds of objects. Not surprisingly, several other methods we can use with arrays, such as reverse, are also not defined for priority queues. The main thing to remember when you are working on this project is that although a priority queue might

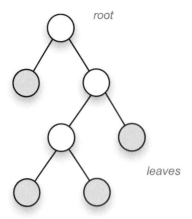

root

leaves

Figure 7.4: *A binary tree is a collection of* **nodes**. *Lines connecting nodes define relationships similar to those in a family tree. At the top of the tree is a single node, the* **root**, *that has no predecessors. Nodes that have no descendants are called* **leaves**.

look like an array, the types of things you can do with it are restricted in order to make sure the queue is always sorted.

The second new data structure is a **binary tree**. A tree can be seen as another type of container, an alternative to a linear structure like an array or queue. The items that make up a tree are called **nodes** (Figure 7.4). When we add an object to a tree, we will make a node to refer to the object and then insert the node into the tree somewhere by attaching it to one or more existing nodes. Every tree has one special node called the **root**. Computer scientists (like genealogists) prefer to draw trees with the root at the top of the picture, as shown in Figure 7.4. Nodes at the bottom of a tree are known as **leaves**.

Trees are very important structures in computer science, used to organize information in a wide variety of algorithms. Most of these algorithms use iterators that have been specially designed for trees. These iterators serve the same purpose as iterators for arrays, allowing us to "walk through" a tree to visit each node in order to perform some operation. For this project, however, we'll simply use a tree as a container and not concern ourselves with how to visit the nodes.

One way to envision how Huffman's algorithm works is to imagine a set of tinker toys, where the circular pieces correspond to nodes, and sticks are used to connect nodes to each other. The algorithm starts with a collection of unconnected circular pieces that will eventually be the leaves of the tree. Each step of the algorithm will find two pieces of the tree that were built on a previous step and connect them to each other, as shown in Figure 7.5. The new piece of the tree will be an interior node that has the existing parts as its children. Since the only pieces available initially are leaf nodes, the first step will connect two leaves. Later steps might connect a leaf to an interior node, or connect two interior nodes. Eventually all the nodes will be connected and the final product will be a tree that looks like the one in Figure 7.4.

The objects in the tree we are going to build with Ruby are another new type of object, called a Node. The leaf nodes in Huffman's algorithm contain two pieces of information: the name of a letter from the alphabet, and the frequency of that letter. For example, the letter *M* has a frequency of 0.03 in Hawaiian words (meaning 3% of the letters in random Hawaiian text should be the letter *M*). An expression that makes a node for this letter is:

```
>> leaf = Node.new("M", 0.03)
=> ( M: 0.030 )
```

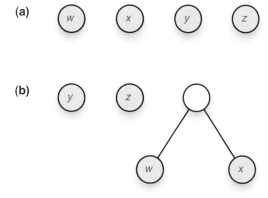

Figure 7.5: *Building a Huffman tree is analogous to making a tree shape with tinker toys.* **(a)** *At the start of the project all the circle pieces that will be leaves of the tree are not connected to anything.* **(b)** *Each step will take two existing pieces and connect them to a new piece. The new piece will be the common ancestor of the two existing pieces.*

The string printed by Ruby is a shorthand notation that tells us the node has a label of *M* and a frequency of 0.03. The parentheses surrounding the label and frequency are intended to look like the sides of a circle in a drawing of a tree.

A second way to make a new node object is to call a method named `combine`. This method is used when the algorithm wants to make a new interior node in the tree (or, to use the tinker toy analogy, when two existing nodes are attached to a new round piece). A call to `combine` must include references to two existing nodes. The result of the call will be a new node object that has the existing nodes as its children. Here is an example. Suppose two existing nodes are named `t0` and `t1`:

```
>> t0 = Node.new("X", 0.1)
=> ( X: 0.100 )

>> t1 = Node.new("Y", 0.2)
=> ( Y: 0.200 )
```

This statement shows how to make a new interior node with `t0` and `t1` as descendants:

```
>> t2 = Node.combine(t0, t1)
=> ( 0.300 ( X: 0.100 ) ( Y: 0.200 ) )
```

The string printed by Ruby to describe the new node is a little harder to understand, but if you look closely you'll see three nodes are included in this string. The outer set of parentheses identify the new interior node, and next to the opening parenthesis is the frequency of the new node, which is the sum of the frequencies of its two children. The two descendants are shown as they were before, printed one after the other, but inside the outer parentheses to indicate they are now below this new interior node.

Now that we have the two main pieces—a new type of object that implements a priority queue and a second new type of object to represent nodes in a tree—the Huffman tree algorithm is very simple. The first step is to make leaf nodes for every symbol in the alphabet and put the resulting nodes into the priority queue. Then remove the first two nodes from the queue, make a new interior node with these two nodes as its children, and insert the new interior node back into the queue. Repeat the step of removing two nodes and inserting one until the queue has been reduced to a single node; this node will be the root of the final Huffman tree.

(a)

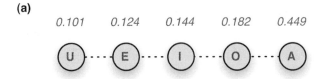

A	0.449
E	0.124
I	0.144
O	0.182
U	0.101

(b)

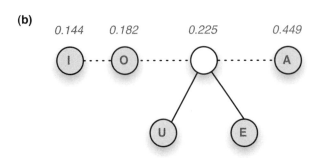

Figure 7.6: *The first two steps in building a Huffman tree for vowels, using their relative frequency in the Hawaiian alphabet. **(a)** The initial queue has only leaf nodes. **(b)** The first step makes an interior node for the two least frequent letters.*

Figure 7.6 shows the first two steps in the construction of a small tree for the five vowels, using their relative frequency in Hawaiian words. The first step initializes the priority queue with a new node for each letter. Note the letters are sorted, with the least frequent letters at the front of the queue. The second row in the figure shows what happens on the first iteration. The two letters at the front of the queue, which are the two least frequently used vowels, are removed from the queue. They are connected by a new interior node, and the new node is put back in the queue. The number above each node is its frequency. For letters, the frequency is taken from the table, but for interior nodes the frequency is the sum of the frequencies of the letters below it. The two important things to remember about this algorithm are (1) at each step the nodes in the queue are ordered, according to their frequency, and (2) the queue grows shorter at each step, so the algorithm is guaranteed to terminate after $n - 1$ steps.

As you work on the project in this section, you will be able to view a drawing of the current state of the priority queue. After you initialize the queue, call a method named view_queue to draw the queue on the RubyLabs canvas. Initially each queue entry will be a leaf node, but at each iteration you will see new interior nodes appearing in the queue.

Tutorial Project

T45. Make a new priority queue object and save it in a variable named pq:

```
>> pq = PriorityQueue.new
=> []
```

As you can see from the result above the new queue is empty.

T46. Priority queues, like arrays in Ruby, can contain any type of object, as long as the objects can be compared (if they can't be compared to one another there is no way to keep them in order). Try adding a couple of strings to your new queue:

```
>> pq << "lemon"
=> ["lemon"]

>> pq << "grape"
=> ["grape", "lemon"]
```

Note how the << method found the right place for the second string. The queue figures out where to insert the object and automatically puts it in a place that makes sure the queue remains sorted.

T47. Add a few more strings to your queue:

```
>> ["kiwi","strawberry","pear"].each { |x| pq << x }
=> ["kiwi", "strawberry", "pear"]
```

T48. Check to make sure the queue now has five items, and that they are stored in the queue in alphabetical order:

```
>> pq.length
=> 5

>> pq
=> ["grape", "kiwi", "lemon", "pear", "strawberry"]
```

T49. Even though the queue looks like an array, it is not an array. If you try to call methods that work for arrays you will get an error message:

```
>> pq.insert("orange")
NoMethodError: undefined method 'insert' for ...

>> pq.reverse
NoMethodError: undefined method 'reverse' ...
```

This makes sense, because if we were allowed to apply these other operations the order of the items in the queue might change, and the items would no longer be stored according to their priority.

♦ Do you think Ruby will allow you to make a priority queue containing numbers? Try some experiments on your own, making PriorityQueue objects and adding numbers to them.

The way to remove an item from a priority queue is to detach the item at the front. The method that does this is named `shift`. It's an unusual name, but it is the name of a similar operation defined for arrays, and is a name that is used in many different programming languages.

T50. Type these expressions to remove the first two strings from your queue:

```
>> s = pq.shift
=> "grape"

>> t = pq.shift
=> "kiwi"
```

T51. Ask Ruby to print the queue, and double-check to make sure there are three items left in the queue after removing the first two:

```
>> pq
=> ["lemon", "pear", "strawberry"]

>> pq.length
=> 3
```

T52. To make a new leaf node (a node that corresponds to a single letter) call `Node.new` and pass it a letter and its frequency:

```
>> t0 = Node.new("A", 0.2)
=> ( A: 0.200 )

>> t1 = Node.new("B", 0.3)
=> ( B: 0.300 )
```

T53. A method named `combine` will create a new interior node by combining two existing nodes:

```
>> t2 = Node.combine(t0, t1)
=> ( 0.500 ( A: 0.200 ) ( B: 0.300 ) )
```

The string printed by Ruby for an interior node is slightly more complicated, but it follows the same pattern for leaf nodes. The outer set of parentheses show that this is a node. The 0.5 after the open parenthesis means the combined node has a frequency of 0.5, which is the sum of the frequencies of the two nodes passed as arguments. The two descendants of this interior node are shown following the frequency.

The next set of exercises will use the letter frequency data for vowels in Hawaiian words to make the Huffman tree shown in Figure 7.6. The frequency data is in a file that is installed along with the RubyLabs gem.

T54. Use a method named `read_frequencies` to make an associative array (a "hash") to hold the data:

```
>> vf = read_frequencies(:hvfreq)
=> {"A"=>0.45, "O"=>0.18, "E"=>0.12, "I"=>0.15, "U"=>0.1}
```

The name `vf` is short for "vowel frequency."

T55. To see the frequency of one of the letters, just access that item in the hash:

```
>> vf["A"]
=> 0.45

>> vf["E"]
=> 0.12
```

Are these the frequency values shown in the table in the figure?

T56. A method named `init_queue` will make a priority queue, make a node object for each item in the hash, and add the nodes to the queue:

```
>> pq = init_queue(vf)
=> [( U: 0.100 ), ( E: 0.120 ), ... ( A: 0.450 )]
```

Does this look accurate to you? Are the items in the queue node objects? Are they sorted? How many are there?

T57. Call `view_queue` to draw the queue on the RubyLabs canvas:

```
>> view_queue(pq)
=> true
```

T58. Type this expression to remove the first node from the queue and save it in a variable named `n1`:

```
>> n1 = pq.shift
=> ( U: 0.100 )
```

Notice that the queue shown on the canvas now has only four nodes.

T59. Call `shift` again, and save the node in a second variable named `n2`:

```
>> n2 = pq.shift
=> ( E: 0.120 )
```

The queue should now be down to three nodes.

T60. Call `Node.combine` to make a new node from the two you just removed, and put the new node back in the queue:

```
>> pq << Node.combine( n1, n2 )
=> [( I: 0.150 ), ( O: 0.180 ), ... ( A: 0.450 )]
```

T61. Can you see how, as a result of the previous three expressions, the nodes for *U* and *E* were removed, and the front of the queue is now the node for *I*? Is the new interior node in the correct location in the middle of the queue?

T62. How long is the queue after this first iteration of the algorithm?

T63. To continue building the tree, repeat the previous three steps. You can put all three operations on a single line for IRB:

```
>> n1 = pq.shift; n2 = pq.shift; pq << Node.combine( n1, n2 )
=> [( 0.220 ( U: 0.100 ) ( E: 0.120 ) ), ... ( A: 0.450 )]
```

T64. Repeat the expression above until there is only one node left in the queue. This single node represents the complete tree for all five vowels.

T65. If you want to watch the steps for the construction of the Huffman tree for the entire alphabet, get the set of frequencies from a data set called `:hafreq`:

```
>> vf = read_frequencies(:hafreq)
=> {"K"=>0.106, "W"=>0.009, ... "U"=>0.059}
```

Now repeat the steps that initialize the queue, draw it on the canvas, and combine the nodes.

7.5 Huffman Codes

After we build a tree using Huffman's algorithm we can use the structure of the tree to define Huffman codes for each letter. Start by attaching labels to the lines that connect interior nodes to their children. At every interior node, including the root, the line going down to the left child should be labeled with a 0, and the line going down to the right child should be labeled with a 1. The labels on the connections in the final tree for the Hawaiian vowels are shown in Figure 7.7.

The labels in the tree can now be used to define codes for letters. Each letter of the alphabet used to make the tree appears somewhere in the tree as a leaf node, because the tree construction algorithm initialized the priority queue with one leaf for each letter, and

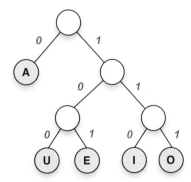

Figure 7.7: *The Huffman tree for the Hawaiian vowels, with labels attached to each connection.*

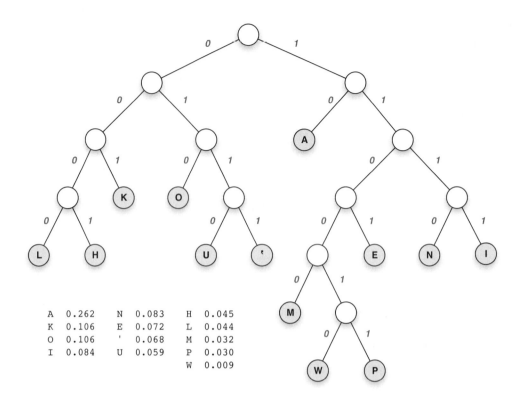

A   0.262   N   0.083   H   0.045
K   0.106   E   0.072   L   0.044
O   0.106   '   0.068   M   0.032
I   0.084   U   0.059   P   0.030
                             W   0.009

Figure 7.8: *Huffman tree for the Hawaiian alphabet. The frequencies in this table are the same as the ones on page 179, but here they are sorted by frequency.*

the leaves were eventually all added to the tree. The code for a letter is defined by the labels on the path from the root to the leaf for that letter. For example, the three connections from the root to the leaf for the letter *U* in Figure 7.7 are labeled 1, 0, and 0, so the code for *U* is 100. The path to *O* goes down the other side of the tree, and the bits on this path give 111 as the code for *O*.

The Huffman tree for the full 13-character Hawaiian alphabet is shown in Figure 7.8. An important thing to notice about the tree is that the most common letters are closest to the root, so the codes for these letters use fewer bits.

It's easy to see why the least frequently used letters are farthest from the root. They correspond to nodes that were toward the front of the queue when it was initialized, and were thus the first to be attached to interior nodes. Now that these letters are below an interior node the path to these leaves will be longer than paths to letters at the end of the queue.

As an example of what happens to infrequent letters, the two letters that were initially at the front of the queue for the complete Hawaiian alphabet were *W* and *P*. The first iteration attached these two letters to a new interior node and put the new node back in the queue. This new combined node still had a low frequency (0.039), so it remained near the front of the queue, where it was picked up and combined with another node, moving the letters even farther down the tree.

The experiments in this section will use the codes defined by a Huffman tree to encode Hawaiian words. The hope is that by using shorter codes for more common letters, the total number of bits to encode a word will be smaller, *i.e.*, using a Huffman code will help compress the text. Words with lots of *A*s will require fewer bits, since the code for *A* is the shortest code defined by the tree. The question is, will the benefit of using a shorter code for *A* be negated by the fact that *W* and *P* have longer codes, so there is no overall benefit from using a Huffman code?

The BitLab module contains a method named `assign_codes` that will iterate over the tree to collect the codes for all the leaf nodes and put them in a hash:

```
>> hc = assign_codes(tree)
=> {"K"=>001, "W"=>110010, "A"=>10,
```

If you want to see the code for a letter, just look it up in the hash:

```
>> hc["H"]
=> 0001
```

Or, if you want to see a complete list of codes, pass the hash to the `print_codes` method you used for projects in Section 7.1:

```
>> print_codes(hc)
0000   L
0001   H
001    K
. . .
```

Now that we know the bits that will represent each letter, encoding a Hawaiian word is simply a matter of writing the bits for each letter in one long string. For example, the encoding of *ALOHA* is

```
                100000010000110
```

The first two bits are the code for *A*, the next four bits are the code for *L*, the next three the code for *O*, then four bits for *H*, and finally two bits for the last *A*.

Decoding a string of bits in order to determine the original word is also defined in terms of the tree. The idea is to use the sequence of bits in a code to choose a direction at each step on a journey that starts at the root of the tree. As an example, suppose we want to decode the string 1101. The first bit is 1, so the first step goes down the right path from the root. From this node, the second step is also a 1, so we go down the right side again. The third step is a 0, so we go left, and the last step is a 1, so we go right again. Now we're at a leaf node, and the label there is *E*, so we're done: the code 1101 is for the letter *E*.

To decode a longer bit string that corresponds to a complete word, we just need to repeat the process described in the previous paragraph. Each time we reach a leaf node we can write the letter at that node, and then go back to the root and continue decoding with the next bit in the input. Suppose we want to decode the bits 10011110. Start walking from the root, and after two steps (10) we're at the letter *A*. Go back to the root, and start a new path from the third bit (since the first two have already been decoded). This path takes us to the okina, or glottal stop symbol, after four steps (0111). Go back to the root again, and the last two bits take us again to *A*. There are no more bits to decode, so this input string contained the encoding of the word *A'A*.

```
# Make a Huffman tree using the frequencies in f

 1:   def build_tree(f)
 2:     pq = init_queue(f)
 3:
 4:     while pq.length > 1
 5:       n1 = pq.shift
 6:       n2 = pq.shift
 7:       pq << Node.combine( n1, n2)
 8:     end
 9:
10:     return pq[0]
11:   end
```

Figure 7.9: *Ruby implementation of the algorithm that makes a Huffman tree.*

For the projects in this section we will be using the Huffman tree shown in Figure 7.8. The letter frequencies for the complete alphabet are in a file named hafreq (for "Hawaiian alphabet frequencies"). A method named build_tree (Figure 7.9) automates the steps of the tree-building algorithm and creates a tree that we can use to encode and decode words. For example, if the letter frequencies are in a hash object named f this statement creates the Huffman tree:

```
>> tree = build_tree(f)
=> ( 1.000 ( 0.428 ( 0.195 ( ... ) ) ) )
```

Earlier in the chapter we saw how to call a method named encode to generate a binary encoding of a string. One of the arguments we passed was a specification of how to do the encoding, either plain ASCII or with a parity bit attached. For this project, we will pass a Huffman tree to encode, and it will encode the string using Huffman codes:

```
>> msg = encode("ALOHA", tree)
=> 100000010000110
```

To go in the opposite direction, and convert a message that was encoded with Huffman codes back into a string of letters, pass the tree to the decode method:

```
>> decode(msg, tree)
=> "ALOHA"
```

The BitLab module has a set of Hawaiian words that can be used to test the encoding process. To load the words into your IRB session, call a method named read_words, and it will return an array containing all the words:

```
>> words = read_words
=> ["A'U", "ALI'I", ... "UA"]
```

To test your understanding of the encoding process, you can try encoding one of the words, using the tree in Figure 7.8, and then check your answer by passing the word to encode.

There is also an array of binary codes, which you can get by calling read_codes:

```
>> codes = read_codes
=> [011010, 001010, 0011111, ... ]
```

For one set of exercises, you will select a code from the list, see if you can decode it by hand using the tree in Figure 7.8, and then check your result by passing the code to decode.

Tutorial Project

T66. Read the letter frequencies for the full Hawaiian alphabet:

```
>> f = read_frequencies(:hafreq)
=> {"K"=>0.106, "W"=>0.009, "L"=>0.044, ... }
```

T67. Make the tree shown in Figure 7.8 by passing the frequencies to `build_tree`:

```
>> tree = build_tree(f)
=> ( 1.000 ( 0.428 ( 0.195 ( ... ) ) ) )
```

T68. Assign binary codes to each node in the tree:

```
>> hc = assign_codes(tree)
=> {"K"=>001, "W"=>110010, "A"=>10,        ... }
```

T69. According to the tree, what is the code for the letter *U*? the letter *P*?

T70. Check your answers by looking up the codes for the same letters:

```
>> hc["U"]
=> 0110

>> hc["P"]
=> 110011
```

Were you able to figure out a code by tracing a path from the root to the leaf for a letter? Try some more examples on your own until you are sure you understand how Huffman codes are defined.

T71. Use the `print_codes` method to get the full list of codes. In order to print them so they are sorted by letter, pass the option `:by_name` as the second parameter:

```
>> print_codes(hc, :by_name)
'    0111
A    10
E    1101
H    0001
...
```

T72. Make an array of the test strings that come with the BitLab module:

```
>> words = read_words
=> ["A'U", "ALI'I", ... "PO", "UA"]
```

T73. Define a variable named `s` to be one of the test strings. You can either pick out a word on your own, or ask Ruby to choose one at random. For example, to work on the word in location 1 of the array, type

```
>> s = words[1]
=> "ALI'I"
```

To choose a random word, type

```
>> s = words[ rand(words.length) ]
=> "MOANA"
```

T74. Using the codes printed by the call to `print_codes`, write down the binary code for the word you selected. Check your answer by passing the word to `encode`:

```
>> encode(s, tree)
=> 100000111101111111
```

The code shown above is for `"ALI'I"` but you will naturally see a different code if you chose to use your own word.

T75. Try encoding some more words, by selecting them from the `words` array, encoding them by hand, and then passing them to `encode` to check your results.

When you are confident you understand how words are encoded in binary, you are ready for the next set of projects, which will decode binary strings to recover the original text.

T76. According to the tree in Figure 7.8, which letter is encoded by the bits 010? By 110011?

T77. You can check your answer to the previous problem by calling `print_codes` again, but this time pass the option `:by_code` so the codes are ordered by code:

```
>> print_codes(hc, :by_code)
0000  L
0001  H
001   K
. . .
```

T78. Make an array of codes using the predefined codes that come with the BitLab module:

```
>> codes = read_codes
=> [011010, 001010, 0011111, 110011010, ...                    ]
```

As you can see, the array is sorted by code length, with the shortest codes first.

T79. Define a variable named x to be the first binary code in the list:

```
>> x = codes[0]
=> 011010
```

T80. Try decoding this binary number, using the codes printed by the call to `print_code`. Check your result by passing the code to `decode`:

```
>> decode(x, tree)
=> "UA"
```

T81. Try decoding some more codes by repeating the previous two exercises with different items from the `codes` array.

The idea behind Huffman codes is that a word should be encoded with fewer bits when the encoding scheme has a short bit pattern for more common letters. In the next set of exercises we'll do some tests to see whether this idea works for Hawaiian words.

T82. The alphabet we are using has 13 letters. This expression will tell us how many bits it would take to encode each letter if we use a fixed-width code, *i.e.*, if we used the standard method of using the same number of bits for each letter:

```
>> log2(13).ceil
=> 4
```

T83. The length of the code for a word would be the number of bits per letter times the number of letters:

```
>> word = "ALOHA"
=> "ALOHA"

>> word.length * 4
=> 20
```

T84. To count the number of bits used by a Huffman code we can ask Ruby to print the number of bits in a message created by a call to `encode`:

```
>> encode("ALOHA", tree)
=> 100000010000110

>> encode("ALOHA", tree).length
=> 15
```

So the Huffman code for this test word saves five bits. That's not surprising, since the word has two A's, and the Huffman code for A is the shortest code.

fixed length code (4 bits per letter):	0100 0010 1010 0010 0011 0110
Huffman code (variable number of bits per letter):	11000 10 0001 10 0000 010

Figure 7.10: *A code with the same number of bits for each letter would use 24 bits to encode "MAHALO," but the Huffman code requires only 20 bits.*

T85. Since we're going to run this same test a few more times let's define a method that will print the number of bits in each encoding:

```
>> def test(s, t) puts s.length * 4; puts encode(s, t).length end
=> nil
```

T86. Pass a string and the tree to this new method:

```
>> test("MAHALO", tree)
24
20
=> nil
```

The first number printed is the number of bits for a fixed-width code, the second is the number of bits for the Huffman code (Figure 7.10).

T87. Try a word that has one of the infrequent letters:

```
>> test("WIKI", tree)
16
17
=> nil
```

Even though the code requires six bits for the W, the K is shorter, so the total number of bits is almost the same.

T88. Try a few more tests:

```
>> test("KAKA", t)

>> test("POHAKU", t)

>> test("HUMUHUMUNUKUNUKUAPUA'A", t)
```

♦ Make a new method named `diff` that is just like `test` (defined in Problem T85) except instead of printing the two lengths, it returns the difference of the lengths. For example:

```
>> diff("MAHALO", tree)
=> 4
```

because the encoding of *MAHALO* with the Huffman tree is 4 bits shorter.

♦ Use a call to `words.each` to iterate over all of the test words, calling `diff` to see how many bits are saved by the Huffman code. What is the total over all the words in the test set?

♦ What do you think will happen if an error occurs in a message encoded by a Huffman tree? Encode a long word or place name, *e.g.*,

```
>> msg = encode("HONOLULU", tree)
=> 000101011100100000011000000110
```

Next enter an expression that decodes a garbled copy of the message:

```
>> decode( garbled(msg, 1), tree )
=> "HEWLULU"
```

Can you explain what happened to this message?

7.6 Summary

The general topic for this chapter was *data representation*: how to encode information as a sequence of 1s and 0s so it can be stored in a computer's memory, saved in a file, or transmitted over a network.

We focused on techniques for encoding text. The simplest method simply assigns a unique pattern for each letter in an alphabet. The number of bits to use in a code depends on the size of the alphabet. For example, to encode the 26 letters used for English words, we need 5 bits for each letter. That's because there are 2^5 different 5-bit numbers, and since $2^5 = 32$ there are enough patterns to assign a different sequence of 1s and 0s to each letter.

The general formula for figuring out how many bits are needed to encode a set of items is $\lceil \log_2 n \rceil$, where n is the number of items, and the notation $\lceil x \rceil$ means "round x up to the next integer." So to include upper and lowercase letters, digits, and punctuation marks in a coding scheme, we would simply add up the number of symbols, compute the base-2 logarithm, and round up. The same idea works for any type of data; an example used in Section 7.1 was a coding scheme for answers on a questionnaire, where there are a fixed number of choices for each response.

We also looked at algorithms that work with binary encodings. An error detection algorithm examines a bit pattern to see if anything has changed since the data was originally encoded. The simplest algorithms are based on the idea of *parity*, which is just a count of the number of 1 bits in a piece of data. When the data is first encoded, an extra bit, called the parity bit, is added to the end of the code. The bit that is added is chosen so the total number of 1s is an even number. Then, if any one of the bits changes, the total will be an odd number, and we know something has gone wrong. More sophisticated algorithms can detect more than one change, and in some cases can even figure out which bits changed and correct the error.

Another type of algorithms that works with encoded information is a data compression algorithm. We looked at a simple technique that is based on the frequency of the letters in an alphabet. The idea is to use only a few bits to encode common letters, and more bits for uncommon letters, and as a result (on average) a complete piece of text will require a smaller number of bits to encode.

The Huffman tree algorithm touches on many of the themes from earlier chapters. Chapter 6 introduced the idea of a data structure, and made the point that how data is organized plays an important role in computation. That's certainly the situation for the Huffman tree algorithm, since the steps are defined in terms of a priority queue. The fact that nodes are always sorted in the queue makes it easy to see how a process that repeatedly removes the first two items always selects the two with the lowest frequency. Being able to save a new node in the queue, knowing that it will find its proper place in line, leads to the definition of a simple, yet elegant, algorithm for building a tree.

Another theme from earlier chapters is the strategy of using iteration to repeat the same basic steps. The goal of the Huffman tree algorithm is to assemble a collection of nodes into a final tree structure. The algorithm starts with a list of disconnected nodes. Each iteration makes definite progress toward the goal, since each step connects two pieces of the tree, and the number of disconnected pieces shrinks by one on each iteration.

Concepts and Terminology Introduced in This Chapter

---

encoding	A method for representing a piece of data as a sequence of bits (binary digits 0 and 1)
decoding	The opposite of encoding, a process for figuring out what is represented by a sequence of bits
error detection	An algorithm that decides whether a binary code was altered since it was first created
parity	A simple technique for error detection, attaches an extra bit to a code
text compression	An algorithm that reduces the number bits needed to encode a piece of data
binary tree	A data structure that represents relationships between objects; each node in a binary tree has at most two descendants
Huffman tree	A binary tree where each leaf node corresponds to a character, and the path from the root to a leaf defines the code for that character
priority queue	A data structure, similar to an array, except objects are always in order

---

Exercises

1. Figure 7.2 shows how responses to a questionnaire can be encoded in binary. Suppose the list of states is extended to include the District of Columbia and the U.S. Territories (*e.g.*, Guam or Puerto Rico). How many additional responses can be encoded before the code needs to be expanded to 7 bits?

2. Suppose the survey asks which country a person lives in, instead of which state. How many bits would be needed to encode a country ID? (There are 192 member states in the United Nations.)

3. If the survey is expanded to include other sports, how many different team IDs could be encoded if the code used 9 bits for a person's favorite team?

4. What are the ASCII codes of the following characters?

   ```
   Q   w   e   r   t   !
   ```
   Note: you can use IRB to check your answer:
   ```
   >> "Qwert!".each_byte { |x| puts x.code(7) }
   ```

5. What characters are encoded by the following binary numbers (based on the ASCII coding scheme)?

   (a) `1111000`   (d) `1101111`   (g) `0101000`
   (b) `0101110`   (e) `1100100`   (h) `0110111`
   (c) `1100011`   (f) `1100101`   (i) `0101001`

6. Extend each of the codes shown in the previous problem with a parity bit that results in an 8-bit code with even parity (*i.e.*, an even number of 1 bits).

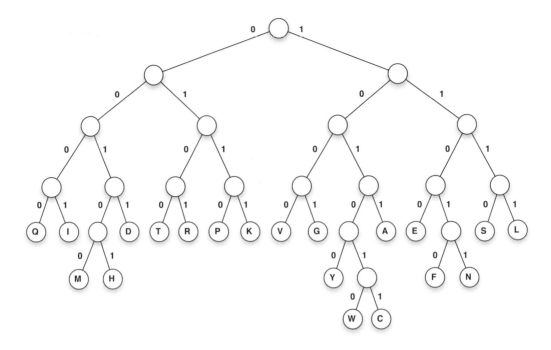

7. The codes shown below were originally made by a transmitter that added an even parity bit. Which codes have a single-bit error somewhere in the code?

   (a) `11100111`   (d) `11100001`   (g) `11000111`

   (b) `11111011`   (e) `11100100`   (h) `11001010`

   (c) `11100100`   (f) `11010000`   (i) `01000010`

The next questions all refer to the Huffman tree in the figure at the top of this page. The tree defines a code for the 20 letters used to abbreviate amino acids, based on their frequency in protein molecules. If you want to check your answers, the letter frequencies can be read by calling `read_frequencies(:aafreq)`.

8. Which letters have the shortest codes?

9. Which letters have the longest codes?

10. What is the code for the letter *M*? For *V*?

11. Use the tree to encode the following sequences of letters:

   a) *MGF*

   b) *GARW*

   c) *VAEYNK*

12. Which letter is encoded by the bits `1001`? By `101011`?

13. Use the tree to decode the following strings of bits (*note:* the decoded words are all common English words):

   a) `10110011`

   b) `011011001001`

   c) `01101100101101011110`

14. DNA sequences can be written as strings made up of the letters *A, T, C,* and *G.* Describe the code you would get if you built a Huffman tree for these characters, assuming each letter has the same probability (0.25).

15. ♦ Find a web site with a Hawaiian-English dictionary, or another source of Hawaiian words. Can you find a word that would be longer if encoded with a Huffman code than with a standard 4-bit code?

16. ♦ Explain what would happen if an error changes a single bit in a message encoded with a Huffman code. Try some experiments that make encoded Hawaiian words and then call `garbled` to change one bit. Can you still decode the resulting bit string?

17. ♦ After working on the previous project, try some experiments that add a single error to some long words with 7 or more letters. Instead of calling `garbled` to add a random error, call `flip` to introduce an error somewhere toward the front of the message. Now pass the result to `decode`. What happens? Is the decoding process hopelessly lost, or does it eventually recover and start correctly decoding letters toward the end of the word?

Chapter 8

The War of the Words

An introduction to computer architecture and assembly language programming

The first machines capable of carrying out the steps of an algorithm were based on the technology that was available when the machines were designed. Early computers used basic mechanical devices like gears, levers, pulleys, and rotating axles.

From the early 1800s, when the first computing machines were conceived, until the mid 1940s, when the first fully automatic computers were working, progress in computing went hand in hand with progress in technology. Improved materials and better manufacturing techniques allowed steady improvement in calculators and computers. The introduction of the vacuum tube, which was basically an "electronic switch" that could take the place of older mechanical switches, was a major step. The ENIAC, the first electronic machine, could carry out its arithmetic operations 1000 times faster than its electromechanical predecessors.

It was another innovation from this era that made it possible for later computers to evolve into the machines we use today. This innovation was more than simply an improvement in technology. It was a conceptual breakthrough in how computers were organized.

Up until 1945, the program that controlled the sequences of operations carried out by a machine was external to the machine itself. The data processed by the machine was stored internally, encoded in the form of decimal or binary digits, using techniques similar to those described in the last chapter. The programs that controlled the machines, however, were typically written on punch cards or paper tapes, and the operations were read one at a time as the machine worked its way through the algorithm. To program the ENIAC, cables were plugged into a panel on the front of the machine (Figure 8.1). These cables connected different parts of the processor, so that output from one component was fed as input to another component.

Figure 8.1: *The ENIAC was programmed by plugging cables into a panel on the front of the machine.*

The major innovation that changed computer science and enabled the phenomenal growth in computing since the 1940s was the idea that the steps in a program could be stored inside the computer along with the data. Computer scientists use the word "architecture" to describe how a set of processors, memories, and I/O devices are interconnected. This new design eventually became known as the **von Neumann architecture**, after John von Neumann (1903–1957), the Hungarian-born American mathematician who wrote the first paper that proposed storing both programs and data in a computer's main memory unit (Figure 8.2).

It would be hard to overstate just how important this idea has been in the development of computer science, since it has had a major impact on almost every area, including computer technology, programming languages, software engineering, and theoretical computer science. In terms of computer technology, modern systems would not be able to execute programs anywhere nearly as quickly if programs were still stored externally. Computer chips today can execute billions of arithmetic operations each second, but all that computing power would be wasted if processors had to read instructions from an external source

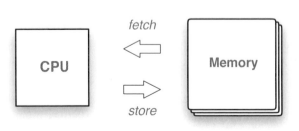

Figure 8.2: *In a von Neumann (stored-program) computer, the memory holds the encoded forms of both instructions and data.*

that couldn't provide commands quickly enough to keep the processor running at full speed. When the program is in the same memory as the data, processors can access instructions just as quickly and easily as data.

The idea of a stored-program computer opened up the possibility of implementing algorithms by using programming languages like Ruby, Java, or C++ to describe the steps in an algorithm. The first stored program computers were programmed by writing out extremely detailed step-by-step instructions, where each step corresponded to one of the operations the machine was capable of executing. This type of program is called called a "machine language" program, because the operations were defined in terms of what the machine could do. Now, however, we write programs in "high level languages" that describe operations in more abstract mathematical terms. Text editors, compilers, and other applications can treat programs as data. A compiler reads instructions written in a high level language and translates them into the encoded form that is stored in memory where a processor can execute them.

In this chapter we will explore the idea that programs are, like numbers and strings, a form of data that can be encoded in binary and stored in memory. The projects are based on an old computer game named **Corewar**. The game is essentially a computerized version of Battleship. Two programs are loaded into a computer's memory, and they take turn executing instructions. The idea is for a program to lob a "bomb" at the other program that will cause it to stop executing. The bombs are the encoded form of the machine language instruction that tells the machine to halt. A program will pick up a bomb and store it somewhere else in memory, hoping to write it over the code of the other program, so that when the second program executes the instruction it halts and loses the game.

The games are run inside a specially constructed machine that is organized along the lines suggested by von Neumann, where a processor is connected to a memory that holds both instructions and data. To the first program, the bomb is simply a piece of data that can be moved around from one place to another in memory. To the second program, however, the bomb is a machine instruction that tells the program to halt. The fact that instructions and

Core Memory

The "core" in "Corewar" comes from the main type of technology used to implement computer memories from the 1950s through the 1970s.

A core was a small "donut" made of a material that could be magnetized in a clockwise or counterclockwise direction. Wires threaded through a core could read the direction or set the field to either direction.

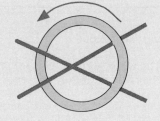

The word "core" is still used as a synonym for main memory in a computer system. An "in-core" algorithm is one that carries out all its steps on data that has been read into memory, while an "out-of-core" algorithm can work on data that is too large to read into memory all at once. A "core dump" is a file created by writing the entire contents of memory out to a disk or CD.

data are both encoded and stored in the same memory unit is the central concept explored in the projects in this chapter.

Computer architects refer to an item that can be stored in a single location in memory as a "word." In a von Neumann architecture, a word can be either an instruction or a simple piece of data like a number or a character. The title of this chapter summarizes what is going on inside the machine during a game of Corewar: two collections of words are throwing data at each other until one of them is forced to halt.

8.1 Hello, MARS

The computer used to play Corewar is named MARS. In order to play a game, we don't need to build a MARS system out of processor and RAM chips, but instead we can write a program in Ruby that mimics the actions of a real system. A program that emulates the actions of a computer is known as a **virtual machine**, or VM..

MARS and other processors based on the von Neumann architecture execute programs according to what is known as the **fetch-decode-execute** cycle. The cycle starts when the machine reads an instruction from memory. This is the "fetch" portion of the cycle.

When the instruction arrives, the processor decodes it in order to figure out what to do. Often this phase performs additional memory operations; for example, if the instruction tells the machine to add two numbers, the decode phase is when the machine reads the values from memory. Finally, after all the necessary information has been gathered, the machine carries out the last phase of the cycle and does the actual instruction execution. When the operation is complete, *e.g.*, after the two numbers have been added and the result saved, the cycle repeats with the next instruction in the program.

In our virtual machine simulator, memory is simply a large array. The items in the array will be a new type of object, called a Word, that will represent either a single MARS machine language instruction or a single piece of integer data. When the virtual machine fetches a word from memory, it specifies the array index of the item it wants. For example, if the memory is named `mem`, and the processor wants to execute the instruction in the first location in memory, it retrieves the word from `mem[0]`. When computer architects talk

---

Virtual Worlds

The word "virtual" was originally used in computer architecture to describe something that was not really present but was implemented instead by software.

For example, some old computers had 64KB of **physical memory**, meaning there were 65,536 locations to store information in the RAM chips. By special techniques implemented in the operating system, a program could behave as if there were several megabytes of memory. This extended memory was known as the **virtual memory**.

The idea of a **virtual machine** is still widely used in computer science. A familiar example is the Java Virtual Machine, or JVM. Web designers write "applets" in Java, and when users connect to a web site, the applet is downloaded and executed by a JVM that is part of a web browser.

```
; test_hello.txt -- a "hello, world" program for MARS

x          DAT #7
y          DAT #4
hello      ADD x, y        ; add the contents of x to y
           DAT #0          ; a DAT instruction halts the machine
           end hello
```

Figure 8.3: *A "hello, world" program in Recode, the assembly language of the MARS virtual machine. This program executes one instruction (the ADD on the line labeled* `hello`*) and then halts.*

about reading or writing items in memory, they use the term **address** to refer to a memory location, so they would say "the processor fetches the word from memory address 0."

To run a MARS program, we will use a text editor to create a file that has a single MARS instruction on each line. But instead of trying to figure out what the binary encoding of each instruction might be, we can use a symbolic notation known as **assembly language**. An assembly language program is a plain text file that has one line for each machine instruction, along with comments and a few other lines that tell the computer how to execute the program. The assembly language for our virtual machine is known as Redcode. The name of the machine, "MARS," is an acronym for Memory Array Redcode Simulator.

Our first project with MARS will be the machine language equivalent of a "hello, world" program. This program doesn't print a message, like a normal "hello, world" program, but it serves the same general purpose: it's a trivial program that illustrates how to use the machine. The MARS version of "hello, world" will just add two numbers and then halt. The Redcode instructions for this program are shown in Figure 8.3.

The main thing to remember when looking at an assembly language program is that each instruction in the program will be translated into a word object that will be stored in memory. The program of Figure 8.3 has four instructions, and when we test it, the program will be loaded into memory locations 0 through 3. The first two lines define the data values that will be added, the third has the instruction that performs the addition, and the fourth is the word that halts the program.

At the front of each line there can be a **label**, which is like a variable name or method name in a Ruby program. Labels are typically simple names, written in lowercase, that remind us what is stored in memory at that location. If a line does not have a label, there need to be spaces at the front of the line so the machine knows there is no label.

The most important part of each line is the **opcode**, which is short for "operation code." It is the opcode that specifies what the processor should do when it fetches this word. The two kinds of instructions used in this trivial little program are ADD and DAT. ADD, as you might expect, tells the machine to add two numbers, and DAT indicates that the word holds a piece of data.

Following the opcode there will be one or two **operands**. ADD instructions have two operands. When the machine executes an instruction of the form `ADD x, y` it goes out to memory to find the values in locations specified by `x` and `y`, adds them, and stores the result back in memory at location `y`. Notice that this instruction updates memory, and that the old value in location `y` will be replaced by the sum.

For our programs, each DAT instruction will have just one operand, which will be a piece of data used by the program. MARS programs can only operate on integer data, but the numbers can be positive or negative.

The last line in the program in Figure 8.3 is a special statement. It just has the word `end` followed by a label. This tells MARS that when the program is loaded into memory the first instruction to execute is the one on the line labeled `hello`.

For the first set of experiments with MARS programs we will make a small "test machine" (called MiniMARS) to run our test programs. These machines have a tiny memory unit that is just big enough to hold a single program. You can think of it as a USB "memory stick" containing a machine language program being inserted into the side of a MARS processor. To make one of these test machines, pass the name of the program to a method named `make_test_machine`. If you have your own program in a file, for example `my_redcode.txt`, you would call

```
make_test_machine("my_redcode.txt")
```

The "hello, world" program and some other small tests are included as part MARSLab, the module we will use for projects in this chapter. To make a test machine using one of these programs, put a colon in front of the name, instead of enclosing it in quotes. This statement in IRB will create a test machine for the "hello, world" program:

```
>> m = make_test_machine(:test_hello)
=> #<MiniMARS ... >
```

If you pass `make_test_machine` a symbol it will look for one of the predefined programs included in the MARSLab module, otherwise it will look for a file in your project directory.

When a program is loaded into memory, the method that translates the Redcode instructions into machine language removes the labels and converts the opcodes and operands into an internal binary representation. But for our projects, in order to be able to follow the execution of a program, the MARS simulator will print instructions and data in a format that looks like the original Redcode. To see what a program looks like in memory, call a method named `dump`. Here is the machine language translation of the "hello, world" program:

```
>> m.dump
0000:  DAT #0  #7
0001:  DAT #0  #4
0002:  ADD -2  -1
0003:  DAT #0  #0
```

Notice how the labels in the ADD instruction have been translated into addresses. Since the word labeled `x` in the Redcode program is two places before the ADD, the assembler replaced the `x` with -2. Similarly, the word labeled `y` is one location in front of the ADD, so `y` was replaced by -1. In later examples we might see positive numbers as operands, in cases where an instruction refers to data that is later in the program. The important thing to remember is that when MARS executes an instruction it uses these numbers to create addresses, not as the data to be added. We'll come back to this point later in the chapter.

Tutorial Project

The MARS virtual machine and other Ruby methods you will use for this project are in a module named MARSLab. Start a Ruby session and include this module:

```
>> include MARSLab
=> Object
```

T1. The "hello, world" program described in this section is in a file named `test_hello`. Make a test machine for this program:

```
>> m = make_test_machine(:test_hello)
=> #<MiniMARS mem = [DAT #0 #7, ... pc = [ *2 ]>
```

This statement asked Ruby to make a MiniMARS test machine object and save a reference to it in the variable named `m`.

T2. Call the `dump` method to get a "core dump" showing the complete contents of the memory of this machine:

```
>> m.dump
0000: DAT #0 #7
0001: DAT #0 #4
...
```

The number at the front of each line is the memory address where an item translated from the original Redcode program was stored.

You may have noticed the `dump` method prints two operands for the DAT instructions. That's because the internal machine language always has two operands, even though some Redcode instructions just have one. In our programs the first operand on a DAT instruction will always be `#0` and we can safely ignore it. The data we use will always be the second item printed after DAT.

T3. When this test machine was initialized it was set up so the first instruction will come from memory location 2. Call a method named `step` to tell the machine to fetch and execute this instruction:

```
>> m.step
=> ADD -2 -1
```

The value returned by this method is the instruction that was just executed. As expected, the machine executed the ADD instruction.

What we want to do now is see if the ADD instruction did in fact add the contents of the two locations specified in the instruction. The ADD instruction should have added the contents of the memory cell two places in front of the instruction to the memory cell one place in front of the instruction. Since the instruction is in memory address 2, the contents of cell 0 should have been added to cell 1.

T4. Call the `dump` method again:

```
>> m.dump
0000: DAT #0 #7
0001: DAT #0 #11
...
```

Do you see how the item in memory location 1 is now the sum of 7 and 4?

Every time we call `step` the machine keeps track of which instruction it just executed and updates the address used to fetch instructions so the next call to `step` will get the next instruction in the program. There is also an "on/off" switch inside the machine. Executing a DAT instruction halts the current program by setting this switch to `:halt`.

T5. Call a method named `status` to get some information about the current state of the machine:

```
>> m.status
Run: continue PC: [ *3 ]
```

This output means the program is still running ("continue" means the MARS program has not halted yet) and the program counter (the address of the next instruction to fetch) is memory location 3.

T6. Call the `step` method again:

```
>> m.step
=> DAT #0 #0
```

As expected, the machine executed the DAT instruction in location 1.

T7. Call the `status` method again:

```
>> m.status
Run: halt
```

The state has changed to `halt` because the DAT instruction terminated the program.

T8. If we try to call `step` again we'll get an error message because there are no more steps to execute:

```
>> m.step
=> "machine halted"
```

8.2 The Temperature on MARS

As a way to introduce some of the other operations that can be used in Redcode programs, let's look at a program that converts temperatures from Fahrenheit to Celsius. When we did this calculation in Ruby we wrote it as a single expression:

```
c = (f - 32) * 5 / 9
```

If we want to do this same computation with MARS we need to write a Redcode program that has DAT instructions for the variables `f` and `c` and a sequence of machine instructions that carry out the steps in the calculation. When the program halts we'll look in the memory cell designated to hold the value of `c` to find the converted temperature value.

MARS is a very simple machine, with a total of only 11 instructions. There are ADD and SUB instructions, to do addition and subtraction, but there are no multiply or divide instructions. We can still do multiplication and division, though. We just have to implement the algorithms that do these operations in the form of a sequence of MARS instructions.

To multiply x times y we can initialize a memory cell to 0, and then add x to it y times. We can implement this operation using three Redcode instructions, as shown in Figure 8.4. The first instruction is one we've seen already. It just tells the machine to add the contents of x to a cell named `acc`, which has presumably been initialized with a `DAT #0` instruction. The label `acc` is short for "accumulator."

The second instruction subtracts 1 from the memory cell that holds the value of y. What is new about this instruction is the # symbol in front of the 1. The # tells MARS to subtract the number 1, instead of using the 1 to create an address to fetch a piece of data from memory. An operand with a # in front is called an **immediate** operand, and it means "use this number as the data" instead of "use this number as an address to fetch the data from."

Figure 8.4: *To multiply x by y, set an accumulator to 0 and then add x to it y times.*

```
; a loop to compute x * y

mult     ADD x, acc
         SUB #1, y
         JMN mult, y
```

The third instruction is what is known as a **branch** instruction, or a "jump." This instruction tells MARS to look at the value in y, and if it is not 0, to go back to the cell labeled mult and continue execution there. The JMN opcode stands for "JuMp if Not zero." The two operands are the label on an instruction to jump to, and the name of a memory location to check to see if it is 0.

These three instructions define an iteration, in this case a simple loop that is executed over and over until y counts down to 0. The number of loop executions depends on the value initially stored in memory location y. As an example, if x contains DAT #6, and y contains DAT #7, this loop will be executed seven times, each time adding 6 to acc. In other words, it will compute 6×7.

The complete code for the Celsius program is in a file named test_celsius (it's also shown in Figure 8.5). The input temperature is placed in the memory location labeled fahr. When the program halts, the output will be in the location labeled cels. The first step is to subtract 32 from the input temperature. Next, we want to multiply this value by 5, so the accumulator is set to 0, and a counter is set to 5. The loop that starts at the instruction labeled mult is iterated five times, and after the last iteration the product of (fahr-32)*5 is stored in the cell labeled acc.

```
; compute cels = (fahr - 32) * 5 / 9 and halt

fahr     DAT #80          ; input temperature
cels     DAT #0           ; store result here
ftmp     DAT #0           ; save fahr-32 here
start    MOV fahr, ftmp   ; (1) subtract 32
         SUB #32, ftmp
mult     ADD ftmp, acc    ; (2) multiply by 5
         SUB #1, count
         JMN mult, count
div      SUB #9, acc      ; (3) divide by 9
         SLT #0, acc
         DAT #0           ; stop here when division done
         ADD #1, cels
         JMP div
acc      DAT #0           ; accumulator
count    DAT #5           ; counter
         end start
```

Figure 8.5: *A Redcode program to convert a temperature value from Fahrenheit to Celsius. The input temperature is the value of the DAT instruction with the label* fahr. *After the program runs, the DAT instruction labeled* cels *will hold the output temperature.*

The division is implemented by a loop that is similar to the multiply loop. The algorithm for dividing x by y is to count how many times y can be subtracted from x. Since we want to divide the value in `acc` by 9, this second loop repeatedly subtracts 9 from `acc`, and keeps track of the number of times the loop is executed by adding 1 to `cels` each time through the loop. The test for the end of this second iteration is a little trickier. The instruction `SLT #0, acc` means "skip the next instruction if $0 < acc$" (SLT is an abbreviation for "Skip Less Than"). So as long as `acc` has a value greater than 0, the machine skips over the halt instruction, adds 1 to `cels`, and jumps back to the top of the loop at the instruction labeled `div`. As soon as `acc` becomes less than 0 the machine does not do the skip, but instead executes the `DAT #0` instruction and halts.

Tutorial Project

Before running the complete Fahrenheit-to-Celsius program, we will do some experiments with the loop that does the multiplication. A small test program is in a file named `test_mult`. The Recode program is shown at the bottom of the page. The first three lines are DAT instructions that define locations for the input values and the place where the result will go, and the next three lines have the instructions for the multiplication loop.

T9. You can ask MARS to print the source code for a program by calling a method named `listing`. Type this expression to see the Recode instructions for the multiplication test program:
```
>> MARS.listing(:test_mult)
```

T10. Make a test machine that has this program loaded into its memory:
```
>> m = make_test_machine(:test_mult)
=> #<MiniMARS ... >
```

T11. Use the `dump` method to look at the machine language. Since all we want to do is look at the DAT instructions in memory locations 0 through 2, pass those two addresses to `dump` so it prints only those locations:
```
>> m.dump(0,2)
0000: DAT #0 #7
0001: DAT #0 #6
0002: DAT #0 #0
```
What these lines show is that the two values that will be multiplied are 7 and 6, and the result is initialized to 0.

```
;redcode
;name test-mult
;strategy    multiplication demo: compute acc = x * y

x         DAT #7          ; multiplicand
y         DAT #6          ; multiplier
acc       DAT #0          ; accumulator
mult      ADD x, acc      ; add x to acc
          SUB #1, y       ; subtract 1 from y
          JMN mult, y     ; repeat if y is not 0
          DAT #0          ; algorithm halts here
          end mult
```

T12. Call the `step` method three times, so that each instruction in the loop is executed once:
```
>> m.step
=> ADD -3 -1
>> m.step
=> SUB #1 -3
>> m.step
=> JMN -2 -4
```
The lines printed by `step` are machine language instructions, where labels have been turned into addresses. `ADD -3 -1` means "add the contents of the word three locations before this instruction to the word one location before this instruction." Do you see what the SUB instruction is supposed to do? And how the JMN instruction means "jump back two instructions if the value in the word four locations back is not zero"?

T13. Call `dump` again to look at the memory cells that hold the data:
```
>> m.dump(0,2)
0000: DAT #0 #7
0001: DAT #0 #5
0002: DAT #0 #7
```
Can you see how the number 7 was added to `acc` (memory location 2), and how the counter (memory location 1) decreased by 1?

T14. Instead of calling `step` three times, we can call another method named `run`. `run` will call `step` for us, and return either when the program stops or it has called `step` the specified number of times:
```
>> m.run(3)
=> 3
```
The return value is the number of instructions that were executed.

T15. Call `dump` again:
```
>> m.dump(0,2)
0000: DAT #0 #7
0001: DAT #0 #4
0002: DAT #0 #14
```
Once again 7 was added to `acc`, and 1 was subtracted from `count`.

T16. This expression will execute the loop four times by repeatedly calling `run` and then `dump`:
```
>> 4.times { m.run(3); puts "--"; m.dump(3,5) }
--
0003: ADD -3 -1
0004: SUB #1 -3
0005: JMN -2 -4
--
0003: ADD -3 -1
0004: SUB #1 -3
0005: JMN -2 -4
...
```

T17. The next instruction should be the DAT instruction that terminates the algorithm:
```
>> m.step
=> DAT #0 #0
```
If you don't see the DAT instruction, call `run(3)` a few times until the return value is 0.

T18. Finally, look at the memory location that holds the final answer. Does `acc` contain 6×7?
```
>> m.dump(0,2)
0000: DAT #0 #7
0001: DAT #0 #0
0002: DAT #0 #42
```

T19. If you want to play around with this algorithm some more, save a copy of the program in a file in your project directory by calling checkout:

```
>> MARS.checkout(:test_mult, "my_test_mult.txt")
Copy of test_mult saved in my_test_mult.txt
```

Now you can edit the file (*e.g.*, to change the value of x or y), save it, make a new test machine for it, and running it again. When you make a test machine for your copy of the program, enclose the name in quotes:

```
>> m = make_test_machine("my_test_mult.txt")
=> #<MiniMARS mem = [DAT #0 #7, ... pc = [ *3 ]>
```

T20. Make a test machine that has the temperature conversion program loaded into its memory:

```
>> m = make_test_machine(:test_celsius)
=> #<MiniMARS ... >
```

T21. Look at the first two locations in memory. The first is the input temperature, and the second is where the output temperature will be placed when the program is done:

```
>> m.dump(0,1)
0000: DAT #0 #80
0001: DAT #0 #0
```

T22. Run the program:

```
>> m.run
=> 124
```

The return value means the machine executed 124 instructions before it halted.

T23. Look at the first two memory locations again:

```
>> m.dump(0,1)
0000: DAT #0 #80
0001: DAT #0 #26
```

Does the second location hold the Celsius equivalent of $80°F$?

You can now move on to the next section, which will explain how MARS runs two or more programs that compete against each other in a game of Corewar, but if you want to explore the division algorithm you can work on the following exercises.

♦ The test program for the division algorithm is in a file named test_div:

```
>> m = make_test_machine(:test_div)
=> #<MiniMARS ... >
```

♦ Use dump to look at the first three memory locations:

```
>> m.dump(0,2)
0000: DAT #0 #20
0001: DAT #0 #9
0002: DAT #0 #0
```

♦ When the program runs, it should compute $20 \div 9$ and store the result in memory location 2. Run the program and look at memory after the program halts. Are the results correct?

♦ Check out a copy of the program, and try some experiments with other data values:

```
>> MARS.checkout(:test_div, "my_test_div.txt")
Copy of test_div saved in my_test_div.txt
=> nil
```

♦ There is a bug in this code: it gives the wrong answer for $x \div y$ when y divides x evenly, *i.e.*, when the remainder is 0. Try the program with $x = 20$ and $y = 10$. Can you see what the problem is?

```
;redcode
;name Dwarf
;author A. K. Dewdney
;strategy Throw DAT bombs at every
;         4th memory cell.

bomb    DAT #0
dwarf   ADD #4, bomb
        MOV bomb, @bomb
        JMP dwarf
        end dwarf
```

Figure 8.6: *The Dwarf program stores DAT instructions in every fourth memory cell.*

8.3 Corewar

In Battleship, a classic board game, players choose where to place a set of ships on a rectangular grid that cannot be seen by the other player. When both players are set up, they take turns calling out board locations, defined by a row and column. For example, player *A* might call out "C-5." If player *B* has a ship on square C-5 it is hit, and when a ship takes enough hits it sinks.

In Corewar, our computer-based version of Battleship, two Redcode programs are assembled and loaded into the MARS machine's memory at random locations. Our VM is set up to have a default memory size of 4096 words, so instructions can be loaded into any location between 0 and 4095. Neither program is given any information about where the other program is located.

When the machine starts it will alternate instructions from the different programs, first executing a single instruction from Program *A*, then an instruction from Program *B*, then *A* again, and so on. The idea is for a program to lob a "bomb" onto a random location in memory, hoping to hit the other program. A bomb is simply a DAT instruction—if the other program tries to fetch and execute the DAT, instead of the instruction that was originally there, it will halt because a DAT serves as a halt instruction in Redcode. When a program runs into a halt instruction it loses, and the other program is declared the winner. If both programs are still running after a predefined number of rounds the game ends in a draw.

A very simple Corewar program, named Dwarf,[1] illustrates one strategy for playing the game. The program is very small: it has one DAT instruction (the bomb) and a three-instruction loop (Figure 8.6). The DAT instruction in the location labeled `bomb` serves two purposes in this program: it is the address of where the bomb should be thrown, and, since it is a DAT instruction, it's also the bomb itself. The main idea is to add 4 to this address each time through the loop, so the bomb is stored in every fourth location in memory.

The first and third instructions in the loop are straightforward: the ADD instruction adds 4 to the address where the bomb will go, and the JMP branches back to the top of the loop. The MOV, which stores the bomb at the specified address, illustrates a new programming technique, called **indirect addressing**. Notice that there is an @ symbol before the second operand. Without the @, an instruction like `MOV x, y` means "copy x to location y." But when the operand includes @, as in `MOV x, @y`, it means "copy x to the location pointed to by y." In other words, y is not the destination, it holds the *address* of the destination.

---

[1]Corewar programs often have names inspired by Dungeons and Dragons or *Lord of the Rings*.

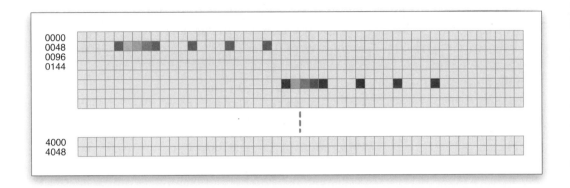

Figure 8.7: *Visual display of MARS memory, showing the progress of a Corewar tournament with two Dwarf programs.*

In this program the word labeled `bomb` holds the address where we want to throw the DAT instruction, and by adding 4 to it each time through the loop we end up storing DAT instructions every four locations in memory.

As the programs start getting more complex, and especially when there are two programs in memory at once, it becomes harder and harder to follow the execution of the programs by simply calling `dump` to print the contents of memory. The MARSLab module includes another method for viewing the contents of memory. This method, called `MARS.view`, opens the RubyLabs canvas (described on page 150) and draws a series of gray boxes, one per memory cell. Since there are 4096 memory cells, they can't all be displayed on a single line, so they are split into a set of rows. The window shown in Figure 8.7 has 48 cells per row. The top row has boxes for memory cells 0 through 47, the next row for cells 48 through 95, and so on.

When the MARS virtual machine is running a set of programs the view will be updated. Each time the machine fetches a word from memory or writes a word into memory, the corresponding box will change color. The viewer uses different colors for each program to make it easier to watch the progress of the different programs.

Figure 8.7 shows what you might see during a tournament being played by two Dwarf programs. A cluster of four colored cells in a row indicates where the instructions are being executed; each time MARS fetches an instruction it colors one of these cells. When you watch the display in action it will be easier to see, but as the programs iterate over the three instructions in the body of the loop these cells will be changing color so you can watch the progress of each program. The other colored cells indicate where bombs are dropped. As the programs run you will see how bombs are placed in every fourth cell.

Figure 8.7 also helps explain how MARS deals with two issues related to memory addresses. The pointer used in indirect addressing to specify where a bomb is written out to memory is initially set to 0, and then on successive iterations it is set to 4, 8, 12, *etc*. As you can see from the pattern of colored squares in the figure, MARS interprets the pointer to be a distance from the start of the program. Just as the number -2 in an instruction like `JMP -2` tells the machine to go to an address relative to the current instruction, a value in a pointer tells the machine to use an address relative to where the pointer is located.

This is a fussy little detail that you will need to understand if you want to write your own Redcode programs, but for the "big picture" it's enough to know that the Dwarf program always tosses bombs at locations that start at the same address the program is loaded into. The leftmost colored cell for each program in Figure 8.7 is the location of the pointer. The first bomb went into a memory cell 4 places after this one, and each new bomb goes into a location 4 places past the previous one.

The other issue is, what happens when a pointer value reaches the end of memory? With a 4096-word memory, the addresses run from 0 to 4095. If a Dwarf program is loaded at location 4000, will it only be able to put bombs in locations 4004, 4008, *etc.*, up to location 4092, before it runs out of room? If that were the case, the contest wouldn't be very fair, since the program loaded closer to address 0 would always have an advantage. To deal with this, MARS memory references "wrap around" to the start of memory again. If you watch the display, you will eventually see that bombs reach the end of memory and then start over again in the low addresses.

Tutorial Project

T24. Create the graphics window that will display the state of the MARS machine by calling the MARS.view method:

>> *MARS.view*

You should see a grid of gray cells similar to the one in Figure 8.7.

T25. To play a game using two programs, call a method named contest, passing the IDs of the programs. This command will start a game with two Dwarf programs:

>> *MARS.contest(:dwarf, :dwarf)*

As the programs run, the squares in the graphics window will start filling in. The contest method will return either when one program halts or after a predefined number of rounds have been executed (the default is 1000).

If both programs are still running and you want to continue the game for another 1000 rounds just type MARS.run. You can pass an argument to run to specify the number of rounds to execute, *e.g.*, MARS.run(50) will run 50 rounds and then return. If you want to stop the program before 1000 rounds have been executed just type ^C (hold down the control key while typing C).

Some things to look for in the display:

- Can you see how each program is in a three-instruction loop?

- Bombs are being dropped on every fourth location, starting at the end of the set of instructions for each program and moving toward higher addresses.

- If the programs run long enough, you should see bombs "wrap around" from high addresses to low addresses.

T26. If you're still not sure how the Dwarf program works, and you'd like to do some more experiments with it, create a test machine. You can use a second parameter to tell the make_test_machine method to make a memory with extra words:

>> *m = make_test_machine(:dwarf, 20)*
=> *#<MiniMARS ... >*

Using the techniques shown in the previous section, call dump to look at the contents of memory, run the machine for a few steps, and then look at memory again.

8.4 Self-Referential Code

The MARS projects up to this point have demonstrated the main idea of the von Neumann architecture. The two major components in a computer are the processor and memory. The memory holds encoded forms of both instructions and data, and programs are executed through a fetch-decode-execute cycle.

One of the implications of this design is that programs can be treated as data. Programs can make copies of other programs—or even themselves—and modify code by writing new instructions. The project in this section will explore this idea in a little more depth.

The Dwarf program tries to force the other program in memory to halt by writing DAT #0 over the code of the other program. But the Dwarf is not limited to writing DAT instructions; it could write any MARS instruction. One possibility is to write JMP 0 to every fourth location. Recall that the operand in a JMP instruction is the distance, relative to the location of the JMP, to the next instruction in the program. The Dwarf itself implements its loop through a JMP -2 that means "go back two instructions." So JMP 0 means "go to the instruction in the same location as this instruction." In other words, any program executing this instruction would be caught in an infinite loop! It's a small loop, containing only one instruction, but it's a loop nonetheless.

In more abstract terms, a program that writes JMP 0 instead of DAT #0 is trying to "stun" the other program instead of kill it. The other program will still remain in the game, but it can't make any progress, since it will be stuck in the location of the JMP instruction. This strategy might be effective in certain situations that will be explained in the next section.

The fact that a program can use any instruction as a type of data opens up another Core-war strategy: a program can copy itself, and then jump to the location of the copy and continue execution there. For example, suppose the Dwarf program has executed 1024 iterations of its loop and the other program is still running. Since the memory size in our MARS virtual machine is 4096 words, and the Dwarf drops a bomb on every fourth location, the next 1024 iterations will just write DATs over the same locations. One way to get the program to write DATs to different locations is to copy the four instructions in the Dwarf code to a new location in memory and then jump to this new location. If the difference between the address of the copy and the address of the original is not a multiple of four, the copy will start writing DATs over new locations.

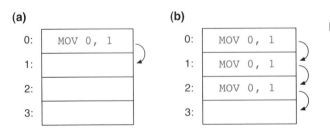

(a)

0:	MOV 0, 1
1:	
2:	
3:	

(b)

0:	MOV 0, 1
1:	MOV 0, 1
2:	MOV 0, 1
3:	

Figure 8.8: *A program named Imp copies itself. (a) When it is first loaded it occupies only one memory cell. (b) After two rounds the program has copied itself twice, and is ready to make a new copy.*

A very simple program that copies itself is named Imp (Figure 8.8). This program has just one instruction: MOV 0, 1. An instruction of the form MOV x, y means "copy the contents of location x to location y." Here again the 0 and the 1 are interpreted as memory locations relative to the address of the instruction. The 0 means "the address of this instruction" and the 1 means "the address following this instruction." When the machine executes the MOV, the instruction copies itself one location ahead in memory. When MARS goes to fetch the next instruction for this program, it sees the newly made copy, and the cycle repeats.

This tiny little one-instruction program simply moves through the entire memory of the machine. Unless it is stopped or modified by the other program, after 4096 cycles the entire MARS memory will be filled with MOV 0, 1 instructions. The natural question is, how effective is this strategy? What will happen when the IMP writes itself over the code of the other program?

The project in this section will be to run a contest between an Imp and a Dwarf. Before you start, make some predictions about what might happen in this game:

- What will happen if the Imp writes its code over the Dwarf?

- Can the Imp ever cause the Dwarf to halt?

- What are the odds of the Dwarf being able to stop the Imp by writing a DAT instruction over it?

Tutorial Project

T27. Load the Imp program into a small test machine with 10 words:

```
>> m = make_test_machine(:imp, 10)
=> #<MiniMARS mem = [MOV 0 1, nil, ... nil] pc = [ *0 ]>
```

T28. Call dump to look at the memory:

```
>> m.dump
0000: MOV 0 1
0001: DAT #0 #0
0002: DAT #0 #0
...
```

As you can see, the program is only one instruction long, and it is loaded into location 0.

T29. Tell the machine to execute one instruction, and print the memory again:

```
>> m.step
=> MOV 0 1

>> m.dump
0000: MOV 0 1
0001: MOV 0 1
0002: DAT #0 #0
...
```

Do you see how the Imp copied itself by simply moving a copy of what is in location 0 to location 1?

T30. Step the machine a few more times, and print the memory again. Each time the machine executed the Imp instruction it made a new copy one location further down in memory.

T31. If you are still in the same IRB session you were in for the previous projects, type this command to clear the Dwarf programs from the MARS virtual machine and clear the display:

```
>> MARS.reset
=> []
```

T32. Start the game that pits an Imp against a Dwarf:

```
>> MARS.contest(:imp, :dwarf)
```

As these programs run it should be easy to see how the Imp moves through memory. If the contest stops before the Imp catches up to the Dwarf, type MARS.run to run for another 1000 cycles.

Did you successfully predict what would happen when the Imp caught up to the Dwarf? If not, can you explain what actually happened? There are two different outcomes, depending on which instruction the Dwarf is executing when the Imp arrives. Try running the contest a few times until you see both situations.

8.5 ◆ Clones

In the previous section we saw how a Redcode program can make a copy of its instructions, and then branch to the copy so that it starts executing from a new location. We can take this idea one step further—we can also split the current program into two parts, so that both copies are active. Not only is the program executing instructions from the new location, but it also continues to execute the original instructions. It's as if the program was able to clone itself, so that two identical copies are running.

The MARS instruction that activates the second copy of the program is named SPL. When the machine executes an instruction of the form SPL x, it starts running a new program located at memory address x. But it also keeps the current program running, and the instruction immediately following the SPL is also executed.

It is not necessary for the program at label x to be an exact copy of the original. Corewar programs often have separate little pieces that each do a specialized task. A piece can be activated by executing a SPL instruction that tells the machine to start that piece running while also keeping the original program going.

In modern computing terminology, we would say that the SPL instructions starts a new **thread**. Writing a program that has two or more threads is more difficult than programming with a single thread, but it can be very useful for complex applications. For example, web browsers typically have multiple threads, with threads for the different windows and for pieces of code that manage menus and operations inside browser windows.

Monitoring a program on the visual display when it executes a SPL instruction is similar to playing an old video arcade game named Centipede. A worm-like creature winds its way down the screen, and when a player shoots the centipede it splits in two, and each piece becomes a new, shorter worm. When you watch a MARS program that contains SPL instructions, you will see a new thread appear in the display. For example, if you were to run a version of Dwarf that copies itself and then uses SPL to start the copy, you would soon see two places on the screen running the 3-instruction main loop of the Dwarf program. To let you know that both copies are from the same player, both will have the same color.

Making new threads of execution comes at a cost, both in real applications and in MARS programs. In MARS, the new thread alternates steps with the old thread. Suppose program *A* executes a SPL, breaking itself into two threads, but program *B* still has a single thread.

The machine will execute one instruction from *A*, then one from *B*, then one from the second thread in *A*, then back to *B*, then back to the first thread in *A* again. In other words, the machine cycles are still split evenly between *A* and *B*, but within program *A*, the machine will switch back and forth between the two threads. Even though *A* has two threads, the programs running in those threads proceed half as quickly as the single program before the split.

What this means in terms of Corewar strategy is that there is an advantage in breaking into multiple threads. A program is still alive as long as any of its threads is still running. But this advantage is offset by the fact that each thread is running more slowly, and as a result it will be easier for the other program to knock it out.

In the previous section there was a discussion of whether it made sense for a program to try to "stun" an opponent by writing `JMP 0` instead of killing it by writing `DAT #0`. One place where this strategy might be advantageous is if a program expects to compete against an opponent that starts several threads. If a thread is killed, the surviving threads will all speed up slightly because there is one less thread to share that program's machine cycles. But if a thread is stunned, it still consumes its share of machine cycles. All of the opponent's threads still run at the same slow speed, and it may eventually be easier to kill them all off.

Tutorial Project

♦ A program named `test_threads` has an example of how to use the SPL instruction. Type this expression to get a listing of the program (or call `MARS.checkout` if you want to get a copy you can view in your text editor):

```
>> MARS.listing(:test_threads)
start    SPL imp
         JMP start
imp      MOV 0, 1
```

♦ Make a test machine to try out this program. Since the program uses the Imp strategy to fill memory with copies of itself, pass a second argument to `make_test_machine` to add extra room in the memory:

```
>> m = make_test_machine(:test_threads, 10)
=> #<MiniMARS mem = [...] pc = [ *0 ]>
```

♦ Get the machine's status:

```
>> m.status
Run: ready PC: [ *0 ]
```

The program counter (PC) shows there is one thread, and that the next instruction for this thread is in location 0.

♦ Call `step` to execute the first instruction in the program:

```
>> m.step
=> SPL 2 #0
```

♦ Now get the status again:

```
>> m.status
Run: continue PC: [ *1 2 ]
```

There is a lot of new information here. First, notice there are two threads. The next instruction to be executed for the first thread is in location 1. That's the JMP instruction that will be executed by the original thread. The new thread is starting at location 2, which is where the Imp is located.

♦ Call `step` again, and then look at the machine status:
```
>> m.step
=> JMP -1 #0
>> m.status
Run: continue PC: [ 0 *2 ]
```
The new status shows that the first thread executed the JMP, so its next instruction will come from location 0. It also shows that the next instruction for this program will come from the second thread. The asterisk in front of an address indicates the address that will has the instruction that will be executed next.

If you keep calling `step` for this test machine, you will see that it continually launches new Imps (`MOV 0, 1` instructions). More and more threads will be added to the list printed for the PC, and you should see that the machine cycles between all active threads.

♦ A program named `mice` uses the strategy of cloning itself, making so many copies that it would be hard for the opponent to stop them all. Run a contest that uses `mice` and one of the other programs, *e.g.*,
```
>> MARS.contest(:mice, :dwarf)
=> nil
```
Don't forget to call `MARS.reset` if you need to clear the machine after a previous contest. Can you see how the `mice` program is making copies of itself?

♦ When `mice` runs, do the copies also make copies? Will the number of `mice` "clones" grow exponentially?

8.6 Summary

The important idea introduced in this chapter is the concept of a stored program computer. Proposed by mathematician John von Neumann in 1945, the plan of connecting a processor to a memory that holds both instructions and data is the dominant architecture used in practically every computer system today. Modern systems have a variety of different kinds of memories, sophisticated "multi-level cache" systems to improve performance, and separate caches for instructions and data, but these are all methods for efficiently implementing the basic plan of the stored program computer.

Being able to encode programs as data has had far-reaching implications beyond the area of computer engineering. In order to store a program in memory, it has to be encoded in the form of binary symbols. The first computer programs were written in a binary machine language, but over the years more abstract notations were developed. An assembly language, like the Recode language we used in this chapter, has symbolic names like ADD and MOV for machine operations, and allows us to assign names to memory locations. Modern programmers use higher level (more abstract) programming languages. Compilers, debuggers, and other applications used for software development are all based on this notion that programs are data that can be operated on by other programs.

The projects in this chapter played a game called Corewar, a computer-based version of Battleship, a classic board game where players try to guess the location of their opponent's ships. In Corewar, the contestants are programs, and the goal is for one program to halt the execution of the other program. The game depends on the fact that a program can overwrite

---

Concepts and Terminology Introduced in This Chapter

---

von Neumann architecture	A design for a computer system in which programs are encoded in binary and stored in memory
processor	The part of a computer system that contains the logic circuits that carry out steps of an algorithm
memory	The component that stores data and instructions
virtual machine	A piece of software that simulates the actions of a processor and memory
assembly language	A programming language where each statement is a single instruction that can be carried out by a processor
opcode	The part of an assembly language statement that tells the processor which operation to perform
operand	Part of an assembly language statement that specifies data to be used in an instruction or where in memory to store a result
thread	A sequence of instructions that implements a single program or method

---

instructions in the other program, copy itself to a new location in memory, or use any of a variety of other strategies derived from having instructions and data in the same memory.

Our Corewar contests were played on a virtual machine named MARS, a simulator written in Ruby that mimics the actions of a real processor. The simulator allows us to test programs in isolation, executing the program one step at a time and watching what each instruction does. We can also load two programs into the machine and let them run until one program stops or until a predetermined number of instruction cycles have been executed.

The technique of using software to implement virtual machines is widely used in computing. One familiar example is the Java Virtual Machine, or JVM, which is a standard part of almost every web browser. When you connect to a web site that contains an "applet" written in Java, the web server sends your browser the encoded form of the JVM instructions for the applet, allowing the program to run on your computer as the browser executes the JVM instructions. A growing new technology known as "cloud computing" is also based on the notion of a software description of a real machine. Computer administrators can configure a machine as if it were a real piece of hardware, but then hand the machine specification off to a server on a network that will eventually find a place to execute programs on an actual piece of hardware that matches the description of the virtual machine.

The advanced section on "clones" introduced the idea of a thread. The SPL instruction splits MARS programs into two pieces that appear to run in parallel. In reality, however, the virtual machine divides its time among the different threads in a round-robin manner, running one instruction at a time. Threads are also widely used in real-world programming, especially in complicated applications where threads execute separate tasks.

The fact that applications are often organized as a group of threads that all cooperate to solve a complex problem brings up an interesting question: can we build a computer system with more than one processor, so that all the threads can run at the same time on different processors? New generations of processor chips do in fact use "multi-core" technology that is able to speed up these sorts of applications. Systems known as "multi-CPU clusters" are built from hundreds, and in some cases thousands, of processors, and are being used to solve very large problems in science and other areas that require high performance computing.

About Corewar

The game of Corewar was first described by A. K. Dewdney in a series of columns written for *Scientific American* in 1984. The MARS virtual machine we used for our projects is based on a 1988 standard, but the most widely used virtual machines today are based on a newer standard that was published in 1994.

If you would like to try your hand at writing your own Corewar programs you can find more information on the web at `http://corewar.co.uk`. There you will find tutorials, articles on the history of the game, and implementations of the newer virtual machine that you can download and install on your own computer.

♦ Proc Objects in Ruby

We've seen several examples of programs as data in projects in earlier chapters. The Ruby-Labs methods that trace the execution of a program are based on that fact that we can make a Ruby object that represents a program. The methods named `trace`, `count`, and `time` are all defined in terms of a single parameter, and they all tell Ruby to execute the program represented by the block object passed to the method.

Here is the definition of `time`, which measures measures the execution time of a program:

```
def time(&f)
  tstart = Time.now
  f.call
  return Time.now - tstart
end
```

The notation `&f` means `time` expects us to pass an object that represents a piece of Ruby code. In the body of the method the expression `f.call` means "invoke the code in the object `f`." All `time` has to do is record the system time (by calling `Time.now`) before and after executing the code defined by `f` and then return the difference.

Here is an example of how we used `time`:

```
>> time { isort(a) }
=> 0.622742
```

When IRB sees a Ruby expression enclosed in braces it creates an object that represents the expression, and it is this object that is passed as an argument to `time`. In Ruby, the object is called a "Proc," which is short for "procedure," an older term for "method."

The idea that programs and data can both be encoded in the form of binary numbers so they can be stored in the same memory can be traced to work in mathematical logic in the 1930s. One of the most important implications of that work is that it even though programs can be encoded and used as data, there are limits to what we can do with this type of data.

We might be able to create an object to represent a piece of Ruby code, but the only thing we can do with this object is tell Ruby to execute the code. Programs often have statements like

```
if x == y ...
```

that compare numbers x and y and carry out some operation if they have the same value. But no programming language allows us to write an expression of the form

```
if p1 == p2 ...
```

where p1 and p2 both refer to programs. It is possible to prove, using the theories in logic that were derived in the 1930s, that there is no way to compare two arbitrary programs to see if they compute the same thing.

Exercises

The first questions are about the multiplication and division algorithms presented in Section 8.2. The multiplication algorithm computes $x \times y$ by repeatedly adding the contents of a word labeled x to an "accumulator." The division algorithm computes $x \div y$ by counting how many times y can be subtracted from x.

1. How many iterations does the multiplication algorithm make if it is asked to compute 2×8?

2. How many iterations are required to compute 8×2?

3. Will the algorithm give the correct result when $x = 0$? Explain what the program named test_mult would do if the instruction labeled x is changed from DAT #7 to DAT #0.

4. Repeat the previous problem, but explain what would happen if $y = 0$, *i.e.*, what would the algorithm do if the instruction labeled y is changed from DAT #6 to DAT #0.

5. Will the multiplication algorithm work if either input is negative? For example, would the algorithm give the correct result if asked to multiply -3×4? What about 3×-4?

6. What would the division algorithm do if it is asked to compute $x \div 0$?

7. When the bombs thrown by the Dwarf program reach the end of memory they "wrap around" and start going into low addresses. Will the bombs ever overwrite the Dwarf's own instructions, so it accidentally "shoots itself in the foot" and causes itself to halt?

8. What would happen if the ADD instruction in a Dwarf program is changed from ADD #4, bomb to ADD #2, bomb? Would this change your answer to the previous question?

♦ Check out a copy of the test_celsius program and save it in a file in your project directory. Modify the program so it converts temperatures in the opposite direction, from Celsius to Fahrenheit.

♦ Check out a copy of test_div and fix the bug that causes the program to give the wrong answer when x is a multiple of y. Hint: add a test that checks to see if $x - y = 0$.

♦ Explain what would happen if a program executes the instruction SPL 0.

♦ Check out a copy of the dwarf program, and modify it so it uses the strategy described in Section 8.4 (page 212), where the program throws 1024 bombs and then copies itself to a new location an odd number of words away.

Chapter 9

Now for Something Completely Different

An algorithm for generating random numbers

Most popular games involve some element of chance. Players roll dice, shuffle a deck of cards, spin a wheel, or use some other method for making a random selection in the game. Computer-based versions of these games are played the same way, but instead of using real dice or using a real deck of cards a computer program manages the game. Somewhere inside the application is an algorithm that generates the next roll of the dice or shuffles the deck of cards.

As you might imagine, a computer-based game would not be very popular if the algorithm that generates moves is not realistic. If you're playing a board game like backgammon, you expect rolls of the dice to be similar to rolls of real dice, and you would start to become suspicious if your opponent rolls doubles far more often than you do. Some people like to practice playing poker against a computer to prepare for tournaments. If the program they train with doesn't deal the same kinds of hands that will be seen in the tournament, the software is not going to be very useful.

The word that best describes our expectations for the computer-based games is **random**. When we play a board game we want the computer to generate a pair of numbers that is just as random as rolling a pair of real dice, and when we play cards we want the computer to generate an ordering for a deck of cards that is as random as what we would get if we carefully shuffle a real deck.

The natural question, of course, is how to define what we mean by "random." Colloquially, random often means "unusual" or "unexpected." But in games, and in other situations where random values are required, something is random if it is **unpredictable**. To be more precise, what we are looking for is an algorithm that generates a sequence of values in which there is no apparent pattern or relationship between one value and the next. If the algorithm is used to simulate rolls of a six-sided die, each new number should be independent of the previous number, and if it is used to deal cards, each new hand should be unrelated to the previous hand. In other words, to use the phrase in the title of this chapter, each value produced by the algorithm should be "completely different" from the previous value.[1]

The goal for this chapter is to explore techniques for using a computer to generate random values. We will take a closer look at the idea that a sequence of values is random if successive values are independent of one another, and explore various ways of trying to determine whether the values are, in fact, random, or whether there are some unexpected connections between them.

9.1 Pseudorandom Numbers

Before you read this section, here is a simple experiment to try. Start your text editor and create a new file. Type 50 numbers between 1 and 6, putting one number on each line. Try to write the numbers as if you were rolling a die; in other words, each number should be unrelated to the one on the previous line.

As you were thinking of numbers to write, were you able to concentrate on each new number, forgetting about what you had written before? Or did you find yourself thinking something like "I haven't written a 3 in a while, I'd better write one now" or "hmmm, that's two 6s in a row, I'd better write something else." If you gave in to the temptation to think about previous values you were starting to add a bias to your results. It is very hard for people to generate a truly random sequence of values.

Engineers, statisticians, and other professionals have used random sequences for many years. Before the algorithms described in this chapter were available, people who needed a random sequence would look in a book of random numbers. A well-known reference book, published in 1955 by the RAND Corporation, was *A Million Random Digits*, a 400-page book with 2500 random digits between 0 and 9 on each page. The authors used what they called an "electronic roulette wheel" to generate random electronic pulses. The electronic circuit was connected to a computer, and a device that measured the pulses converted them to digital form. Another way to generate random signals is to use a white noise generator, an audio device that produces something that sounds like static.

To play a game with a computer we do not need to connect to a roulette wheel, white noise generator, or other physical device that behaves randomly. Instead, applications use an algorithm that produces a different value each time it is called. For example, there is a method built into Ruby called `rand`. If we pass a number n to `rand`, it will return a value between 0 and $n - 1$:

```
>> rand(100)
=> 54
>> rand(100)
=> 39
```

---

[1]The title is a catch-phrase from the BBC comedy series, *Monte Python's Flying Circus*.

There is an interesting paradox here. According to the definition given in Section 1.3, an algorithm has a well-defined set of inputs, and a precise and unambiguous sequence of steps that leads to the output. If that's the case, how can an algorithm produce a different result each time it is run? Or, in terms of Ruby programs, how is it possible for a method to return a different value each time we call it?

The answer to these questions is the method *appears* to be random, but in fact it follows a predefined set of rules, just like any method. The general idea is to produce a long sequence of numbers, and if we look at a small set of numbers in the middle of the sequence, they will appear random. Even though they are not truly random, many applications, such as computer games, can use them in place of actual random values. Because the numbers are created by an algorithm, and not an external source of random data like the RAND corporation's roulette wheel, we refer to them as **pseudorandom**, and the algorithm that produces the sequence is a **pseudorandom number generator**, or PRNG.

As an introduction to how a PRNG might work, imagine a situation where an event needs to be scheduled at regular intervals throughout a day. Perhaps a nurse in a hospital needs to periodically administer medications to a patient, or a researcher needs to collect data from an experiment, and it is our job to schedule these events. If the events occur every eight hours the schedule is simple and easy to remember. One plan would be to schedule the first three events at midnight, 8 A.M., and 4 P.M. The next event would be at midnight again, and the schedule would repeat. Since the schedule is the same every day it is easy to remember, and we can simply tell people what the schedule is (Figure 9.1).

However, if the events need to occur every seven hours the schedule is more complicated. If the first event is at midnight, the next two events would be 7 A.M. and 2 P.M. But now the fourth event will be at 9 P.M., and the second day would not be like the first.

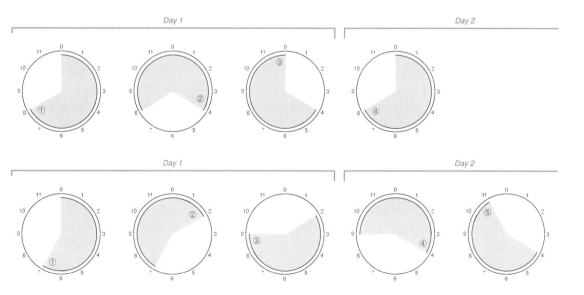

Figure 9.1: *If an event occurs every eight hours (top row), a schedule is very predictable, and events occur at the same time every day. However, if events are seven hours apart (bottom row) each day is different, and it is more difficult to keep track of when events are scheduled to occur.*

It's very straightforward to use Ruby to figure out the schedule. Start by making a list named t and initializing it so it has the time for the first event:

```
>> t = [0]
=> [0]
```

For now we will use a 12-hour clock. Let's assume we can tell by context whether "0" is midnight or noon, and whether a "4" means 4 A.M. or 4 P.M. (one of the exercises at the end of the chapter will be to modify the schedule to use a 24-hour clock).

To add the time for the next event, the list needs to be extended with a time that is seven hours later than the one currently at the end of the list. Since we're using a 12-hour clock, the expression is

```
>> t << (t.last + 7) % 12
=> [0, 7]
```

Recall from previous chapters that an expression of the form a << x means "attach x to the end of array a," and that the % symbol is Ruby's mod operator, *i.e.*, the expression x % 12 means "the remainder after dividing x by 12."

If the above operation is repeated several more times, this is what the list would look like:

```
[0, 7, 2, 9, 4, 11, 6, 1, 8, 3, 10, 5, 0, 7, 2]
```

That's a pretty difficult schedule to remember, but it matches the specification, and we could either put it on a poster, or ask people to store it in the calendar on their cell phone.

To get back to the subject of generating random numbers, at first it looks like the list above might be random. If we gave the list to a person who did not know the numbers were times from a 12-hour clock, they would probably have a hard time guessing the list was generated by a simple rule, or what that rule was. On closer inspection, however, we can see some regularity in the list. For one thing, every hour appears exactly once in the first 12 places in the list (assuming we interpret 0 as 12 o'clock). When we are rolling a 6-sided die, it would be very rare to see a sequence of rolls in which every number between 1 and 6 showed up before any number appeared again. In fact, it's not at all uncommon to have the same number appear twice in a row. The fact that we used a rule that forces every number to occur once before the first number appears again means the list is not truly random.

If we add a few more items to the list, using the same rule, we will see conclusive evidence that the list is not random. As soon as the first number is generated again, the pattern will start to repeat itself. In other words, as soon as the rule attaches a 0 to the end of the list, all the values that followed 0 the first time will occur again, and in exactly the same order. If we want to know what follows 2 in the list shown above, we can either apply the rule again and evaluate $(2 + 7) \bmod 12$, or we can find the 2 earlier in the list and look at the number came after it.

The rule of adding 7 and taking the remainder mod 12 does not create a very useful list of pseudorandom numbers, but it is a good starting point. A more general formula for adding a new value x_{i+1} to the end of a list is

$$x_{i+1} = (a \times x_i + c) \bmod m$$

where a, c, and m are all constants and x_i is the previous item in the list. The "add seven to the current time" rule follows this general pattern, since it has $a = 1$, $c = 7$, and $m = 12$.

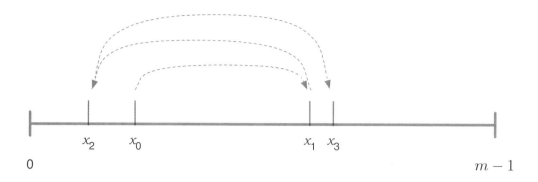

Figure 9.2: *A pseudorandom sequence defined by the rule $x_{i+1} = (a \times x_i + c)$ mod m. If values of a, c, and m are chosen carefully, every value between 0 and $m - 1$ will appear in the first m numbers in the sequence and the number line will be completely filled in.*

If a and c are defined properly, this rule will place every value from 0 to $m - 1$ in the list before it repeats (Figure 9.2). As we saw previously, with $a = 1$ and $c = 7$, the rule makes a list of all 12 values between 0 and 11. But if we use $c = 8$ (which is what we did originally, when the schedule called for events every eight hours) the list is much shorter:

```
[0, 8, 4, 0, 8...]
```

The number of items in the list before it starts to repeat is called the **period**. The period when $c = 8$ is 3, because the list only has 3 items before it starts repeating. The period when $c = 7$ is 12 because all 12 numbers are in the list before it repeats.

If we use a large value of m, and values of a and c that work well for that m, we can make a very long list of unique numbers. Furthermore, if we look at small portions of the list, it will be very difficult to figure out what rule is used to generate the numbers. In practical terms, we will have a list of random numbers. Even though they were produced by a pseudorandom number generator, and will not be truly random, for many applications they might be "random enough."

The formula shown above was used in some of the earliest pseudorandom number generators, and is still widely used because it is very easy to implement and does a reasonable job for games and other casual applications. In current implementations, m is typically 2^{32}, or roughly 4×10^9, so this technique will generate a list of over four billion numbers before it repeats. The PRNG built into Ruby is based on a newer and more sophisticated algorithm that has a period of 2^{19937}.

In the next section, we will see how to put this formula to use in making random numbers for games and other applications. Later in the chapter we will come back to the claim that the sequence is "random enough" by looking at different ways to measure randomness and using these techniques to assess the randomness in the sequence of values produced by a PRNG.

Tutorial Project

The first exercises will explore the formula for making pseudorandom numbers. Start an IRB session and load the module that will be used in this chapter:

```
>> include RandomLab
=> Object
```

T1. Make the schedule for events that occur every eight hours. Initialize the schedule with the time for the first event:

```
>> t = [0]
=> [0]
```

T2. Apply the rule that adds a new event that will occur eight hours after the previous event:

```
>> t << (t.last + 8) % 12
=> [0, 8]
```

T3. Apply the rule three more times. Enclose the previous expression in braces and call the `times` method to make a schedule that repeats the basic pattern of 0, 8, 4:

```
>> 3.times { t << (t.last + 8) % 12 }
=> 3

>> t
=> [0, 8, 4, 0, 8]
```

See the sidebar below for an explanation of the `times` method.

T4. Start a new schedule by typing the first expression again:

```
>> t = [0]
=> [0]
```

T5. Repeat the expression that adds new events, but make 16 events scheduled 7 hours apart:

```
>> 16.times {t << (t.last + 7) % 12}
=> 16

>> t
=> [0, 7, 2, 9, 4, 11, 6, 1, 8, 3, 10, 5, 0, 7, 2, 9, 4]
```

Can you see how the period for this new schedule is 12, *i.e.*, how all 12 numbers between 0 and 11 appear in this list before it starts to repeat?

The `times` Method

The projects in this chapter use a Ruby method named `times`. This method is used to repeat an operation a specified number of times. The first example right tells IRB to print a string three times.

```
>> 3.times { puts "hello" }
hello
hello
hello
=> 3
```

We can also put a variable name to the left of the method, as shown in the second example, which appends 10 random numbers to an array. The value returned by `times` is just the number of times the operation was executed.

```
>> n = 10
=> 10
>> n.times { a << rand(100) }
=> 10
```

The RandomLab module defines a method named `prng_sequence` that implements the general form of the equation that creates a list of pseudorandom numbers. The three parameters to the method are the values of a, c, and m to plug into the equation. The list returned by this method will have m numbers, and with the right combination of a and c all values from 0 to $m - 1$ will be in the list.

T6. As a first test, use the `prng_sequence` method to make the schedule of events that occur every eight hours:

```
>> sched8 = prng_sequence(1, 8, 12)
=> [0, 8, 4, 0, 8, 4, 0, 8, 4, 0, 8, 4]
```

As expected, this sequence is not very random.

T7. Ruby arrays have a method named `uniq` that will return a list of all the unique values in an array, *i.e.*, the method returns a copy of the array with all the duplicates removed. Apply this method to the schedule you just made:

```
>> sched8.uniq
=> [0, 8, 4]
```

Do you see how the `uniq` method confirms the fact that only 0, 8, and 4 are used in this array?

T8. Now make a second schedule for events that occur every seven hours:

```
>> sched7 = prng_sequence(1, 7, 12)
=> [0, 7, 2, 9, 4, 11, 6, 1, 8, 3, 10, 5]
```

T9. Get a list of unique numbers in this list:

```
>> sched7.uniq
=> [0, 7, 2, 9, 4, 11, 6, 1, 8, 3, 10, 5]
```

T10. One way to see whether every number between 0 and $m - 1$ is in the list is to count the number of unique values. If an array has no duplicates, the length of the list returned by `uniq` will be the same as the length of the original list:

```
>> sched8.uniq.length
=> 3

>> sched7.uniq.length
=> 12
```

T11. This combination of parameters will generate a list of numbers between 0 and 999:

```
>> seq = prng_sequence(3, 337, 1000)
=> [0, 337, 348, 381, 480, ... 97, 628, 221, 0]
```

Look at the list of 1000 numbers in your terminal emulator window. Does it look "random" to you?

T12. Let's see if this combination of a, c, and m made a list with a period of m:

```
>> seq.uniq.length
=> 100
```

So this combination of a, c, and m yields a list that has the same 100 numbers repeated 10 times—not very random at all.

T13. Here is a better combination. Change the first argument to 81 instead of 3:

```
>> seq = prng_sequence(81, 337, 1000)
=> [0, 337, 634, 691, ... 749, 6, 823]
```

T14. How many unique numbers are in this list?

```
>> seq.uniq.length
=> 1000
```

Just what we were looking for. Does this list look random?

Figure 9.3: *The first 100 values produced by a PRNG with a = 81, c = 337, and m = 1000.*

The RandomLab module includes methods that will draw pictures based on the values produced by a pseudorandom number generator. The next set of exercises will plot the values returned by a PRNG on a number line, so you can get a sense of how the values are scattered over a specified range (Figure 9.3).

T15. Create a window with a number line for values between 0 and 999:

```
>> view_numberline(1000)
=> true
```

You should see a canvas with a line running through the middle of it.

T16. A method named `tick_mark` will display a mark at a specified point on the line. Type this expression to display a tick mark in the middle of the line:

```
>> tick_mark(500)
=> nil
```

T17. This command will draw 100 tick marks, at locations 0 through 99:

```
>> 100.times { |i| tick_mark(i) }
=> 100
```

T18. Reinitialize the display by calling `view_numberline` again (Exercise T15).

T19. Type this expression to plot the points from the pseudorandom sequence that contained only 100 different numbers:

```
>> prng_sequence(3, 337, 1000).each { |i| tick_mark(i) }
=> [0, 337, 348, 381, ... 97, 628, 221]
```

Can you see how only about 1/10 of all the points are filled in? So even though the call to `prng_sequence` made a list of 1000 numbers, the list has 100 different values repeated 10 times.

T20. Reinitialize the drawing, and repeat the call to `prng_sequence`, but change the 3 to 81 so you get the sequence with all 1000 numbers:

```
>> view_numberline(1000)
=> true

>> prng_sequence(81, 337, 1000).each { |i| tick_mark(i) }
=> [0, 337, 634, 691, 308, 285, ... 749, 6, 823]
```

Now your number line should be completely filled in. There are 1000 numbers in the list, and each value from 0 to 999 will occur exactly once.

The expression in Exercise T20 filled in each point in the number line. But we could have done that by repeating Exercise T17 and telling it to draw 1000 tick marks. When you were watching the display, were the numbers created by `prng_sequence` added in a random order?

9.2 Numbers on Demand

Although the sequences produced in the exercises for the previous section may appear to be random, appearances can be deceiving. Later in the chapter we'll look at some techniques for evaluating sequences to test randomness, but first we'll see how the random number generator can be implemented in a method intended for games and other applications that need random numbers.

As a practical matter, applications do not generate a list of all numbers in a pseudorandom sequence. A game may need only a few hundred rolls of the dice, and it would be a big waste of time and space to generate the full list of billions of pseudorandom numbers defined by the best random number generators. Instead, games and other applications use a programming technique that creates random numbers "on demand."

As an analogy for how this might work, imagine a scenario where a statistician has a lab assistant who is in charge of random numbers. The situation where the full sequence is generated ahead of time would correspond to the lab assistant carrying around the RAND book of random digits. Each time the statistician needs a random value, the assistant would look up the current digit in the book, and then move his bookmark one place to the right. However, when the numbers are generated by an algorithm, the assistant just needs to keep track of one number on a small piece of paper. When the statistician needs a random number, the assistant plugs the number into the equation to compute the next value in the sequence and then erases the old number and replaces it with the new value.

The RandomLab module has the definition of a new type of object called a PRNG that uses the one-number-at-a-time strategy to implement a pseudorandom number generator. To make a PRNG object, pass the a, c, and m parameters to the method that makes a new object. For example, in the last section we made a pseudorandom sequence of $m = 1000$ numbers using $a = 81$ and $c = 337$:

```
>> seq = prng_sequence(81, 337, 1000)
=> [0, 337, 634, 691, ... 749, 6, 823]
```

To make a PRNG object based on this sequence, we pass these same values to PRNG.new:

```
>> p = PRNG.new(81, 337, 1000)
=> #<RandomLab::PRNG a: 81 c: 337 m: 1000>
```

A method named advance moves to the next pseudorandom number in the sequence. If we call advance a couple of times, it should return the first two values in the sequence:

```
>> p.advance
=> 337
>> p.advance
=> 634
```

Each time advance is called it acts like the assistant with the piece of scrap paper. It sets the value of x to the current value in the sequence. It then evaluates $(a \times x + c)$ mod m. The value of this expression is saved as the new state of the sequence, and it's also returned as the value of the call to advance.

Tutorial Project

T21. If you started a new IRB session since you worked on the project in the previous section, type this expression again to make a pseudorandom sequence:

```
>> seq = prng_sequence(81, 337, 1000)
=> [0, 337, 634, 691, ... 749, 6, 823]
```

T22. Print the first 10 numbers in the sequence:

```
>> seq[0..9]
=> [0, 337, 634, 691, 308, 285, 422, 519, 376, 793]
```

T23. Make a PRNG object using the same values of *a*, *c*, and *m*:

```
>> p = PRNG.new(81, 337, 1000)
=> #<RandomLab::PRNG a: 81 c: 337 m: 1000>
```

T24. A method named `state` will show us the current number in the sequence (the number written on the "scrap paper"):

```
>> p.state
=> 0
```

So the PRNG object starts out with the same value as the array made by `prng_sequence`.

T25. If you call the `advance` method you should get back the next value in the sequence:

```
>> p.advance
=> 337
```

T26. Call `advance` a few more times. Do you get the same numbers you see at the front of the list named `seq`?

T27. Call `p.state` to see what the current number is, and then call `p.advance` again. Do you see how each call to `advance` simply computes the next value in the pseudorandom sequence?

9.3 Games with Random Numbers

The previous two sections showed how it is possible to create a list of numbers that appear to be random. We looked at the equation that defines how each element in the list is derived from the previous one, and then saw how we might implement this technique in a way that lets us get new numbers "on demand" instead of creating the entire list at once. The goal for this section is to show how a pseudorandom sequence can be put to work by an application that uses random numbers.

If the three parameters (named *a*, *c*, and *m*) that define the relationship between successive numbers are chosen carefully, the pseudorandom sequence will have values from 0 up to $m - 1$. However, if we are writing a program to play a game like backgammon, which is based on rolling a pair of dice, we want a series of numbers between 1 and 6.

A simple approach to simulating random rolls of a six-sided die is to get a value from the pseudorandom sequence, find its remainder after dividing by 6 (which will yield a number between 0 and 5), and then add 1. Here is a Ruby expression that transforms a number from a PRNG object named p into a number between 1 and 6:

```
>> (p.advance % 6) + 1
=> 2
```

Figure 9.4: *If pseudorandom numbers are used by a program that plays a game with dice, the values in the pseudorandom sequence are mapped to numbers between 1 and 6.*

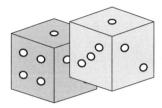

If we evaluate this expression several more times and save the results in a list, we can see something interesting:

```
>> rolls = []
=> []
>> 10.times { rolls << (p.advance % 6 + 1) }
=> 10
>> rolls
=> [2, 5, 2, 3, 4, 3, 4, 5, 2, 1]
```

As expected, all the numbers in the list are between 1 and 6. But notice the numbers that follow the 2s in this list: the first 2 is followed by 5, the second by 3, and the third by 1.

The fact that a number can be followed by different values adds to the illusion that this sequence is truly random, and not just pseudorandom. Because 1000 was passed as the value of m when the PRNG was created, the pseudorandom sequence generated by p will have every number between 0 and 999. But when we divide these numbers by 6, the remainders will be between 0 and 5. Each remainder will show up hundreds of times, and the value that follows a remainder can be any one of the 6 numbers between 0 and 5.

PRNG objects have a method named random that does this conversion for us. This method takes two parameters, min and max. The method gets the next value from the pseudorandom sequence, and then it uses the mod operator to turn that value into an integer between min and max. For example, a program that wants to use the PRNG object to simulate rolls of a die would call p.random(1,6) to get numbers from 1 to 6:

```
>> p.random(1,6)
=> 1
>> p.random(1,6)
=> 2
```

There is one more issue that needs to be addressed if we want to use a PRNG in an application, and that is how to initialize the sequence. Each time a new PRNG object is created, the initial state is set to 0. Since the application alway uses the same values of a, c, and m to create the PRNG, it will always get the same sequence of numbers at the start of each game. Our backgammon program will always start off with the same rolls of the dice: 2, 5, 2, 3, 4, 3, 4, 5, 2, *etc.* Players will soon recognize that every game starts the same way, and they will lose interest.

One way to address this problem is to set the state of the PRNG to a particular value. We can create the new object, letting the `new` method set the state to 0, but then we can change the state to any value we choose. This process of changing the state of a PRNG is called "seeding." If we want to seed one of our PRNG objects we just have to call a method named `seed`, *e.g.*,

```
>> p.seed(226)
=> 226
```

The next call to `advance` will get the number that follows 226 in the underlying pseudo-random sequence.

But now the question becomes, which value do we use for a seed? If we're using the simple PRNG that has $m = 1000$ we can choose any number from 0 to 999. But if we just choose our favorite number, and write that number into the program, we will have the same problem. The game will start out with a different set of rolls than if the sequence started with 0, but since the PRNG always starts in the same state players are again going to see the same set of rolls at the start of each game.

A common solution is to use the system clock. Ruby has a module named `Time` that has a collection of methods for dealing with dates and times. Calling `Time.now` will return the current date:

```
>> Time.now
=> Mon Dec 28 09:30:34 -0800 2009
```

It's also possible to turn a date into an integer by calling the `to_i` method:

```
>> Time.now.to_i
=> 1262021437
```

This value is the number of seconds since January 1, 1970 (the date of the "Big Bang" when time started in the Unix world).

To initialize each new game, an application typically gets the current time, converts it to an integer, and uses that value to seed the random number generator. Although the time value is increasing in a predictable way, the values that follow in the pseudorandom sequence will be very different, and the games will start out with a different sequence of rolls each time.

Tutorial Project

T28. Make a PRNG object with the same values of a, c, and m used previously:

```
>> p1 = PRNG.new(81, 337, 1000)
=> #<RandomLab::PRNG a: 81 c: 337 m: 1000>
```

T29. Get 10 rolls of a die by calling the `random` method and asking it for values between 1 and 6:

```
>> 10.times { puts p1.random(1,6) }
2
5
2
3
. . .
```

T30. Make a second pseudorandom number generator using the same parameter values:

```
>> p2 = PRNG.new(81, 337, 1000)
=> #<RandomLab::PRNG a: 81 c: 337 m: 1000>
```

T31. Now print the first 10 rolls of the die produced by this PRNG:

```
>> 10.times { puts p2.random(1,6) }
2
5
2
...
```

Do you see why they are the same rolls? Each new PRNG object is initialized to start at the same place in the pseudorandom sequence, so each will create the same sequence of values.

T32. Call the method that gets the current date and time:

```
>> Time.now
=> Mon Dec 28 10:33:57 -0800 2009
```

Obviously the result you get will be different, but you should see a string that contains a date formatted according to the conventions you set for your operating system (this string is in the format seen by users in the United States).

T33. Type the expression again after adding to_i to convert the time into an integer:

```
>> Time.now.to_i
=> 1262025240
```

T34. If you want to use the current time to set the seed for the PRNG, convert the time to a number between 0 and $m - 1$:

```
>> rs = Time.now.to_i % 1000
=> 574
```

T35. Create a third PRNG, and then set its seed using the current time:

```
>> p3 = PRNG.new(81, 337, 1000)
=> #<RandomLab::PRNG a: 81  c: 337 m: 1000>

>> p3.seed(rs)
=> 574
```

T36. Now get 10 rolls of the die from this new PRNG:

```
>> 10.times { puts p3.random(1,6) }
4
1
4
...
```

T37. Set the seed for p3 again, but this time use a value that is one greater than the previous seed, and then get 10 rolls of the die:

```
>> p3.seed(rs+1)
=> 575

>> 10.times { puts p3.random(1,6) }
1
6
3
...
```

After doing these last two exercises, do you see how a slight change in the seed leads to a big change in the pseudorandom sequence? Just a slight difference in the time used to set the seed will result in a completely different sequence of random numbers at the start of each game.

9.4 Random Shuffles

If we want to write a program to play bridge, poker, or some other card game we are going to need a method that makes random collections of cards. Simulating the roll of six-sided die just involved computing a number between 1 and 6, but here the problem is slightly different. It would be easy enough to label each card in the deck with a number between 0 and 51, and then use a pseudorandom number generator to compute a number between 0 and 51 as a way of choosing a card at random. But if we want to deal a hand there is a chance that we could end up with two copies of the same card. In bridge, each hand has 13 cards, and if we make a hand by just calling `random(0,51)` 13 times, odds are we will have a collection where the same number appears twice.

One way to solve this problem is to use a method that "shuffles" a deck of cards. We will start with a collection of all 52 cards, and then each time we play a game we can shuffle the collection by rearranging the items in a random order. The goal for this section will be to develop a method that rearranges objects in an array. The method will use values from a pseudorandom number generator to produce a random ordering.

To make the project more realistic, the RandomLab module defines a type of object named Card. A card object will have a rank (ace, king, queen, *etc.*) and a suit. We will represent a deck of cards as an array of 52 objects that each represent a different card from a standard deck. As with other kinds of objects, an expression with the name of the type followed by the word `new` will create an object of that type. If you don't pass an argument to `new` you will get back a random card:

```
>> Card.new
=> 10♥
>> Card.new
=> 2♣
```

When we make a deck of cards, we don't want 52 random cards, but instead we want one of each possible card. You can pass an integer between 0 and 51 to the `new` method to tell it which card to make:

```
>> Card.new(0)
=> A♠
>> Card.new(1)
=> K♠
>> Card.new(50)
=> 3♣
>> Card.new(51)
=> 2♣
```

As a convenience, we can get a complete deck by calling `new_deck`, which makes each of the 52 card objects for us:

```
>> d = new_deck
=> [A♠, K♠, Q♠, ... 3♣, 2♣]
```

Mathematicians refer to an ordering of items as a **permutation**. The goal for the project in this section is to define a method named `permute!` that we can use to make a new random ordering of the items in an array. For example, after making a full deck of cards as shown above, we can call `permute!` to shuffle the deck. Each time we call `permute!` it will make a new random permutation:

```
>> permute!(d)
=> [9♠, A♣, 4♣, Q♣, 7♠, J♣, 4♥, ... ]
>> permute!(d)
=> [10♣, Q♣, 7♥, A♦, 6♣, 2♥, 8♣, ... ]
```

Unlike the sorting methods we developed in Chapters 4 and 5, which returned sorted copies of the input arrays, each call to `permute!` changes the order of items in the array it is passed.

A program that plays poker would probably "deal" the cards the same way we would in a real game. After calling `permute!` to shuffle the deck, it would then give `d[0]` to the first player, `d[1]` to the next player, and so on. For our experiments with poker hands, however, we'll just need one hand, and we can make this hand by using the first five cards in the deck. This Ruby expression shuffles the deck and makes an array containing the first five cards in the new deck:

```
>> permute!(d).first(5)
=> [2♣, A♠, 5♦, 10♣, 7♣]
```

We've seen the method named `first` before: the expression `a.first` returns the item at the front of the array `a`. This time we're passing the number 5 as an argument so it returns an array of the first 5 items.

One straightforward algorithm for permuting the items in an array is based on an iteration that exchanges two items at each step. Begin by picking up the first item and exchanging it with a random item somewhere to the right. Then exchange the second item with a random item somewhere to its right, then exchange the third item with a random item somewhere to its right, and so on until you reach the end of the array. This algorithm is reminiscent of the insertion sort algorithm: on each iteration the array has two regions, where items in the left part of the array have been exchanged and items to the right are waiting to be moved.

It's easy to exchange the values of two items in Ruby by using a construct known as **parallel assignment**. A normal assignment statement has a single variable on the left side of the assignment operator, *e.g.*,

```
>> x = 7
```

A parallel assignment has two or more variable names on the left, as in

```
>> x, y = 3, 4
```

The variables on the left are assigned the corresponding values from the right, so in this example `x` is set to 3 and `y` is set to 4. It's called a "parallel" assignment because we can think of the two assignments happening at the same time.

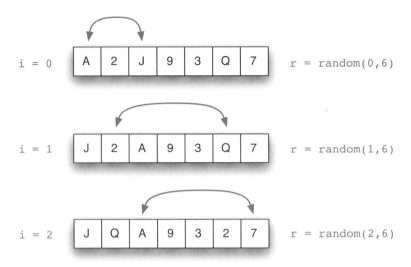

Figure 9.5: *On each iteration, the item at location* i *changes places with an item at a random location to the right.*

To use a parallel assignment to exchange the value of two variables, write the names of the variables in different orders on the left and right sides of the assignment operator:

```
>> x, y = y, x
```

Ruby executes this statement by fetching the current values of y and x and then writing them to the specified locations. It might help to think of a juggler picking up two balls and tossing them to different hands in the same motion: while the value of y is being tossed to x the value of x is being tossed to y.

Parallel assignment also works for locations in an array. For example, to exchange the values in the first two locations of an array a we can just write

```
>> a[0], a[1] = a[1], a[0]
```

Now that we know how to exchange two items in an array, writing the method is straightforward. The algorithm is an iteration in which a variable named i takes on values from 0 to $n - 2$, where n is the length of the array. At each step, we just need to set a variable r to a random value between i and $n - 1$, which is the last location in the array. We then exchange the items at locations i and r using a parallel assignment. One small detail to notice here: it's possible that i and r will have the same value, in which case the exchange operation has no effect.

An example showing the progress of the `permute!` method as it scrambles an array of 7 items is shown in Figure 9.5. On the first iteration, i is 0, and the program sets r to a random value between 0 and 6. After assigning r the value 2, the program exchanges r[0] and r[2]. In the second iteration, r is a random value between 1 and 6, and this time the program exchanges r[1] and r[5]. Note that it is possible for an item to be moved several times. The "A" that originally started out in location 0 is going be moved again on the third iteration (unless the call to `random(2,6)` returns 2).

Figure 9.6: *A Ruby method that makes a random permutation of the items in* x *(which can be a string or an array).*

```
# Rearrange the items in x in a random order
1:  def permute!(x)
2:    for i in 0..x.length-2
3:      r = random(i, x.length-1)
4:      x[i], x[r] = x[r], x[i]
5:    end
6:    return x
7:  end
```

In the exercises below we will use the `trace` method to monitor the progress of a permutation and see how successive items from an array are exchanged with random items. The listing of the `permute!` method is shown in Figure 9.6. The parallel assignment is on line 4, so if we attach a probe here we will be able to see the state of the array just before the current item is exchanged with another one to its right.

Tutorial Project

The Ruby code for the `permute!` method is shown in Figure 9.6. If you would like to print a version in your IRB session you can call the `listing` method:

```
>> Source.listing("permute!")
```

You can also "check out" a copy if you want to view it in your text editor:

```
>> Source.checkout("permute!")
```

T38. Make a small array of strings:
```
>> a = TestArray.new(5, :colors)
=> ["plum", "thistle", "khaki", "chocolate", "hot pink"]
```

T39. Type a parallel assignment expression that exchanges the values in the first and third locations, and then print the array again:
```
>> a[0], a[2] = a[2], a[0]
=> ["khaki", "plum"]

>> a
=> ["khaki", "thistle", "plum", "chocolate", "hot pink"]
```
Can you see how Ruby exchanged the strings at `a[0]` and `a[2]`?

T40. Make an array of numbers to use when tracing the `permute!` method:
```
>> a = Array(0..9)
=> [0, 1, 2, 3, 4, 5, 6, 7, 8, 9]
```

T41. A method named `brackets` (similar to the one used in the searching and sorting labs) will print the contents of an array. Attach a probe to the `permute!` method so `brackets` is called just before line 4 is executed:
```
>> Source.probe( "permute!", 4, "puts brackets(x,i,r)" )
=> true
```

T42. Trace the execution of a call to `permute!` as it scrambles your test array:

```
>> trace { permute!(a) }
4: [0  1  2  3  4  5  6  7  8  9]
2:  4 [1  2  3  0  5  6  7  8  9]
6:  4  2 [1  3  0  5  6  7  8  9]
3:  4  2  6 [3  0  5  1  7  8  9]
...
=> [4, 2, 6, 3, 8, 9, 7, 1, 5, 0]
```

Each line above shows the current state of the array. The number at the front of the line, before the colon, is the value of `r`, which is the location where an item will be moved. The left bracket is printed just in front of `a[i]`, which means the item to the right of the bracket is the one that will be moved. The first line shows that the 0 in `a[0]` is going to be swapped with the item in `a[4]`. You can see the effect of this change in the array printed on the second line.

The actual result you get will be different, since the location used for the exchange is chosen at random. But you should be able to follow the steps of the algorithm by looking at the front of each line to see which location was chosen, and then noticing on the following line how the item next to the bracket was swapped with the item at the chosen location.

T43. Call `Card.new` a couple of times to make some random cards:

```
>> Card.new
=> 8♥
>> Card.new
=> 3♦
```

T44. Type this expression to make an array named `a` with 13 random cards, and then print the array after sorting it:

```
>> a = []; 13.times { a << Card.new }; a.sort
=> [K♠, J♠, J♠, 4♠, ... ]
```

There is a chance that by simply calling `Card.new` 13 times to deal a hand we will get a duplicate card. Does your array have any duplicates? Repeat the expression a few times. How often do you get duplicate cards?

T45. Make a full deck of cards containing each of the 52 cards:

```
>> d = new_deck
=> [A♠, K♠, Q♠, J♠, ... 3♣, 2♣]
```

T46. Call the `permute!` method to see if it shuffles the deck:

```
>> permute!(d)
=> [2♥, Q♠, 9♦, K♣, 4♥, 4♣, 2♠, ... ]
```

Repeat the expression a few times. Does it look like a random shuffling after each call?

T47. Shuffle the deck, and save the first five cards in an array named `h`:

```
>> h = permute!(d).first(5)
=> [9♦, 3♦, A♣, 5♠, 7♠]
```

Repeat this expression a few times. Do you get a random poker hand each time?

T48. The following expression is like the one above, except it deals a bridge hand (13 cards) and it sorts the hand before it is displayed:

```
>> h = permute!(d).first(13).sort
=> [J♠, 10♠, 9♠, 5♠, 3♠, 2♠, K♥, 10♥, 9♥, 6♥, 9♦, 10♣, 9♣]
```

Repeat the expression a few times. Since the hand is a random shuffle of a full deck you should never see any duplicates.

9.5 Tests of Randomness

So far we've been relying on our intuition that the output from our pseudorandom number generators looked random. In this section we'll perform some tests on the sequence of numbers produced by a PRNG to investigate the question of whether the sequences are random or not.

The two techniques we will use are both informal tests that use graphical displays. Visualization is a very powerful method for seeing whether there are any biases or hidden patterns in the data. These informal tests will not give us a definitive answer, of the form "yes, this sequence is random" or "no, this sequence is not random," but the tests are simple to do. If a sequence of numbers is not random, the nonrandomness often shows up in the form of patterns in the display.

The first type of display is a **histogram** (often called a "bar chart"). To test a sequence of simulated rolls of a die, we can just count the number of times each number from 1 to 6 occurs in the sequence. The histogram will have one bar for 1s, another bar for 2s, and so on, where the height of a bar indicates how often that number showed up in the random sequence (Figure 9.7).

The algorithm we have been using for making pseudorandom sequences should create what is known as a **uniform distribution**. In an experiment with 1000 rolls of the die, we expect each number will occur roughly $1000 \div 6 = 167$ times. But suppose there is a problem with the PRNG, and it gives us 300 6s. If all the other rolls occur equally often, the bar for 6s will be over twice as high as the others, and we can tell at a glance that the sequence is biased. But if all six bars are roughly the same height we could then do some further statistical tests if we wanted to know for certain whether the sequence is random.

The RAND corporation used a similar approach to test the numbers in their book of random digits. One of their tests was called the "poker test." If we want to apply this test to our random number generator, we can use the PRNG to deal 1000 poker hands. According to the laws of probability we should see 420 hands that have two cards with the same rank (*i.e.*, a pair), 21 hands with three cards of the same rank (three of a kind), and so on. If we plot the results with a bar chart, and see a lot more straights and full houses than pairs, then something is clearly wrong.

To draw a histogram on the RandomLab canvas, we first need to create a set of "bins," with one bin for each result in the experiment. For example, we need six bins for the

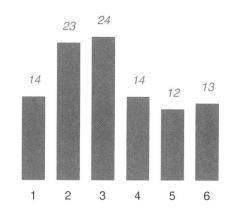

Figure 9.7: *A histogram (bar chart) showing the frequency of random rolls of a die based on the first 100 numbers from the pseudorandom sequence.*

experiment with simulated dice, one for each of the numbers from 1 to 6, so the command to initialize the histogram is

```
>> view_histogram([1,2,3,4,5,6])
=> true
```

To add one to the count in a bin, call a method named update_bin, passing it the ID of the bin to increment. This expression adds one to the count for bin 6:

```
>> update_bin(6)
=> true
```

When Ruby evaluates the expression you should see the bar for bin 6 grow slightly.

Another approach to test whether the sequence of values generated by a PRNG is random or not is to look for patterns or correlations between successive numbers in the sequence. It turns out there is a pattern in the sequence produced by the PRNG we used earlier in this chapter. Here is the expression we typed to make a list of 1000 numbers:

```
>> seq = prng_sequence(81, 337, 1000)
```

and these are the first 10 numbers in the sequence:

```
>> seq[0..9]
=> [0, 337, 634, 691, 308, 285, 422, 519, 376, 793]
```

From this short sample it's not clear what the pattern is, but it starts to become more apparent if we convert each of these numbers into a value between 1 and 6, which is what we would do if we're using the numbers in a game based on rolling dice. If you want to try to figure out the pattern yourself, cover up the paragraph below the IRB output shown below. Here are the first 20 numbers in the sequence, converted to values between 1 and 6. Do you see the pattern?

```
>> seq[0..19].map { |x| (x % 6) + 1 }
=> [1, 2, 5, 2, 3, 4, 3, 4, 5, 2, 1, 4, 3, 2, 3, 4, 5, 2, 3, 2]
```

Based on this short list it looks like the numbers alternate between even and odd. If the computer used this pseudorandom sequence to play a game, every time it rolled an even number it would follow with an odd number, and *vice versa*. It wouldn't take long for a person playing a game based on this PRNG to suspect something was wrong. In a game with two die, the computer would never roll "doubles."

A visual display that makes this type of pattern easy to see is known as a **dot-plot**. As the name implies, the drawing will be a set of dots on a canvas. To make a dot-plot on the RubyLabs canvas, call a method named view_dotplot:

```
>> view_dotplot(500)
=> true
```

The argument is the number of pixels for the height and width of the plot.

To display a dot on the canvas call plot_point, passing it the *x* and *y* coordinates of the dot. For example, this expression will draw a dot in the middle of the canvas:

```
>> plot_point(250,250)
=> nil
```

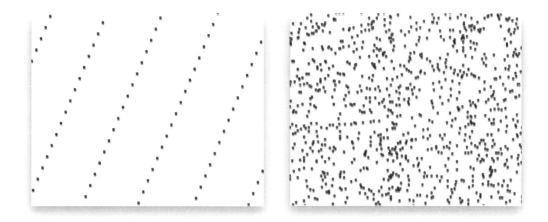

Figure 9.8: *In these images the x and y coordinates of a dot are determined by two successive numbers from a pseudorandom number generator (PRNG). The image on the left is from a PRNG that alternates between even and odd numbers. The image on the right is based on Ruby's own* rand *method.*

One way to look for patterns in values from a pseudorandom sequence is to plot a set of points in which the x and y coordinates of a dot are determined by getting two successive numbers from the PRNG. If we make a PRNG named p, this loop will paint 1000 points using p to set the x and y coordinates:

```
>> 1000.times {
      x = p.random(0,499);
      y = p.random(0,499);
      plot_point(x,y)
   }
```

If there is no correlation between successive numbers, the points will be scattered at random all over the canvas. But if there are any hidden patterns in the data, even if they are very subtle, the dots will line up or form some other distinctive shapes.

The dot-plot for our pseudorandom number generator with $a = 81$, $c = 337$, and $m = 1000$ is shown on the left in Figure 9.8. It's obvious from this plot that the sequence of numbers is not at all random. Even if we had not noticed that the sequence alternated between even and odd numbers, this plot would have told us there was some sort of pattern, and we would not use this set of parameters in an application that requires random numbers.

As was the case with histograms, the visualization is convenient for getting an initial impression. When there are hidden patterns in the data they often jump out when the data is displayed with a dot-plot. But to truly check to see whether the data is random it would be necessary to perform a detailed statistical analysis of the x and y coordinates. For this chapter, however, we will simply do informal tests using pictures.

Tutorial Project

The first set of projects will plot a histogram for rolls of a die produced by the pseudorandom number generator we used in previous projects.

T49. Type this expression to clear the canvas and draw an empty histogram to count the rolls of a die:

```
>> view_histogram( [1,2,3,4,5,6] )
=> true
```

You should see six lines in the middle of the screen, each representing a bin with zero items.

T50. Type this command to increment bin number 5:

```
>> update_bin(5)
=> nil
```

Do you see how the rectangle for this bin got slightly taller?

T51. Repeat the previous command a few times, using different bin numbers, until you're confident you know how this expression updates the histogram.

T52. Make a new pseudorandom number generator using the same parameters from earlier in the chapter:

```
>> p = PRNG.new(81, 337, 1000)
=> #<RandomLab::PRNG a: 81 c: 337 m: 1000>
```

T53. This command will get a random number between 1 and 6 from the PRNG and then pass it to update_bin to increment the count for that number:

```
>> x = p.random(1,6); update_bin(x)
=> nil
```

T54. Repeat the command several more times. Do you see how a different bin is updated each time the expression is evaluated?

T55. Clear the histogram by entering the expression from problem T49 again:

```
>> view_histogram( [1,2,3,4,5,6] )
=> true
```

T56. Put the command that draws a random number and then plots it inside a call to times so it is executed 100 times:

```
>> 100.times {  x = p.random(1,6); update_bin(x) }
=> 100
```

What does your histogram look like? Are the numbers fairly evenly distributed?

If any of the bins reaches the top of the canvas the drawing methods will automatically rescale the histogram to make it smaller. So if you see the histogram suddenly shrink, don't worry. It just means the drawing methods are "zooming out" and making room for more data.

T57. In an earlier exercise we saw how this PRNG generates every number between 0 and 999. That means there should be the same number of 1s, 2s, 3s, *etc.*, when those numbers are converted to values between 1 and 6. To verify this claim, repeat the previous loop 900 more times, so a total of 1000 numbers are in the histogram:

```
>> 900.times {  x = p.random(1,6); update_bin(x) }
=> 900
```

What did you see? Are the results what you expected?

The next set of exercises make another histogram, this time using poker hands.

T58. The classification of a poker hand is often called its "ranking." To get an array of names of poker hands, call a method named `poker_rankings`:

```
>> poker_rankings
=> [:high_card, :pair, ... :four_of_a_kind, :straight_flush]
```

Note that the rankings are ordered, with the most common hands on the left.

T59. Pass the array of names to the method that creates a new histogram:

```
>> view_histogram(poker_rankings)
=> nil
```

Since there are nine names in the list, you should see nine empty bins.

T60. Make a new deck of cards:

```
>> d = new_deck
=> [A♠, K♠, Q♠, ... 4♣, 3♣, 2♣]
```

T61. This expression will shuffle the deck and copy the first five cards to an array named h:

```
>> h = permute!(d).first(5)
=> [Q♥, K♦, 5♠, J♥, K♣]
```

It's OK to cheat if you want to repeat that expression until you get an interesting hand.

T62. A method named `poker_rank` will figure out what sort of hand is in the array:

```
>> poker_rank(h)
=> :pair
```

T63. Since the result from `poker_rank` is one of the histogram labels, we can increment the count for that type of hand by passing the label to `update_bin`:

```
>> update_bin(poker_rank(h))
=> true
```

T64. The next command combines all the operations into a single expression: it shuffles the deck, makes a hand from the first five cards, figures out what sort of hand it is, and updates the corresponding bin in the histogram:

```
>> h = permute!(d).first(5); update_bin(poker_rank(h))
=> nil
```

T65. Repeat that command several times. Are the bins updating? Are they mostly on the left, for the most common hands?

T66. If you want to repeat the experiment for a larger number of hands, put the expression in the body of a loop:

```
>> 1000.times {h = permute!(d).first(5); update_bin(poker_rank(h))}
=> 1000
```

♦ The histogram methods keep track of the number of each type of item. If you want to see the final counts, call a method named `get_counts`:

```
>> get_counts
=> {:straight_flush=>0, :high_card=>524, :two_pair=>41,
    :three_of_a_kind=>29, :full_house=>1, :straight=>5,
    :four_of_a_kind=>1, :flush=>4, :pair=>395}
```

This output shows there were no straight flushes, 524 hands classified as "high card," 41 with two pair, and so on. Find a table of poker probabilities on the Internet, and compare your results with the expected frequency for each type of hand. Do you think there is any bias in these hands, or does it seem like `permute!` did a good job of shuffling the deck?

9.6 Summary

Making lists of random values is a subtle problem. It's hard for humans to do, and it turns out it's not quite so easy for machines, either.

In this chapter we saw how to make a long list of numbers where short stretches appear to be random. Given the right combination of parameters, the equation that defines how to add a new value to the list as a function of the previous value will make sure every number between 0 and $n - 1$ eventually appears in a list of n numbers.

But even though the list has all n numbers, and they appear to be in a random order, there can be some hidden patterns. For example, when we took a closer look at one of the sequences we studied, we saw that it alternated between even and odd numbers. Even more subtle patterns might be hidden in the data. We explored two methods for testing randomness by drawing pictures based on values produced by the random number generator. These informal methods can help us tell, at a glance, whether there is a hidden bias or pattern.

Random number generators are among the most important and widely used algorithms in computer science. They are used in games, of course, but there are many other application areas where random values play a key role. We used random sequences in Chapters 4 and 5 to test searching and sorting algorithms. Many scientific algorithms also use random numbers. For example, some algorithms that reconstruct the evolutionary history of a set of genes use a method based on taking a random sample of all possible family trees to find the one that is most likely given the similarities between the genes. Later in this book, Chapter 12 describes a type of problem known as optimization, where the goal is to find the optimal solution to a problem, and one common approach also uses random samples. E-commerce, Internet banking, and other network traffic depends on encryption algorithms to turn a piece of text into what appears to be a random sequence of letters; many of the important concepts used to design effective random number generators are also used in encryption algorithms.

Concepts and Terminology Introduced in This Chapter

random	A sequence of items is random if the values are independent of one another
pseudorandom	A pseudorandom sequence is one produced by an algorithm; pseudorandom sequences are not truly random, but small subsequences may appear to be random
PRNG	Pseudorandom number generator, an algorithm that creates a sequence of pseudorandom values
permutation	A reordering of the items in a list or array
uniform distribution	The result of generating random values where each value is equally likely
histogram	Also known as a bar chart; a visual display that shows the number of times various events occur

---

Exercises

1. The schedule in Section 9.1 for events that occur every seven hours had a period of 12, *i.e.*, every number between 0 and 11 appears once in the schedule before it repeats. Can you find other intervals besides 7 hours that also lead to a period of 12?

2. Suppose the schedule for events that occur every seven hours starts on a Monday at 12 A.M., so the next events are Monday at 7 A.M., 2 P.M., and 9 P.M. Will this pattern ever repeat? That is, will there ever be another Monday with this same schedule? If so, can you determine how many days will pass before this same schedule is used?

3. Redo the schedule for events that occur every seven hours using a 24-hour clock. Is the period still 12? Or is it now 24?

4. What are some intervals that lead to schedules with a period of less than 24 when a 24-hour clock is used?

5. What are some intervals that lead to a full schedule with 24 times and every time between 0 and 23 occurring exactly once?

6. What do the answers to the last problem have in common? Is there a common attribute for intervals that lead to schedules with a period of 24, *vs.* those the lead to a period of less than 24?

7. The first day of each month usually falls on a different day of the week. Is the pattern of weekdays random? Pseudorandom?

8. Use the Ruby expressions starting on page 241 to create a dot-plot on the RubyLabs canvas. Then make a plot of 3000 points drawn at random from Ruby's own random number generator (the built-in method named `rand`) instead of numbers from a pseudorandom sequence. Do the points appear to be spread uniformly all over the canvas?

9. The Ruby expression given in Exercise T14 used a call to `uniq` to see if all of the numbers between 0 and m−1 were in a pseudorandom sequence. Another way to check the sequence would be to sort it so you can compare it to an array that has every number from 0 to m−1. Can you write a Ruby expression that will do this test?

10. If you took the challenge at the beginning of Section 9.1 and entered a series of random digits in a file, make a histogram from your numbers. One way to make a histogram is to load the numbers into a spreadsheet application and use its "chart" command. Were you able to make a uniform distribution? Or are some numbers a lot more frequent than others?

11. ♦ Here are some values of a, c, and m that should generate much better pseudorandom sequences than the ones we used in this chapter:[2]

a	c	m
1255	6173	29282
171	11213	53125
421	54773	259200

Repeat some of the experiments in this chapter to evaluate the quality of the pseudorandom sequences. Make PRNG objects, and use values from the objects to make histograms (not just of values from 1 to 6, but for other ranges, too) and dot plots.

---

[2]From *Numerical Recipes in C*, by W. H. Press, *et al.*

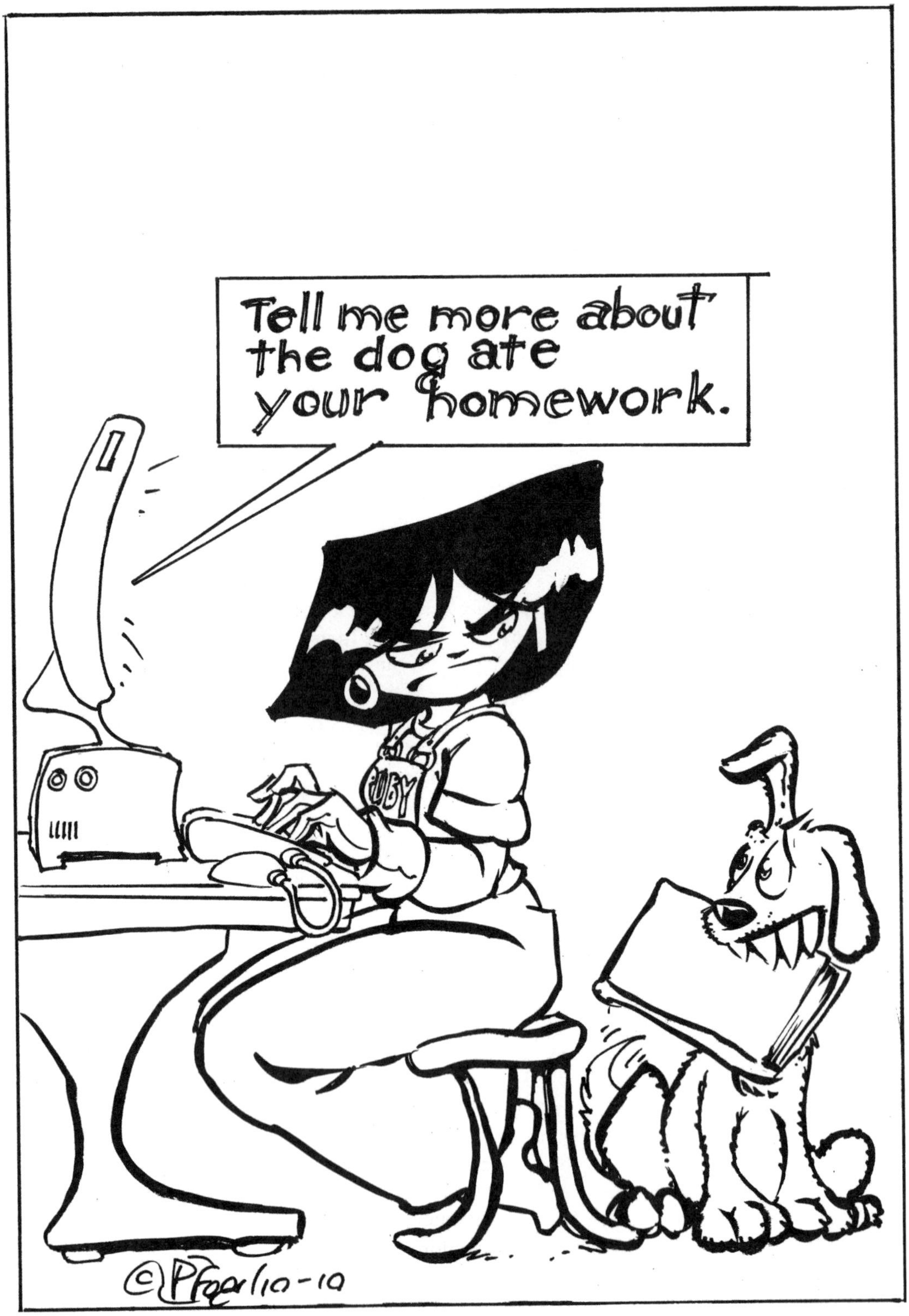
Tell me more about the dog ate your homework.

Chapter 10

Ask Dr. Ruby

A program that understands English (or does it?)

In 1950, Alan Turing (1912–1954), one of the founders of modern computer science, published a paper with the title "Computing Machinery and Intelligence." Electronic computers were just starting to be used outside of math and science, and they were being adopted by businesses and other organizations. There was widespread interest in this new technology, and people began to wonder just what these machines were capable of doing. The topic of Turing's paper was the nature of human intelligence, specifically whether a machine could ever be able to handle problems routinely solved by humans.

To answer the question of whether a computer was intelligent or not, Turing proposed a simple criterion, based on language. He argued that if a computer could carry on a conversation with a person, without the person ever suspecting they were interacting with a machine, then the computer should be considered intelligent.

Building a robot that looks and sounds like a human is, by itself, a daunting prospect. Turing suspected that if a person talked with a robot, no matter how well the robot could converse in English, the person would be biased, and would never accept the machine as being intelligent.

To separate the question of intelligence from questions of looking and sounding human, Turing proposed a simple "thought experiment." In this experiment, which is now known as the **Turing Test**, a person, called the judge, is asked to carry on a conversation using a computer terminal. In modern terminology, we would say the judge would use a chat application, running on their laptop or desktop personal computer. The judge chats with two other participants over a local network connection (Figure 10.1). The other two participants will be in two different rooms, behind closed doors, labeled *A* and *B*. In one room there is a human, and in the other there is a computer, but the judge will not know which participant is in which room.

247

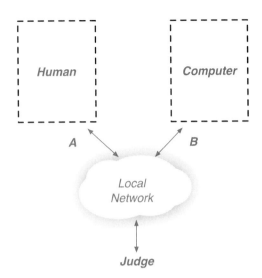

Figure 10.1: *In the Turing Test, three participants are using "chat" software over a local network to carry on a conversation. The judge knows there is a human in one room, and a computer in the other, but does not know which room holds the computer. The object is for the judge to ask questions of both participants, and then decide which one is the computer.*

The judge is allowed to direct questions at either of the other participants. For example, the judge might type "*A*, what color is a fir tree?" or "*B*, should the NCAA establish a playoff system for all levels of college football?" *A* and *B* each try to convince the judge they are human, and that the participant in the other room is a computer. If, at the end of the conversation, the judge can't decide which room holds the computer, we can conclude the machine possesses a high degree of human intelligence.

It is interesting to note that 60 years after Turing's paper was published, his test is being applied in a real way. If you use an instant messaging service or participate in on-line chats at any of the popular social networking web sites, you might have seen a posting by a "chatbot," a computer program that generates posts. Malicious chatbots generate spam messages that fill chat rooms with advertising, and some chatbots have reportedly tried to pose as real users to fool people into revealing credit card numbers and other financial information. Chatbots don't need to be able to converse about subjects in general, and they don't even need to fool a high percentage of people. If just a few users are tricked, even temporarily, the chatbot will have served its purpose.

Issues related to writing computer programs that can converse with humans are part of an active area of research, with many unsolved problems and open questions. The goal is to develop methods for representing everyday knowledge, concepts we take for granted (*e.g.*, that fact that most trees are green), and to design algorithms that use this knowledge to carry on a conversation. The name of this research area is **natural language processing** to distinguish it from another area of computer science, which is concerned with computer programming languages.

One of the first programs to attempt to converse in English was named ELIZA . It was written by Joseph Weizenbaum (1923–2008), a computer scientist at MIT, in 1966. ELIZA was a remarkable program, not only because it was able to generate realistic English sentences, but also because of how people reacted to it. Weizenbaum was amazed by how often people were willing to open up and converse at length with ELIZA, even when they knew they were getting responses from a computer and not another human being. In *Computer Power and Human Reason*, a book he published in 1976, Weizenbaum compared this "ELIZA effect" to going to the theater: we know the people on stage are actors, but we suspend our disbelief for a few hours and think of the characters as real people with real lives. When users were typing sentences into ELIZA, they seemed perfectly happy to carry on a conversation, even when they knew it was a computer on the other end.

This chapter is an introduction to natural language processing. We will use a version of ELIZA, written in Ruby and included as part of the RubyLabs software package, to explore some of the issues faced by programs that attempt to use natural language. Unlike the other chapters, where the goal is to show how a problem can be solved by computation, the project in this chapter will raise new questions instead of illustrating various solutions. As we look at how our version of the program works, we will start to see how difficult it is to even define the problem, much less figure out how to solve it computationally. Natural language processing is a huge challenge, and there is still a long way to go before computers routinely converse in English or any other human language.

10.1 Overview of ELIZA

Weizenbaum's idea for ELIZA was to make a program that would respond by simply rearranging the words in an input sentence. The rearrangements were defined by a set of **rules**. For example, if a user types "I don't like rain," the computer could simply pick out important words, apply a rule that reorders them, and respond with "Why don't you like rain?" To test this idea, Weizenbaum developed rules that would have his system mimic a Rogerian psychologist, so the rules were designed to look for words or phrases a person might use during a therapy session.

The simplest type of rule is to have the program recognize key words and reply with one of several predefined responses for that word. For example, if the input contains "mother" or "father" or other family relationships, the response could be one of several generic sentences about families. Here is a part of a transcript from a session with Eliza, the RubyLabs implementation of ELIZA:

```
H: My father wouldn't buy me a puppy.
C: Tell me more about your family.
```

Whenever Eliza sees any sentence with a word that refers to a family member, it can respond with a generic sentence like the one above. The word "dream" is another key word, and the program will respond to any sentence containing the word "dream" with something like "What does that dream suggest to you?" or "Do you dream often?"

More complicated rules extract a large chunk of the input and use it as part of the response. As an example of how this works, consider statements like "I am sad because my team lost the big game" or "I am sad I never got a puppy." In general, sentences that start out with "I am sad..." continue with a description of a situation or a reason the speaker is sad. The rule Eliza uses to respond to this type of sentence requires the program to find the words following "sad" and use them as part of the response:

```
H: I am sad I never got a puppy.
C: Can you explain what made you sad you never got a puppy?
```

In this second example, the key word is "sad." The program took the words following "sad,"
in this case "I never got a puppy," made a simple transformation that changed "I" to "you,"
and used the result with its predefined response of "Can you explain what made you..."

Of course not all sentences that start out "I am sad..." fit this general pattern, and this
rule can easily construct a response that looks pretty strange:

```
H:  I am sad we never had a chance to play fetch.
C:  I am sorry to hear you are sad we never had a chance
       to play fetch.
```

The problem, of course, is that the word "we" can be interpreted two different ways. Did
the speaker mean "my puppy and I" or "you and I"? A human psychologist would assume
the speaker had the first meaning in mind, but Eliza, using very simple rules to transform
input sentences, generated a sentence that looks like it is based on the second meaning.

The idea of applying rules to input sentences is the central concept in ELIZA. In the
sections below we will explore the details of what defines a rule and how to create new rules,
but for now the main idea is to realize that ELIZA's ability to participate in a conversation
depends entirely on the set of rules it uses. With only a few rules, where ELIZA always
responds with set phrases associated with key words, a person interacting with ELIZA would
soon realize they are conversing with a machine. But with more complex rules that analyze
several parts of the input, it's possible to have ELIZA generate responses that give more of
an appearance that it "understands" the person typing the sentences.

The set of rules used by ELIZA to transform an input sentence is called a **script**. Weizenbaum chose the name "Eliza" for his program to emphasize the fact that the rules in a script
determine how the program will respond. Just as Eliza Doolittle, the fictional character
from George Bernard Shaw's *Pygmalion*, became more refined by learning more rules of
language and etiquette, ELIZA the program should become better at conversing as more
rules are added to its script.

Weizenbaum envisioned a situation where different scripts could be written for different
applications; for example, one might make a script with rules based on key words used in
cooking and baking to make a system that would give the illusion of conversing with a chef.
The original script that Weizenbaum wrote, and the one that made the program famous,

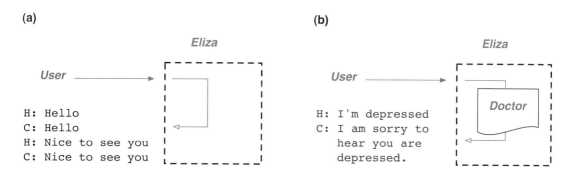

Figure 10.2: *(a) When the* `Eliza` *module is first loaded it does not have any transformation rules, so any sentences typed by a user are simply echoed. (b) The rules in the* DOCTOR *script extract key words and phrases and transform an input sentence into a response.*

was named DOCTOR (Figure 10.2). The DOCTOR script that played the role of Rogerian psychotherapist contained rules based on words like "sad," "depressed," "dream," "wish," and "family."

To begin the project for this chapter we will just run Eliza, using rules based on those from Weizenbaum's DOCTOR script. When you read the responses the "doctor" makes to your statements, see if you can get a sense for what sort of rules the program is applying. In the next section we will start learning how to write our own sentence transformation rules.

Tutorial Project

T1. Start IRB, and load the module that will be used for this project:
```
>> include ElizaLab
=> Object
```

T2. Call the method named `Eliza.run` to start a conversation. When Eliza is running the prompt changes to `H:` to let you (the human) know Eliza is waiting for you to type a sentence:
```
>> Eliza.run
   H:
```

T3. Since Eliza does not have a script yet, it can only apply a default rule that does not change the input. Type a few sentences. Everything you type will simply be echoed right back:
```
>> Eliza.run

   H: Hello.
   C: Hello

   H: How are you?
   C: How are you

   H: Stop repeating what I say
   C: Stop repeating what I say
```

T4. Unlike an annoying sibling, Eliza will quit when you tell it to. Type the single word `bye`, or `quit`, or hit ^D (hold down the control key while typing D) to end a conversation:
```
   H: quit
=> nil
```

Note that the prompt has changed back to the normal IRB prompt. The `Eliza.run` method has returned `nil`.

T5. Call the Eliza module's `load` method to read the "doctor" script:

```
>> Eliza.load :doctor
=> true
```

If you pass `Eliza.load` a symbol it will look for one of the predefined scripts included in the ElizaLab module, otherwise it will look for a file in your project directory.

T6. Call `Eliza.run` again. Now the rules defined in the "doctor" script will be used to transform every sentence you type into a response:

```
>> Eliza.run
How do you do. Please tell me your problem.

  H: I need a puppy.
  C: What would it mean to you if you got a puppy?

  H: I remember my cousin had a puppy.
  C: Do you often think of your cousin had a puppy?
```

T7. Try a few more sentences on your own. Don't forget to type `quit` when you are ready to continue with the next problem in the tutorial.

Eliza uses a set of key words to generate responses. In the example described earlier in this section, a sentence containing the word "mother" or "father" is transformed by a rule defined for family members to produce "Tell me more about your family." If a sentence does not contain a key word Eliza just responds with something noncommittal, like "Go on."

T8. To see a complete list of key words in the current script call `Eliza.info`:

```
>> Eliza.info
Script: ./../data/eliza/doctor.txt
    28 rules with 51 sentence patterns
    78 key words: alike, always, am, are, because, ...
```

T9. Start a new conversation by typing `Eliza.run` again, and make up some input sentences on your own. You can type anything you would like, but you are more likely to keep up the illusion of chatting with a psychologist if your statements include key words from the list printed by the `info` method.

10.2 Sentence Patterns

To implement a program like Eliza in Ruby, one of the first decisions to make is how to look for key words in sentences. The program needs to scan sentences like "My father wouldn't buy me a puppy" to look for words like "father," and to scan sentences like "I'm sad I don't have a dog" to break it into the parts before and after the word "sad."

The easiest way to look for words in a sentence is to do a **string search**. Ruby has several methods that scan a string to look for a specified word or substring. For example, suppose we have a sentence named `s1` defined as

```
>> s1 = "I was afraid of the cow."
```

A script with rules about farm animals would probably want to see if words like "cow," "horse," and "pig," are in a sentence. The `include?` method, which we first saw in Chapter 4, is all we need. A call of the form `s.include?(w)` does a linear search from the beginning of the string `s` to see if another string `w` appears as a substring. The method returns `true` if the letters in `w` can be found anywhere in `s`, or `false` if `w` is not in `s`.

To see if a sentence contains a word, just pass the word as an argument in a call to include?. This example asks if "cow" is in the sentence s1:

```
>> s1.include?("cow")
=> true
```

There are two problems with this simple plan, however. We don't want Eliza to respond to "My cat is scowling at me" with something like "Tell me more about your farm," which it might do with this sentence because it will find "cow" in the middle of the input string:

```
>> s2 = "My cat is scowling at me."
=> "My cat is scowling at me."

>> s2.include?("cow")
=> true
```

What we want is for Eliza to look for "cow" as a complete word, but to not match the letters "cow" in the middle of "scowl."

The second problem is that we are going to end up with a lot of patterns if we have to create a new one for each word. The farm script should respond to sentences with "cow," "pig," "horse," and so on, and it would be nice if we didn't have to write a separate pattern for each word.

Ruby provides an alternative type of search that addresses both of these problems. This more advanced type of search uses what is known as a **regular expression**. A word search based on regular expressions is basically a search for a string that matches a pattern. Instead of looking for a specific string, the search method will look for any string that fits the description specified by the regular expression.

Regular Expressions in Ruby

The pattern-matching operations in Eliza are based on **regular expressions**. In Ruby, a regular expression is an object, just like a string or integer. Regular expressions look like strings, but they are enclosed in slashes, not quotes.

Inside a regular expression, characters like periods and asterisks have special meaning. For example, /.ow/ defines a pattern that means "any letter followed by an *o* and a *w*." As the examples at right show, Ruby can use regular expressions when it does a string search, either to look for an individual word or for all substrings that match the pattern.

```
>> s = "how now brown cow"

>> r = /cow/
>> r.class
=> Regexp
>> s.index(r)
=> 14

>> r = /.ow/
>> s.scan(r)
=> ["how", "now", "row", "cow"]
```

For the Eliza project we will use only a few of the special symbols. The main thing to know is that strings enclosed in slashes are regular expressions that will be used to see if an input sentence matches a particular pattern.

You may have encountered regular expressions when you visited a web site that asked you to enter information in a form. A web page that expects a user to type a phone number, date, social security number, or other information with a specific format will often check to see if the text entered into the form matches the general pattern for that type of data. A web interface with a "search box" is another place regular expressions are commonly used. For example, you might want to search an on-line crossword puzzle dictionary to find a 6-letter word that starts with *s* and ends with *e*. If you know how to write regular expressions, you can enter a pattern that describes the words you're looking for. When you click the "search" button, the web application will find all words in the dictionary that match the pattern.

For this project we will be using a new type of object, implemented as part of the ElizaLab module, to represent sentence patterns. The objects are called Patterns (short for "sentence patterns"). Making a regular expression can be quite involved, but the method that creates new pattern objects will do most of the work for us. All we have to do is pass a simple string in a call to `Pattern.new`, and the method will transform the string into a regular expression.

The simplest type of sentence pattern is one that specifies a key word. Suppose we want a pattern that tells Eliza to reply to any sentence that contains the word "cow" with either "Tell me more about your farm" or "Go on." This expression makes a new pattern object and saves it in a variable named p:

```
>> p = Pattern.new( "cow",
         ["Tell me more about your farm", "Go on"] )
=> cow: ["Tell me more about your farm", "Go on"]
```

The first argument is the string we want to look for in an input sentence; in this case, it's a single word, and it means the object can be applied to any sentence that contains the word "cow." The second argument is a list of strings that will be used as responses when this word is found in an input sentence. The `Pattern.new` method creates the regular expression that will find "cow" as a complete word, so the pattern will not match sentences that contain "scowl" or other words with the letters "cow" in the middle.

```
H: We had a cow.                    /cow/
C: Tell me more about                  "Tell me more about your farm."
your farm.                             "Go on."

H: I never liked that cow.
C: Go on.

H: The cow jumped at me.
C: Tell me more about
your farm.
```

Figure 10.3: *A pattern object (shown in the gray box) has a regular expression and a list of strings. The* `apply` *method will see if an input sentence matches the regular expression. If so,* `apply` *returns one of the responses. The object cycles through the list of responses, so that each match gets a different response.*

Recall that our goal is to devise a set of rules for transforming input sentences into responses. A pattern object implements a very basic type of rule, where strings that match the regular expression lead to responses defined by the list of strings. A method named `apply` will carry out this operation. If `p` is a pattern object, a call to `p.apply(s)` will try to match `s` with the regular expression, and if the match is successful, return one of the response strings. Here is an example of our new pattern being applied to a sentence that contains the word "cow":

```
>> p.apply("I milked the cow")
=> "Tell me more about your farm"
```

When there is more than one response, Eliza will cycle through them (Figure 10.3). The next call to `p.apply` will use the second response in the list:

```
>> p.apply("That cow was scary")
=> "Go on"
```

If a sentence does not match the pattern, the call to `apply` will return `nil`. This sentence does not contain the word "cow" so the call to `apply` fails:

```
>> p.apply("There were pigs, too")
=> nil
```

Because the pattern object makes sure the regular expression matches only complete words, our pattern will not apply to sentences where "cow" appears in the middle of a word:

```
>> p.apply("The cat was scowling at me")
=> nil
```

The second advantage of using regular expressions to match input sentences is the ability to make a single pattern for a set of related words. Instead of having to make separate patterns for different animals, we can make one pattern, using a regular expression that effectively says "match an input sentence that contains any of the following words." All we need to do is list the words, separated by a | character.

Here is an example of a pattern that can be applied to any sentence that contains "cow," "pig," or "horse":

```
>> p = Pattern.new( "cow|pig|horse",
           ["Tell me more about your farm", "Go on"] )
=> cow|pig|horse: ["Tell me more about your farm", "Go on"]
```

Now a call to `apply` will generate a response if any of the three words is in the input sentence:

```
>> p.apply("The cow slept in a barn")
=> "Tell me more about your farm"

>> p.apply("The horse jumped the fence")
=> "Go on"
```

A set of words separated by vertical bars is known as a **group**. As seen in this example, groups make it easy to define rules that can be applied to any one of a set of alternative words. Later in the chapter we will see how groups also let us extract part of the input sentence so words can be echoed to the user when Eliza constructs a response.

Tutorial Project

Before you start this section of the tutorial, remove the DOCTOR script. Either start a new IRB session, or call a method that erases all the sentence patterns currently in the system:

```
>> Eliza.clear
=> true
```

T10. Make a pattern that will apply to sentences that contain the word "cow":

```
>> p = Pattern.new( "cow", ["Tell me more about your farm."] )
=> cow: ["Tell me more about your farm."]
```

Note there is only one response in this example, but it is still enclosed in brackets.

T11. If you want to test the pattern, a method named match will return true if a string matches the sentence pattern:

```
>> p.match("Don't have a cow, man")
=> true

>> p.match("He scowled")
=> false
```

T12. Try applying the pattern to see how Eliza will respond to sentences containing the word "cow":

```
>> p.apply("I milked the cow")
=> "Tell me more about your farm."
```

T13. You can add more sentences to the list of responses by calling add_response:

```
>> p.add_response("What made you think of cows just now?")
=> ["Tell me more ...", "What made you think ..."]
```

T14. Now when you enter sentences that match the pattern Eliza will cycle through the responses. Type a few sentences with the word "cow":

```
>> p.apply("The cow slept in the barn")
=> "Tell me more about your farm."

>> p.apply("She had a cow")
=> "What made you think of cows just now?"

>> p.apply("No, not a real cow, she had a fit")
=> "Tell me more about your farm."
```

T15. Make a new pattern that will apply to more than one word:

```
>> p2 = Pattern.new("Ruby|Perl|Python" )
=> Ruby|Perl|Python: []
```

T16. Add a response string to the new pattern:

```
>> p2.add_response("Is that your favorite language?")
=> ["Is that your favorite language?"]
```

T17. This new pattern should match any sentence that has one of the three words:

```
>> p2.apply("We wrote programs in Ruby")
=> "Is that your favorite language?"

>> p2.apply("Some people wanted to use Python")
=> "Is that your favorite language?"

>> p2.apply("No, I like Java")
=> nil
```

Try some more tests on your own to explore the limits of these sentence patterns. Does the regular expression inside the pattern care about capitalization? What about plurals? If you make a pattern for a word like "cow" will the pattern match sentences that contain "cows"?

10.3 Building Responses from Parts of Sentences

To add to the illusion that the computer is actually carrying on a conversation, Eliza should be able to create responses that use parts of the sentence typed by the user. For example, suppose the input is "I'm afraid of cows." Using the technique we saw in the last section, we could make a pattern that looks for the word "afraid" and responds with generic question like "Why are you scared?" But it would be more realistic if the reply was something like "What is it about cows that worries you?" because the response contains "cows," a word from the input sentence. In order to create these kinds of responses, Eliza needs to be able to extract substrings from the input sentence and then use those substrings as part of the response. In this example, we need to be able to extract the word that follows "afraid of" so we can use it in the response.

If a pattern contains a group—defined in the previous section as a set of words separated by vertical bars—the part of the input sentence that matches the group is saved so it can be used later to make the response. The diagram in Figure 10.4 shows how this process works.

First note that the regular expression in the pattern object (the gray box) has a group of three words, one of which is "cow.". When the `apply` method finds the word "cow" in the input sentence, Ruby will save the word. Words are saved in special variables that have names of the form $n, where the *n* is the group number (since in general a pattern can have more than one group). In this example, the word "cow" is saved in $1.

Notice that the response string also has a variable name. When a response has a variable name, the `apply` method "plugs in" the variable's value at that location. In this example, since the letters "cow" were saved in $1, the response becomes "You had a cow?" A variable name in the response string is basically a "placeholder." Before the response is printed, the `apply` method fills in the output sentence by substituting the value of a variable for the placeholder.

Here are the lines from the IRB session that created the pattern shown in Figure 10.4. The first step is to make the pattern object, supplying it with a string that has a group of words:

```
>> p = Pattern.new( "cow|pig|horse" )
=> (cow|pig|horse): []
```

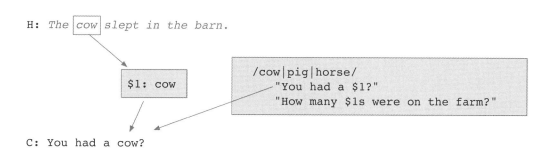

Figure 10.4: *The animal names in this pattern define a group. When an input sentence contains one of these words, the* `apply` *method saves the word in a variable so the word can become part of the response.*

Next, add some response strings that have placeholders:

```
>> p.add_response("You had a $1?")
=> ["You had a $1?"]

>> p.add_response("How many $1s were on the farm?")
=> ["You had a $1?", "How many $1s were on the farm?"]
```

Now when the `apply` method creates a response, it will insert the word it found in the input sentence:

```
>> p.apply("The cow slept in the barn.")
=> "You had a cow?"

>> p.apply("The horse jumped the fence.")
=> "How many horses were on the farm?"

>> p.apply("The pig wallowed in the sty.")
=> "You had a pig?"
```

In this example there was only one group, but in general there can be any number of groups. When a pattern has more than one group, however, there is some ambiguity in how the groups are defined. Here is an example:

```
>> p = Pattern.new("hamster|guinea pig|gerbil")
```

Should Ruby treat this as two groups, one for the words "hamster" and "guinea" and the other for the "pig" and "gerbil"? Or is it one group, with "guinea pig" considered to be an alternative to "hamster" and "gerbil"? The answer is that Ruby considers this pattern to have a single group with three alternatives.

If you really do want two different groups, you need to use parentheses to surround the words in each group. Parentheses in a pattern are just like parentheses in an arithmetic expression: they alter the default precedence, so that Ruby evaluates the expression the way you intend and not by using its default. Here is a pattern that has two groups of words:

```
>> p = Pattern.new("I (like|love|adore) my (dog|cat|ducks)")
=> I (like|love|adore) my (dog|cat|ducks): []
```

Now when we make response strings we can use two placeholders, since there are two groups in the pattern:

```
>> p.add_response("Why do you $1 your $2?")
=> ["Why do you $1 your $2?"]

>> p.add_response("Your $2?")
=> ["Why do you $1 your $2?", "Your $2?"]
```

Notice how the words from each part of the input are inserted into the right place in the response:

```
>> p.apply("I adore my cat")
=> "Why do you adore your cat?"

>> p.apply("I love my dog")
=> "Your dog?"
```

Let's go back to the idea at the beginning of the section, where we want to define a pattern that responds to a sentence like "I'm afraid of ..." with "What is it about ... that worries you?" Now that we know about word groups, we can make a pattern that plugs in the word from the input sentence:

```
>> p = Pattern.new("I'm afraid of (cows|dogs|ghosts)")
=> I'm afraid of (cows|dogs|ghosts): []

>> p.add_response("What is it about $1 that worries you?")
=> ["What is it about $1 that worries you?"]
```

Unfortunately, this pattern only works for the words specified in the group. If we are writing a script to play psychologist, we have to anticipate everything a person could be afraid of.

Another very useful feature of regular expressions allows us to write patterns that save more than just a single word in a placeholder. A regular expression can contain **wild cards** that match any piece of text. A wild card is written with two characters: a period followed by an asterisk. Here is how we would write the same pattern, using a wild card:

```
>> p = Pattern.new("I'm afraid of .*")
=> I'm afraid of (.*): []
```

All the words following "of" in an input sentence are saved in a variable, and they can be inserted into the response:

```
>> p.add_response("Why are you afraid of $1?")
=> ["Why are you afraid of $1?"]
```

Metacharacters (Special Characters in Regular Expressions)

Many symbols have special meanings when they are used in regular expressions.

From a previous sidebar, we saw that inside a regular expression, a period means "any character can be in this location."

The asterisk means "any number of the things to the left," so the pattern .* means "any number of any character" — in other words, an arbitrary string of characters that may appear in an input sentence.

The vertical bar means "or." It is used in situations where we want a pattern to apply to a group of words.

.	match any single character
.*	match any number of characters
x|y|z	match x or y or z
(x|y)s?	x or y, optionally followed by an s

♦ A question mark means the preceding item is optional. One place to use it is when making a pattern that matches a singular or plural form of a word. For example, /(cow|horse)s?/ matches "cow" or "horse" or "cows" or "horses".

Here's what happens when we apply this pattern to some sentences:

```
>> p.apply("I'm afraid of the dark")
=> "Why are you afraid of the dark?"

>> p.apply("I'm afraid of little green men")
=> "Why are you afraid of little green men?"
```

Note that the wild card doesn't just match a single word; the regular expression saves an arbitrarily large part of the input sentence in the variable.

To recap what was introduced in this section:

- If the regular expression that defines a sentence pattern has a group (a set of words separated by vertical bars) or a wild card the regular expression saves the part of the input sentence that matches this part of the pattern.

- The pieces of the input that match a group are saved in variables named 1, *etc.*

- A response can include variable names, and as a result, the output string will contain parts of the input string.

Tutorial Project

T18. Create a pattern that has a group of words:
```
>> p = Pattern.new("green|yellow|red|blue")
=> (green|yellow|red|blue): []
```

T19. Check to see if the pattern matches a sentence that contains one of these words:
```
>> p.match("The sky is red")
=> true
```

T20. Add a response string with a variable, so that the matching word is included in the response:
```
>> p.add_response("That's my favorite color, $1")
=> ["That's my favorite color, $1"]
```

T21. Now call `apply`, to see what sentence is constructed in response to the input:
```
>> p.apply("The sky is red")
=> "That's my favorite color, red"
```

T22. If the pattern is given any sentence with one of the four colors listed you should see a response that contains the color name:
```
>> p.apply("The green bird flew away")
=> "That's my favorite color, green"
```

T23. Make a pattern that has two groups, and add a response that uses both saved substrings:
```
>> p = Pattern.new("The (dog|cat|frog) (ran|jumped)")
=> The (dog|cat|frog) (ran|jumped): []

>> p.add_response("Are you sure the $1 really $2?")
=> ["Are you sure the $1 really $2?"]
```

T24. First see if the pattern matches a sentence that has one word from each group:
```
>> p.match("The dog ran after the car")
=> true
```

T25. Try applying the pattern to some sentences that contain words from both groups:

```
>> p.apply("The cat ran away")
=> "Are you sure the cat really ran?"

>> p.apply("The dog jumped over the fence")
=> "Are you sure the dog really jumped?"
```

T26. Next make a pattern that has a wild card, and add a response that uses the saved text:

```
>> p = Pattern.new("I'm afraid of .*")
=> I'm afraid of (.*): []

>> p.add_response("What is it about $1 that scares you?")
=> ["What is it about $1 that scares you?"]
```

T27. Try some sentences that match the pattern:

```
>> p.apply("I'm afraid of clowns")
=> "What is it about clowns that scares you?"

>> p.apply("I'm afraid of speaking in public")
=> "What is it about speaking in public that scares you?"
```

T28. Make a pattern that has three wild cards:

```
>> p = Pattern.new("I'm .* my .* was .*")
=> I'm (.*) my (.*) was (.*): []
```

Notice that Ruby put parentheses around the wild cards for you.

T29. The response can use the matching pieces of the sentence in any order:

```
>> p.add_response("How did the fact your $2 was $3 make you $1?")
=> ["How did the fact your $2 was $3 make you $1?"]
```

T30. Try out this new pattern:

```
>> p.apply("I'm glad my plane was on time")
=> "How did the fact your plane was on time make you glad?"

>> p.apply("I'm totally disappointed my midterm exam
      was rescheduled for a week later")
=> "How did the fact your midterm exam was rescheduled for a
      week later make you totally disappointed?"
```

T31. Make a new pattern with a single wild card:

```
>> p = Pattern.new("I like .*", ["Why do you like $1?"])
=> I like (.*): ["Why do you like $1?"]
```

T32. Try out the new pattern on some sentences:

```
>> p.apply("I like to solve crossword puzzles")
=> "Why do you like to solve crossword puzzles?"

>> p.apply("I like your hat")
=> "Why do you like your hat?"

>> p.apply("I like her attitude")
=> "Why do you like her attitude?"

>> p.apply("I like my new computer")
=> "Why do you like my new computer?"
```

Some of the replies in the last exercise seem realistic, but others aren't quite right. Can you tell what the odd ones have in common, and how they differ from the realistic ones?

10.4 Substitutions

The exercise at the end of the last section showed how the simple strategy of just "cutting and pasting" entire fragments from an input sentence to build up a response doesn't always work. The pattern object from Exercise T32 responded to "I like my new computer" with "Why do you like my new computer." If we want the "doctor" to respond to statements of the form "I like ..." with "Why do you like ...?", the program cannot simply echo the parts that match the wild card.

The problem with the strange-looking replies is that they have personal pronouns. When the program just plugs parts of the input containing pronouns into the response template, the result can be silly. If the only way Eliza could respond to sentences was through the cut-and-paste operations described in the previous section, Eliza would respond to "I am happy to see you" with "Why are you happy to see you?" In a normal conversation, words like "I" or "you" are typically (but not always) replaced by their opposite. If a patient says "I am happy to see *you*" a real doctor would reply "Why are you happy to see *me*?" The "you" in this sentence was replaced by "me" in the response, but the other words ("to see") are unchanged.

In order to handle situations like this, Eliza uses an operation called **postprocessing**. After breaking the input into pieces with a regular expression, but before reassembling the pieces into a response, Eliza does an additional pattern matching operation on each of the variables $1, $2, *etc.* This second pattern matching step does single-word replacements. Every "I" is replaced with "you," every "my" with "your," and so on. The result isn't perfect, but with a large enough set of replacement strings Eliza can maintain the illusion of carrying on a conversation.

Words substituted during the postprocessing step are kept in a list of **associations**. During the postprocessing phase, Eliza looks in the association list for each word in $1, $2, *etc.*, and if a word is found it is replaced by the corresponding string.

Here are some examples of how postprocessing leads to more realistic responses. The pattern is:

```
>> p = Pattern.new("I am (.*)", ["Are you really $1?"])
=> I am (.*): ["Are you really $1?"]
```

Without postprocessing, responses can be very strange:

```
>> p.apply("I am out of my mind")
=> "Are you really out of my mind?"

>> p.apply("I am sorry I dropped your computer")
=> "Are you really sorry I dropped your computer?"
```

To add a new word to the association list used for postprocessing, just use an assignment statement. These two assignments tell Eliza to replace "me" with "you" and "my" with "your":

```
>> Eliza.post["me"] = "you"
=> "you"

>> Eliza.post["my"] = "your"
=> "your"
```

The list is an associative array named `Eliza.post`.[1] The word to substitute is on the left side of the assignment, between square brackets, and the word it should be replaced by is on the right side.

Now if you ask Ruby to show you the contents of the list you'll see it has saved these two associations:

```
>> Eliza.post
=> {"my"=>"your", "me"=>"you"}
```

After defining a few more postprocessing rules the responses are more realistic:

```
>> p.apply("I am out of my mind")
=> "Are you really out of your mind?"

>> p.apply("I am sorry I dropped your computer")
=> "Are you really sorry you dropped my computer?"
```

Eliza will also do substitutions to a sentence before it tries to apply a pattern. These substitutions, performed during a **preprocessing** phase, are typically used to replace contractions with their more formal counterparts. Suppose we have a pattern for sentences that start "you are":

```
>> p = Pattern.new("You are (.*)", ["I am not $1"])
=> You are (.*): ["I am not $1"]
```

As expected, the pattern works for sentences that contain these two words:

```
>> p.apply("You are crazy")
=> "I am not crazy"
```

But the pattern will not apply to sentences that have the less formal version that uses a contraction:

```
>> p.apply("You're kidding")
=> nil
```

We can fix this problem by defining a preprocessing rule:

```
>> Eliza.pre["you're"] = "you are"
=> "you are"
```

The variable named `Eliza.pre` is, like `Eliza.post`, an associative array that tells Eliza how to replace certain strings. Eliza scans every input sentence, and each time it finds a string from the left side of a rule in `Eliza.pre`, the string is replaced the by the word(s) on the right side of the rule. According to the rule shown in the example, "you're" will be replaced by "you are." After adding this item to `Eliza.pre` the pattern will apply to sentences that have the contraction:

```
>> p.apply("You're kidding")
=> "I am not kidding"
```

---

[1]In Ruby, an associative array is called a Hash. Hashes were introduced in the Huffman Code project in Chapter 7 (see the sidebar on page 166). They are also described in the Ruby Reference section on page 357.

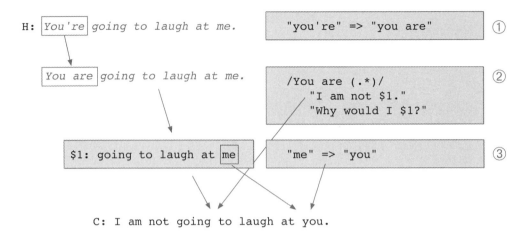

H: You're going to laugh at me.   "you're" => "you are"   ①

You are going to laugh at me.

/You are (.*)/   ②
  "I am not $1."
  "Why would I $1?"

$1: going to laugh at me   "me" => "you"   ③

C: I am not going to laugh at you.

Figure 10.5: *Preprocessing and postprocessing help transform an input sentence into a response. The first step in responding to an input sentence is to expand contractions in the input, using rules in* Eliza.pre. *Next, see if the input matches the regular expression in a sentence pattern, and if so, save the parts of the input that match wild cards or word groups. Finally, postprocessing replaces pronouns according to rules in* Eliza.post.

Note that substitutions defined in Eliza.pre are performed *before* a pattern matching operation is tried, in order to transform an input into a form that will match a pattern. Substitutions defined in Eliza.post are applied *after* a pattern matching operation has succeeded, but before the pieces are reassembled into a response.

The complete process used to transform a sentence is illustrated in Figure 10.5. The first step is preprocessing, where the apply method sees the string "you're" in the input and replaces it with "you are." Now the sentence matches the pattern, and the words following "are" are saved in $1. The postprocessing step replaces the word "you" with "me." After all the pattern matching steps are complete, the pieces are put back together, and the contents of $1 are substituted into the response string to create the output.

Tutorial Project

T33. Make a pattern to use for testing preprocessing and postprocessing:
```
>> p = Pattern.new("you are (.*) ", ["I am not $1"])
=> you are (.*): ["I am not $1"]
```

T34. Since the pattern starts with "you are" it will not match a sentence that uses "you're":
```
>> p.apply("You're going to laugh at me")
=> nil
```

T35. Define a preprocessing rule that tells Eliza to replace "you're" with "you are" before it tries to apply the pattern:
```
>> Eliza.pre["you're"] = "you are"
=> "you are"
```

T36. Tell Ruby to print the preprocessing rules to verify the one you just defined is there:
```
>> Eliza.pre
=> {"you're"=>"you are"}
```

T37. Apply the pattern to some sentences that have the string "you're":
```
>> p.apply("You're right")
=> "I am not right"

>> p.apply("You're going to laugh at me")
=> "I am not going to laugh at me"

>> p.apply("You're killing me")
=> "I am not killing me"
```

The pattern now works equally well for sentences with "you are" or "you're." We've fixed one problem, but now we need to define some postprocessing rules to make the output convincing.

T38. Define a postprocessing rule that tells Eliza to change "me" into "you" when it assembles an output string:
```
>> Eliza.post["me"] = "you"
=> "you"
```

T39. Again it's a good idea to make sure the new rule is there:
```
>> Eliza.post
=> {"me"=>"you"}
```

T40. Try your example sentences again:
```
>> p.apply("You're going to laugh at me")
=> "I am not going to laugh at you"

>> p.apply("You're killing me")
=> "I am not killing you"
```

Try some more sentences on your own. Can you figure out how to define pre- and postprocessing rules that will allow the pattern
```
p = Pattern.new("I am (.*)", ["Are you really $1?", "$1?"])
```
to respond to "I'm sorry I dropped your computer" with "Are you really sorry you dropped my computer?"

10.5 An Algorithm for Having a Conversation

To summarize what we've seen so far:

- a pattern object is defined by a regular expression and a set of response strings;

- when we call an object's `apply` method, it uses the regular expression to see if a sentence matches the pattern;

- pattern matching variables ($1, $2, *etc.*) along with preprocessing and postprocessing help the `apply` method extract parts of the input sentence and reuse them as part of the response.

As you might imagine, there are many more enhancements we might make to give the `apply` method more flexibility in the types of sentences it can transform. At this point, however, we will turn our attention to the problem of how to use patterns as part of an algorithm that will carry on a conversation. Any sentence typed by the user is likely to match several different patterns, and we need to define an algorithm that will choose among all the patterns that might be applied.

A straightforward algorithm would be to simply try patterns until we find one that matches an input sentence. We could make an array of pattern objects, and then write a program that uses an iterator to try the patterns in order. The first time we find a pattern that applies to the input sentence, we would use the response generated by that object. The problem with this approach is that patterns will always be applied in the same order. If any of the patterns at the front of the array have regular expressions that match common words, they will be used fairly often, and responses will become predictable.

Weizenbaum's solution to this problem was to assign a **priority** to each word. He expected people would ask the program questions like "Are you a computer?", or "Am I talking to a machine?", so the original DOCTOR script had patterns to respond to inputs with the words "computer" or "machine." To make sure ELIZA responded directly to these questions, Weizenbaum developed an algorithm that tried to match inputs to high priority words like "computer" before trying any patterns for sentences containing more common words like "are" or "you."

Another issue that needs to be addressed is that some common words, like "I", are going to appear in several different patterns. For example, there might be different sentence patterns to respond to inputs like "I am worried . . . ", "I remember . . . ", or "Why can't I . . . ". We want to make sure Eliza tries all the patterns for a word before moving on to a lower priority word.

A new type of object, called a Rule, is simply a list of pattern objects. When we say Eliza has a "rule for *x*," what we mean is there is a rule object that has a set of patterns that all pertain to sentences containing the word *x*. When *x* is found in an input sentence, Eliza will try to match the sentence with each of these patterns.

Given this definition of a sentence transformation rule, the algorithm for responding to a sentence typed by a user is straightforward (Figure 10.6). The algorithm uses a priority queue to manage the words in the input sentence. A priority queue is a container, similar to an array, but it has the special property that items in the container are always sorted according to some priority. Each sentence is broken into individual words, and if there is a rule for a word, the rule is added to the queue. After all words in the input sentence have been scanned, Eliza starts applying rules according to their order in the queue. If a

---

The ELIZA Algorithm

To respond to a sentence typed by a user:

1. Break the sentence into words.

2. If there is a rule for a word, add the rule to a priority queue.

3. Try the rules in order of decreasing priority.

4. If a rule applies to the input sentence (*i.e.*, the rule has a pattern that matches the sentence) apply postprocessing rules to placeholder variables and return the response.

Figure 10.6: *An algorithm for having a conversation.*

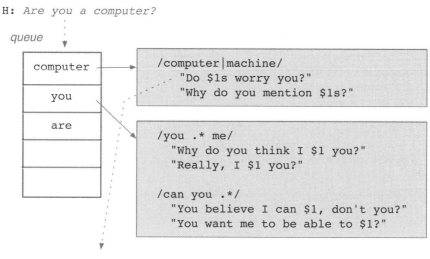

Figure 10.7: *The rules for each word in an input sentence are saved in a priority queue. Eliza tries the rules in order, starting with the highest priority word at the front of the queue. In this example, the word "computer" has the highest priority of all the words in the input sentence, and the response comes from a pattern associated with the word "computer."*

rule applies to the sentence, the response made by that rule is returned, otherwise Eliza continues with the next rule in the queue.

An example of how the algorithm processes a sentence is shown in Figure 10.7. The input is "Are you a computer?" The DOCTOR script has patterns for sentences with the words "computer," "you," and "are," so rules for these words are saved in the priority queue. Since "computer" has the highest priority, it will be at the front of the queue, and on the first iteration Eliza will try to match the input to the regular expression for this rule. Since the regular expression matches the input, the response generated by this rule ("Do computers worry you?") is the sentence that will be printed as the output. If Eliza did not have a priority queue, it might have tried other patterns first, and the output could have come from a pattern for sentences with the word "are" or "you."

For the project in this section we will see how Eliza uses its priority queue to select from a set of several different rules that might apply to an input sentence. After a script has been loaded, you can call a method named `rule_for` to see whether there is a rule for a particular word. For example, to see if the script has a rule for the word "remember" you would type

```
>> Eliza.rule_for("remember")
=>  [5] --> [
  /I remember (.*)/
    "Do you often think of $1?"
  ... ]
```

```
# Transform a sentence using rules in a script,
# return nil if no rule applies

  def Eliza.transform( s )
    queue = PriorityQueue.new
    Eliza.scan( s, queue )
    response = nil
    while queue.length > 0
      rule = queue.shift
      response = Eliza.apply( s, rule )
      return response if response != nil
    end
    return nil
  end
```

Figure 10.8: *The RubyLabs implementation of the ELIZA algorithm.* `Eliza.scan` *splits a sentence into individual words and adds rules to a priority queue. PriorityQueue objects were introduced in the Huffman tree project; see Exercises T45 to T51 in Section 7.4.*

The output means there is a rule, and it has a priority of 5 (the higher the number, the higher the priority). Following the priority, you should see a set of one or more regular expressions, and for each expression, a set of one or more response strings. If a script does not have a rule for a word, the `rule_for` method returns `nil`:

```
>> Eliza.rule_for("cow")
=> nil
```

To see how Eliza will respond to a string, call the `transform` method:

```
>> Eliza.transform("I'm afraid of cows")
=> "Is it because you are afraid of cows that you came to me?"
```

This method will break the sentence into words, look to see if there is a rule for each word, and add those rules to the priority queue. It then starts to apply the sentence patterns in order of the rule priorities. So even though there isn't a rule for "cow," Eliza was able to generate a response using a rule for one of the other words in the sentence, in this case a rule for "afraid."

A Ruby implementation of the complete algorithm is shown in Figure 10.8. The method named `Eliza.scan` applies the preprocessing rules, *e.g.*, to expand contractions like "I'm" into complete words like "I am." It then splits the sentence into individual words, and adds rules for the words to the priority queue. The call to `Eliza.apply` will see if a rule has a pattern that matches a word, and if so do all the necessary postprocessing and return the response string.

If we want to trace the execution of the algorithm, we can see all the details of how Eliza processes a sentence by calling a method that turns on "verbose mode":

```
>> Eliza.verbose
=> true
```

Now when we call `transform`, Eliza will print out the details of every step, showing which words are added to the priority queue, and then which sentence patterns are tried.

Tutorial Project

T41. Load the sentence patterns in the DOCTOR script:

```
>> Eliza.load :doctor
=> true
```

T42. See if the script has a rule for sentences that contain the word "if":

```
>> Eliza.rule_for("if")
=>   [3] --> [
   /if (.*)/
      "Do you think it's likely that $1?"
      "Do you wish that $1"
      ... ]
```

T43. Call the `transform` method to see how Eliza responds to a sentence that contains the word "if":

```
>> Eliza.transform("If wishes were horses")
=> "Do you think it's likely that wishes were horses?"
```

T44. If you call `transform` with another sentence containing the word "if" Eliza should use the second response string:

```
>> Eliza.transform("If wishes were horses")
=> "Do you wish that wishes were horses?"
```

Do you see how the patterns for "if" tell Eliza to extract all the words following "if" and plug them into the response strings?

T45. Print the rule for transforming sentences that contain the word "remember":

```
>> Eliza.rule_for("remember")
=>   [5] --> [
   /I remember (.*)/
      "Do you often think of $1?"
      ...
   /do you remember (.*)/
      "Did you think I would forget $1?"
      ...
```

You should see there are two patterns, one for sentences that match "I remember" and one for sentences that match "Do you remember."

T46. Call the `transform` method with a sentence that matches the first pattern:

```
>> Eliza.transform("I remember that cow")
=> "Do you often think of that cow?"
```

T47. Now try a sentence that matches the second pattern for "remember":

```
>> Eliza.transform("Do you remember why I'm afraid of them?")
=> "Did you think I would forget why you are afraid of them?"
```

T48. Turn on Eliza's "verbose mode":

```
>> Eliza.verbose
=> true
```

T49. Transform a sentence that has several words that are key words in the DOCTOR script:

```
>> Eliza.transform("I remember my sister had a computer")
preprocess: line = 'I remember my sister had a computer'
...
=> "Do computers worry you?"
```

You will see several output lines as Eliza transforms this sentence. The first set of lines will tell you that Eliza found rules for "I," "remember," "my," and "computer." Then Eliza will print the contents of the priority queue, and since "computer" is the highest priority word in this sentence, Eliza tries that pattern first. Can you see why the final output was a question about computers, and not any of the response strings for the word "remember"?

Try calling `transform` with a few more sentences of your own. In each case, you should see one or more words being added to the priority queue, and then see how Eliza tries the rules for those words in order, from highest priority to lowest. If you want to see what the rule is for any of the words in the queue, just call `Eliza.rule_for` (or you can use your text editor to open the text file for the DOCTOR script).

T50. To turn off verbose mode, and go back to the normal way of interacting with Eliza, call a
     method named `quiet`:
```
>> Eliza.quiet
=> false
```

10.6 ◆ Writing Scripts for ELIZA

Making patterns and experimenting with them by typing expressions in IRB is a good way to learn both how Eliza uses regular expressions to break an input sentence into smaller parts and how it creates an output sentence by transforming and reassembling the pieces. But to get Eliza to carry on a conversation, there will be too many patterns to type in interactively. What we need to do is write down the patterns in the form of a script. Then, each time we want to try out the transformation rules, we just have to load the script, and all the patterns defined in the script will be available for Eliza to use.

One way to get started on your own script is to make a copy of the DOCTOR script that comes with Eliza. This command will save a copy of the script in a file named "doctor.txt" in your project directory:

```
>> Eliza.checkout(:doctor)
Copy of doctor saved in doctor.txt
```

Once you have a copy, you can either add new rules to it or just use it as a guide for writing your own new script with a completely new set of rules. To have Eliza use your updated copy of the script, pass the file name to the `load` method. For example, if you check out a copy of the DOCTOR script, as shown above, and then rename it "farm.txt" after adding some rules that respond to sentences with farm animals, use this command to load the modified script:

```
>> Eliza.load "farm.txt"
=> true
```

Script files are plain text files that can be created with the same text editing application you use for Ruby methods. The main part of the file will be a set of rule definitions, but there can also be other information, for example, a specification of which words are expanded during the preprocessing phase and the substitutions to perform during postprocessing.

Scripts have a very simple syntax. There are only two kinds of things in a script: **directives**, which are commands for Eliza, and **rules**, which are specifications for how to transform an input sentence.

```
# A portion of the "Doctor" script for the RubyLabs version of Eliza in Ruby

:pre you're "you are"

:post me "you"
:post myself "yourself"
:post my "your"

was 2
  /was i (.*)/
    "What if you were $1?"
    "Do you think you were $1?"
  /i was (.*)/
    "Were you really?"
    "Why do you tell me you were $1 now?"

i
  /i (want|need) (.*)/
    "What would it mean to you if you got $2?"
    "Why do you $1 $2?"
  /i am (.*)(sad|unhappy|depressed|sick)(.*)/
    "I am sorry to hear you are $2$3."
    "Do you think coming here will help you not to be $2$3?"
  /i am (.*)(happy|elated|glad|better)(.*)/
    "How have I helped you to be $2$3?"
    "Has your treatment made you $2$3?"

remember 5
  /I remember (.*)/
    "Do you often think of $1?"
    "Does thinking of $1 bring anything else to mind?"
    "What else do you remember?"
  /do you remember (.*)/
    "Did you think I would forget $1?"
    "Why do you think I should recall $1 now?"
```

Figure 10.9: *An excerpt from the "doctor" script. Lines beginning with # are comments, lines beginning with a word that starts with a colon are directives, and the remaining lines are all parts of rules. A rule starts with a line that has a single word. Regular expressions (sentence patterns) begin and end with a slash character. Response strings are surrounded by double quotes. Although it's not strictly necessary, if you indent regular expressions by two spaces, and response strings by four spaces, as shown here, it is easier to tell which strings go with which patterns, and which patterns go with which words.*

Directives are words that start with a colon. The two directives you should know about are `:pre`, which tells Eliza to add a new word to the set of words it expands during the preprocessing step, and `:post`, which tells Eliza to add a new item to the association list it uses during postprocessing (described in Section 10.4). Some examples from the DOCTOR script are:

```
:pre don't "do not"
:post your "my"
```

The `:pre` command tells Eliza that whenever the letters `don't` are seen in an input sentence they should be replaced by `do not`. The `:post` command says that "your" should be replaced by "my" in the parts of the input that are echoed back to the user, *e.g.*, when converting "I like your hat" into "Why do you say you like my hat?"

Directives like those shown above are specified on a single line, but rules need several lines, as shown in Figure 10.9. The first line has the word that triggers the rule. So, for example, if you are writing a rule that will process sentences related to cows, the first line in the rule will just have the word "cow." The default priority for a rule is 1, but if you want the rule to have a higher priority put the priority number right after the word.

Following the line with the key word there can be one or more sentence patterns. The first line in a sentence pattern is the regular expression that defines the pattern, and following that are the response strings Eliza will use when an input sentence matches the regular expression. There can be any number of response strings for each regular expression and any number of patterns for any word.

A complete description of how script files are organized and a few more examples can be found in the Eliza section of your Lab Manual.

Project Ideas

If you would like to experiment with Eliza, here are some suggestions for scripts.

♦ Use the `checkout` method to get a copy of the DOCTOR script, and then add some rules for new topics. For example, you can add rules for farm animals, movies, books, or any other subject.

♦ Write a new script for "small talk" at a party. Add rules for a few specific items, like your favorite sports team, but if the input does not match one of the script patterns have it reply with something like "strange weather we've been having" or "did you watch the game last night?"

♦ Search the Internet for a transcript of the old Abbott and Costello routine called "Who's on First?" See if you can devise a set of rules so that when a person types something like "Tell me, what is the name of the first baseman?" your script replies "What's on second." How many other formulaic responses can you generate?

♦ Search for a transcript of the Monte Python skit called "The Argument Clinic." In this skit, the "clinician" tries to carry on an argument by negating every sentence spoken by the "client." If the input is "Yes it is" the response is "No it isn't" and *vice versa*. Can you make a script that argues with the user?

10.7 ELIZA and the Turing Test

From the projects in the previous sections, it may seem that all Eliza needs to pass the Turing Test is a script with rules to respond to a wider variety of questions, and perhaps a few more features, like preprocessing and postprocessing, to deal with new special cases. Human languages are very complex, however, and there are limits to the sorts of things we can "teach" Eliza to respond to.

The first barrier for Eliza is that it does not take into account the structure of a sentence. It simply breaks each input into individual words, without trying to figure out whether a word is an important part of a sentence. When faced with any sentence consisting of more than a few words, Eliza is not likely to make a good choice:

> *You said I wouldn't be afraid of cows if I visited a dairy, but when I was there I was chased by a duck, and now I'm having nightmares about birds.*

It's unlikely that any ordering of the words in this sentence will lead to a realistic response. A human would recognize there are four separate topics (a suggestion for overcoming a fear, a trip to a dairy, an episode with a duck, nightmares) and try to make an appropriate response based on the topics. Eliza just dumps all the words in a queue, and the highest priority rule is applied, regardless of the context of the rest of the sentence. If you typed any long sentences when you were first experimenting with Eliza you no doubt realized it did not do a very good job with complex sentences.

There are algorithms, called **parsers**, that will use rules of English grammar to break sentences into their constituent parts. But even if we replace Eliza's word-oriented strategy with a more complicated algorithm, one that uses a parser to break a sentence into noun and verb phrases, we would soon run into another problem. Consider this short conversation:

> *The cows were lying down in the field.*
> Go on.
>
> *They were quiet.*
> Why do you say the cows were quiet?

In order for the computer to make this second response, it would need to have some sort of memory, to keep track of the fact that the conversation was about cows. Eliza has no way to connect the word "they" in the second sentence with "cows" in the first sentence. Every input is the start of a new conversation for Eliza, and all it can do is make a response from the words in the current sentence.

Here is an example of another conversation that is not possible for a computer that uses Eliza's rule-based algorithm:

> *That cow was huge.*
> How big was it?

This response requires the computer to understand the *meanings* of words. English speakers know the word "huge" means "very large," and a response that substitutes "big" for "large" is natural. Eliza, of course, knows nothing about the meanings of words. It has no data structure to correlate words with possible meanings, and no algorithm for considering whether or not to replace a word with a similar one.

Researchers in natural language processing refer to the general problem of assigning meanings to words as **semantics**. Semantics is about much more than meanings of individual words, however. It is not simply a matter of giving Eliza a dictionary so it can look up synonyms. Meanings are based on what computer scientists refer to as "real-world knowledge" or "common sense knowledge." Here is another simple example that illustrates the challenge:

> *The cow fell into the canal.*
> Was it able to get out?

A dictionary that simply associates "canal" with "channel used to convey water for irrigation" would not help a program make this response. The reply is based on knowledge that canals have water, the cow was in danger because it was in the water, and the cow might have had some difficulty because canals usually have steep walls. A duck is also an animal commonly found on a farm, but a human's response to "the duck fell into the canal" would be very different. All of this knowledge is something we, as humans, have built up over many years of learning and using language, and without it a computer would not be able to carry on a natural sounding conversation.

Even if we were able to put all the pieces described above into a program, so we have an application that can parse complex sentences, can keep track of statements in order to remember the context of a conversation, knows the meanings of words, and has a substantial amount of real-world knowledge, there is yet another difficult challenge. When we see or read a sentence, we are continually making **inferences**. Inference is such a common phenomenon that we take it for granted, and don't even realize that we are doing it. Consider this simple statement:

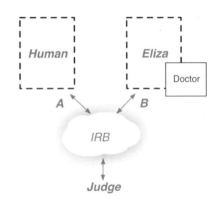

> *The cows walked across the pasture while my dad waited at the barn.*

When you read that sentence, did you immediately think that the cows were walking toward the barn? A perfectly reasonable response to that statement might be "What happened when the cows got to the barn?" But there is nothing in the statement that said anything about where the cows were headed. We simply inferred that the cows were walking in the direction of the barn, because the phrases "cows walked" and "at the barn" were both part of the same sentence.

Depending on the larger context, there could be several different interpretations. Dad could have been waiting for a taxi, or waiting for someone to bring his lunch. Even if the program knows the meanings of the words, and has built up a context from earlier statements, it will have to be able to make inferences in order to respond like a human. People are constantly "connecting the dots" when they hear or read a sentence, and a response from a human will use inferences based on the sentence and the context.

Tutorial Project

The conversation shown below is from a hypothetical program that pretends to converse in English. The responses from this program are each examples of replies that Eliza is unlikely to or unable to generate. Can you explain why Eliza would not make these responses? If it helps, you can type a sentence into the RubyLabs version of Eliza (after loading the DOCTOR script), or use the `rule_for` method to see if DOCTOR has a rule for a particular word. Some comments about these responses can be found in the section with Answers to Selected Exercises on page 344.

```
H: I remember when I lived in the country and there were
     cows and chickens everywhere.
C: Why do you mention cows just now?
H: They were always looking at me.
C: The cows were looking at you?
H: Our parents got mad when we tipped the cows.
C: Did you really push over a cow?
H: The one with the big horns was the scariest.
C: Are you referring to a bull?
H: Yeah, that guy was pretty aggressive.
C: Did the horns seem dangerous?
```

10.8 Summary

In his 1950 paper on the nature of intelligence, Alan Turing was optimistic about the prospects for natural language processing. He wrote:

> I believe that in about fifty years time it will be possible to programme computers ... to make them play the imitation game so well that an average interrogator will not have more than 70 per cent chance of making the right identification after five minutes of questioning.

Another famous "prediction" that natural language processing would be well established by the year 2000 was Arthur C. Clarke's *2001: A Space Odyssey*. The movie, which was released in 1968, featured a computer named Hal that spoke flawless English with the crew of a spaceship on a journey to the planet Jupiter.

Actual achievements in natural language processing have fallen far short of these predictions. There have been some notable successes, but only in very limited areas. Reservation systems for airlines and call centers for banks are examples where natural language interfaces are being used effectively. None of these systems are able to converse on a wide variety of subjects, however.

One area where there has been some success is in parsing sentences. If you do an Internet search for "English parsers" you will find several web sites at research labs that are working on methods for analyzing complex sentences. Many of these sites allow you to type a sentence into a form in your browser and submit it for analysis. The parser will return a page that shows the overall sentence structure, with noun phrases, verb phrases, and other parts of speech.

The Turing Test itself has become a source of controversy in recent years. One criticism is that programs can do well in a contest simply by appealing to the judges, and do not actually have to carry on a conversation. If a programmer knows judges are influenced by quirky

---

Concepts and Terminology Introduced in This Chapter

natural language processing	The area of computer science research concerned with algorithms for understanding human language
Turing Test	A proposal for assessing machine intelligence; a program would pass the test if it could carry on a natural language conversation
ELIZA	The first program to attempt to converse in English; it used simple rules to transform input sentences into responses
regular expression	A type of expression used for pattern matching; used by the RubyLabs implementation of Eliza to process input sentences
semantics	The meaning of a word or sentence; a difficult challenge for natural language processing

---

responses, even to the point of changing topics at random, then a program that spits out non sequiturs every now and then has a better chance of winning. Is this the sort of thing people do, during a conversation, or simply a strategy to win a contest? Should the test be amended, to require programs to try to emulate how people think and talk? Or should Turing's original rules still apply, so that any program that carries on a realistic conversation for five minutes will earn the distinction of being called "intelligent"?

A computer program that truly does converse with a human would be one that remembers what was said earlier in the session, knows the meanings of words, and is capable of making inferences based on what it hears. It will be a long time before such programs are available.

Exercises

The first set of questions ask "what would Eliza do?" Try to answer these questions based on your experience with Eliza in the tutorial projects. To check your answers, you can either look at the text for the DOCTOR script, or start an IRB session with Eliza.

1. Here are some sentences typed by a user, and the responses generated by Eliza (after loading the DOCTOR script). Explain why Eliza responded the way it did.

   ```
   H: I can't remember the date.
   C: How do you know you can't remember the date?

   H: I can't remember her name.
   C: I am not interested in names.

   H: There is no beer left.
   C: Are you saying 'no' just to be negative?
   ```

2. Does Eliza handle plural forms of nouns? Would it generate the same response for "My brother lives in Hawaii" and "Both of my brothers are younger"?

3. Does Eliza work equally well with sentences that have the past, present, and future forms of a verb?

4. Some of the reasons a program like Eliza would have difficulty passing the Turing Test were presented in Section 10.7. Can you think of an example sentence, or a sort of conversation, that illustrates each situation?

   a) A sentence structure that is too complex, *i.e.*, that would require a parsing algorithm.

   b) A conversation that depends on remembering something from a previous sentence.

   c) A response that requires Eliza to know the meanings of words.

   d) A response that requires real-world knowledge.

   e) ♦ A response based on an inference.

5. Here is the definition of a sentence pattern in a session with ElizaLab:
   ```
   >> p = Pattern.new("I (hope|wish|want) (.*)")
   >> p.add_response("Do you really $1 $2?")
   >> p.add_response("Have you always $1ed $2?")
   ```
   Show what Ruby will print for the following expressions, assuming no postprocessing rules have been defined yet:
   ```
   >> p.apply("I want a new computer")
   >> p.apply("I wish you would stop putting words in my mouth")
   ```

6. What postprocessing rules are necessary for Eliza to respond the way it does in the following examples (using the pattern object defined in the previous exercise)?
   ```
   H: I want my computer to understand me.
   C: Do you really want your computer to understand you?

   H: I wish you would be more sympathetic.
   C: Have you always wished I would be more sympathetic?

   H: I hope you can help me understand why our cow used to scare us.
   C: Do you really hope I can help you understand why your cow
      used to scare you?
   ```

7. Define sentence patterns that would "teach" Eliza how to respond to the following types of sentences:

   a) Reply to sentences containing the word "coffee" with "No thanks."

   b) Reply to sentences containing "coffee," "tea," or "latte" with "No thanks, I don't drink *x*," where *x* is one of the three key words.

   c) Reply to sentences of the form "I drink *x*" with "Tell me why you like *x*," where *x* can be any word or phrase.

   d) Reply to sentences containing both drinks and food, for example, respond to "I drink coffee with a scone in the morning" with "Do you really have a scone with your coffee?" Make up your own list of drinks and foods.

   e) Reply to sentences with the words "comics," "books," or "web site" with "Why do you mention *x*?" or "Do you often read *x*?" or "I don't like *x*, myself."

8. Suppose there is a program that will translate spoken words into text, so Eliza could respond to verbal sentences. Discuss some of the issues that would have to be resolved before Eliza could be used to handle telephone calls from customers, for example,

   • credit card customers inquiring about their current balance, or about specific transactions;

   • airline customers calling to make reservations, or ask about itineraries;

   • customers who purchased software who have questions about how the software works or are reporting errors.

9. The Ruby implementation of Eliza's algorithm for transforming a sentence was given in Figure 10.8 on page 268. Could this algorithm be considered a type of search? How does it compare to a linear search?

Chapter 11

The Music of the Spheres

Computer simulation and the N-body problem

Before the seventeenth century, comets were thought to be random, unrelated occurrences. Astronomers who studied the motions of comets thought they moved in a straight line through space, and there was no reason to suspect a strange object appearing in the sky was the same body that had last shown up dozens of years earlier. But by 1687, when Isaac Newton (1643–1727) published *Principia Mathematica*, astronomers realized comets, like planets, were celestial bodies that orbited the Sun. Newton, using his new theory of gravitational attraction, showed that comets followed highly elliptical orbits, so they were only visible from Earth for short periods when they were close to the Sun.

A few years later, in 1705, Edmund Halley (1656–1742) analyzed records from previous comet sightings, and proposed that a body most recently observed in 1682 was a comet that orbited the Sun every 76 years. If this was the case, Halley needed to account for large differences in the lengths of previous orbits. The time between appearances in 1531 and 1607 was 76.1 years, but the period between the sightings in 1607 and 1682 was only 74.9 years (Figure 11.1). Being familiar with Newton's new theory, Halley proposed that these differences were the result of the gravitational pull of Jupiter and Saturn. Comets are very small, so their orbits would be easily influenced by the two largest planets in the solar system. If Halley was right, the comet would appear again some time around 1758. It was the first time anyone had tried to predict a future appearance of a comet.

One of the landmark achievements in the era before there were machines to carry out computations was the calculation of the orbit of Halley's Comet by the French astronomer Alexis-Claude Clairaut (1713–1765). In the summer of 1758, before the comet was sighted, Clairaut decided to test Newton's equations and Halley's prediction by computing the exact orbit of the comet. He began by assuming the comet travelled mainly along a smooth elliptical orbit, and then performed a series of adjustments to account for the gravitational effects of the two large planets. Clairaut spent five months, working with his colleagues Joseph Jérôme Lalande (1732–1807) and Nicole-Reine Lepaute (1723–1788), on detailed corrections to the orbit, each step depending on the result of earlier calculations.

The group began their work in June, and by November Clairaut and his colleagues had determined that Jupiter and Saturn would cause the current orbit to be 618 days longer than the previous one. On November 14, 1758, Clairaut predicted Halley's Comet would be closest to the Sun in the middle of April, with a margin of error of one month. When the comet finally did reappear, astronomers were able to record the actual date it was closest to the Sun, which was March 13, 1759.

The calculations by Clairaut are interesting not only because they are an example of a significant computation from a time well before there were electronic computers, but because they were an important scientific result, as well. Newton's equations of motion were still not universally accepted, and the computation of the gravitational effects of the outer planets was one of the first major confirmations of Newton's theory.

Before Newton, astronomers had attempted to describe the motions of planets by geometric equations. Most ancient philosophers believed the planets rotated around the Earth, following circular paths. In the Middle Ages astronomers began to question this view, and gradually adopted the position of Nicolaus Copernicus (1474–1543), who argued that the planets revolved around the Sun. Johannes Kepler (1571–1630) made an important contribution when he proposed that the orbits were elliptic, rather than circular. This allowed him to simplify the equations for the orbits, and as a result the apparent movements of the planets were described with more accuracy.

Newton's theory, that planets and other bodies simply move according to the gravitational effects of the objects around them, was a major break from previous descriptions of the motions of the planets. According to Newton, the planets appear to move with such regular circular or elliptical orbits because the Sun is so much larger than all the planets. The mass of the Sun is about 10^{30} kg, which is more than 300,000 times the mass of the Earth and 1000 times more than Jupiter, the largest planet. This large mass causes the planets to circulate in orbits that are, for all practical purposes, ellipses.

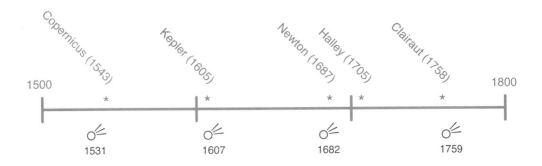

Figure 11.1: *This time line shows the dates of important results in the study of comets. In 1705 Edmund Halley predicted an object last seen in 1682 would reappear in 1758. Alexis-Claude Clairaut used Isaac Newton's law of gravitational attraction to compute the effects of Jupiter and Saturn on the orbit of Halley's comet. Before the comet was visible Clairaut successfully predicted the date it would be closest to the Sun.*

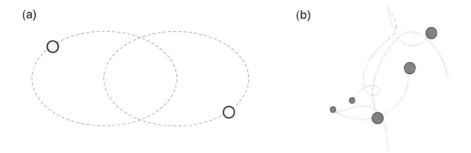

Figure 11.2: *(a) In a binary star system, with two bodies that are approximately the same size, one can solve a set of equations to predict the future location of each star. (b) The more general case is an N-body system. There is no set of equations that will describe the motion of each body, and the only way to predict the future location of each body is through computation.*

The fact that bodies do not follow a prescribed orbit, but simply move according to the force of gravity, means there is no equation that will tell us precisely where any body will be some arbitrary point in the future. One of the first formulas students learn in a science class is $d = v \times t$, which describes the distance d an object will travel when it moves with constant velocity v for a period of time t. If we know how fast an object is moving, and we assume it keeps moving at the same constant speed, we can use this equation to predict where the object will be at any point in the future by simply multiplying the velocity by the time. But for planets, moons, and comets moving under the influence of gravitational forces there is no such formula.

For certain special situations—a binary star system where two similarly sized objects orbit each other, or a small falling object near the Earth's surface—there are equations that describe the motions, and it is possible to solve these equations to predict the future locations of the objects. But in the general case there are no equations to predict exactly where each body in a system will be at any future date. Astronomers use the term **chaotic** to describe this sort of motion. When used in this technical context, chaotic basically means "unpredictable." One of the consequences of a system being chaotic is that very small changes in the current conditions may lead to very different positions after a period of time.

As soon as Newton published his theory of gravity, he and others began an effort to develop a set of equations to describe the relative motions of the Earth, Moon, and Sun, in what became known as the "3-body problem." Eventually mathematicians were able to prove that it would not be possible to derive simple equations that would predict exactly where any one of $N \geq 3$ bodies would be at any arbitrary point in time. Physicists today refer to the problem of trying to describe the motion of a collection of bodies as the **N-body problem** (Figure 11.2).

While it may not be possible to solve an equation to determine the precise locations at any point in the future, it is possible to estimate where a set of bodies will be by doing a series of calculations. The idea is to start with the current locations and headings of each body. We can compute the effect of gravity on each of the objects and adjust their headings accordingly. We then assume the bodies will move in the direction defined by their updated heading, which will tell us where the objects will be a short time later. By repeating this

The Music of the Spheres

Johannes Kepler, like most astronomers before him, believed the planets traveled in regular orbits that could be described by mathematical equations. Kepler's main contribution was the realization that the orbits were elliptical and not circular. In his work *The Harmony of the Worlds* (1619), Kepler described a correspondence between harmonic ratios in musical notes and ratios of the orbital speeds of planets. But unlike vibrating strings on a musical instrument, the motions of planets cannot be described by equations that will predict a location at an arbitrary point in time. Instead, computer simulations, based on calculations of motions over very small time steps, are used to track the positions of planets, comets, and asteroids.

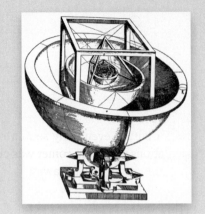

process, recomputing the gravitational attractions from each new position and computing new positions a short time later, we can estimate where the bodies will be at some future point in time. Mathematicians refer to this type of operation as **numeric integration**.

Clairaut used this technique when he computed the orbit of Halley's comet in 1759. He began by assuming the comet would follow an orbit that was basically an ellipse, and when the comet was far away from any planet he computed its trajectory using formulas for elliptical paths. When the comet was relatively close to Jupiter or Saturn he applied numeric integration to make a series of corrections based on the gravitational attraction of the planet.

Our project for this chapter will explore computations that predict the motions of the planets in the solar system. SphereLab, the Ruby module with the software we will use, defines a new type of object, called a Body, that can be used to represent planets and other celestial bodies. When we make a body object, it will encapsulate all the necessary information, such as the planet's mass and position. We will then be able to call methods that compute the gravitational attraction between bodies and update their positions in response to these forces. By repeatedly computing the forces acting on the planets, and moving them for a small amount of time, we will see how an algorithm can paint an accurate picture of the orbits of the solar system.

Using computation to solve problems like predicting the future positions of planets is an example of **computer simulation**. Simulation is an important area in applied computer science, used by professionals in a wide variety of areas, including science, medicine, engineering, and business. Computer-generated imagery (CGI), a growing part of the entertainment industry, is also a form of computer simulation.

Today astronomers use computers to calculate the orbits of several thousand bodies, including comets and asteroids. Simulations are used to predict the locations of satellites, to help plan launches or return trips from the International Space Station, and to track asteroids that are potential threats to Earth. The project in this chapter is a scaled down and simplified version of the algorithms used by astronomers, but it is nonetheless a good introduction to some of the issues in computer simulation.

11.1 Running Around in Circles

One of the decisions Clairaut had to make when planning his computation of the orbit of Halley's Comet was how many times to adjust the comet's orbit to account for the gravitational attraction of Jupiter and Saturn. A single adjustment half way through the orbit might have improved the estimate of the full orbit time, but the calculation would be more accurate if several small adjustments were made at points evenly spaced throughout the orbit. On the other hand, planning to do a calculation for every 24-hour period would lead to far too many steps in the computation. That level of detail might not be necessary, since the effects of Jupiter and Saturn are negligible when the comet is at the farthest reach in its orbit. Clairaut eventually settled on a combination of the two strategies, making detailed calculations when the comet was closest to Jupiter and Saturn, but using the elliptical orbit when the comet was far away.

To begin the project for this chapter, we will explore the tradeoff between how often positions are calculated and the overall accuracy of the computation. For the first set of experiments, imagine a situation where a robot has landed on a distant planet, and we want to send it out on a mission to explore its surroundings. We want the robot to travel a path that is a perfect circle, and we want it to end up at the same location where it started. We can transmit instructions to the robot to tell it to move straight ahead, or to turn so it is heading in a new direction. When we tell the robot to move, it will advance straight ahead for a specified amount of time, and then wait for further instructions.

Figure 11.3 shows how we can get the robot to make a trip that will move along a circular path that brings it back to its starting point. With only six corrections the path would be a hexagon. However, if the goal is to make a smooth circle, we need to correct the path more often. The tradeoff is that the more corrections we make, the longer it will take the robot to make the circuit, since it has to pause to wait for a new heading at the end of each leg.

For the lab project, we will monitor the progress of the robot in the RubyLabs graphics window. To initialize the canvas, call a method named `view_robot`. The method initializes the window to show a map of a square piece of terrain, 400 meters on each side. The robot is initially on the west (left) edge of the map.

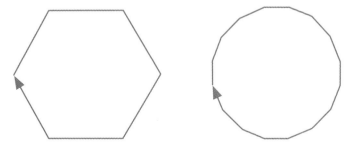

Figure 11.3: *If we tell the robot to correct its heading six times, the path it travels will be a hexagon, as shown on the left. With more frequent updates the path will look more like a smooth circle. The path on the right was made by sending 16 corrections.*

Command	Action
advance(t)	Move ahead for *t* hours
turn(a)	Turn clockwise by *a* degrees
plant_flag	Place a marker at the current location
orient	Rotate until the marker is 90° to the right
heading	Return the current direction (in degrees) the robot is headed
speed	Return the robot's velocity (in meters per hour)
track(:on)	Turn on the tracking option
track(:off)	Turn off tracking

Table 11.1: *Commands that can be transmitted to the robot explorer.*

The Ruby expressions that tell the robot what to do are all of the form `robot.x`, where x is the command to execute. For example, if we type

```
robot.turn(30)
```

the robot will change its current heading by turning clockwise 30°. The complete list of commands is given in Table 11.1.

When we send the command that tells the robot to move forward, we need to tell it how long to move before it stops to wait for the next command. The distance an object moves in an amount of time *t* is defined by the simple formula $d = v \times t$, where *v* is the velocity at which the object moves. Our robot will move at 10 meters per hour. So if we want it to move 50 meters, we have to send the command to tell it to advance straight ahead for 5 hours:

```
robot.advance(5)
```

After the robot advances for the specified amount of time it will stop. At this point, if we want it to move in a circle, we need to send it a command to turn clockwise before it advances again. The question is, how far should we tell it to turn?

One way to answer this question is to imagine the robot has an arm sticking straight out from its right side, at a 90° angle from the direction it is heading. If the robot moves in a perfect circle, the arm will always be pointing at a flag at the center of the circle. After advancing in a straight line for a while, the flag will be behind the robot. To get the robot back on track, we want it to turn until the arm is pointing at the center again.

Figure 11.4 illustrates how the robot moves and turns. At the start of its journey, the robot is facing north, and its arm is pointing due east at the flag. After moving straight for a while, the flag is behind it. Before continuing, the robot should turn until the arm is once again pointing at the flag. The robot actually turns a little further, so the flag is slightly in front of where the arm is pointing; if you're curious about the exact calculation, refer to the SphereLab documentation in your Lab Manual.

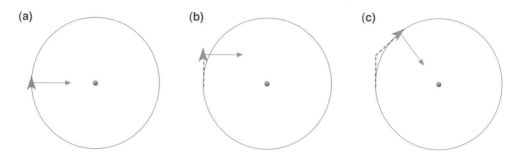

Figure 11.4: *(a) The initial position of the robot explorer. (b) The robot's position after moving straight ahead. (c) After reorienting and moving ahead again, the robot is back on the circular path and the arm is again pointing at the flag.*

The `orient` command tells the robot to turn until it is facing a direction that will put it back on the correct path:

```
>> robot.orient
=> 41.112090439167
```

The return value is the number of degrees the robot turned.

The first step in the project is to plant a flag to mark the center of the circular path. When the robot first lands, it will be facing due north. We want it to move some distance to the east, plant the flag, turn around, and move back to the starting position. Then we'll send it out to the north, and periodically have it reorient as it moves clockwise around the flag. If we repeatedly tell the robot to advance and reorient, it will eventually move in a complete circle and end up back where it started.

Tutorial Project

T1. Start IRB and load the module that will be used for this project:

```
>> include SphereLab
=> Object
```

T2. Initialize the RubyLabs canvas to display the map and robot:

```
>> view_robot
=> true
```

You should see a new window, with the robot on the left side, pointing north (straight up).

T3. Call the `location` method to get the *x* and *y* coordinates of the robot:

```
>> robot.location
=> [40, 200]
```

The coordinates represent distances from the point at the southwest corner of the map. The robot starts out at a point 40 meters in from the left, and 200 meters up from the bottom.

T4. The `speed` and `heading` methods will tell you how fast the robot will move (in meters per hour) and the direction it is currently headed (360° is north):

```
>> robot.speed
=> 10.0
```

```
>> robot.heading
=> 360.0
```

To plant the flag in the center of this territory, we need the robot to turn so it is heading east, travel 160 meters (since the robot's current x coordinate is 40, and the center is at $x = 200$), plant the flag, turn back to the west, and travel the same amount of time to get back to the starting point.

T5. Point the robot to the east:

```
>> robot.turn(90)
=> 90
```

You should see the arrow that represents the robot on the screen turn so it points toward the center of the map.

T6. Since the robot moves at 10 meters/hour, we have to tell it to travel for 16 hours to get to the center of the map:

```
>> robot.advance(16)
=> 16
```

T7. Get the new location of the robot:

```
>> robot.location
=> [200.0, 200.0]
```

Can you see how it has moved 160 meters to the east? The previous x coordinate was 40, and the new one is 200.

T8. Plant the flag:

```
>> robot.plant_flag
=> 0
```

T9. Turn the robot around, send it back to the starting point, and tell it to turn to the north again:

```
>> robot.turn(180)
=> 180

>> robot.advance(16)
=> 16

>> robot.turn(90)
=> 90
```

T10. It will be easier to follow the robot's progress on its circular trip if it leaves a track behind as it moves. Type this command to turn on the tracking option:

```
>> robot.track(:on)
=> :on
```

T11. Tell the robot to move for 3 hours in the direction it is currently heading:

```
>> robot.advance(3)
=> 3
```

You should see the robot icon move a short distance toward the top of the map, leaving a line segment on the canvas to show where it has been.

T12. Tell the robot to turn 90°:

```
>> robot.turn(90)
=> 90
```

T13. You can turn the robot counterclockwise by passing a negative number to turn:

```
>> robot.turn(-30)
=> -30
```

T14. After turning 90° clockwise, and then 30° counterclockwise, the new heading should be 60°:

```
>> robot.heading
=> 60.0
```

Is the robot icon on your screen pointing to 60° (toward the northeast)?

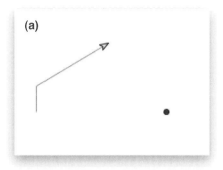 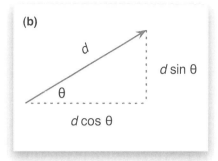

Figure 11.5: *(a) The location and orientation of the robot after moving for three hours, turning 60°, and moving another 10 hours.* **(b)** ♦ *The second part of the path traveled by the robot is the hypotenuse of a right triangle. The distance moved in the x and y dimensions corresponds to the cosine and sine of the angle adjacent to the hypotenuse.*

T15. Move the robot for 10 hours along its new heading:
```
>> robot.advance(10)
=> 10
```
Your canvas should now look something like the screen snapshot in Figure 11.5a.

♦ Get the robot's new location:
```
>> robot.location
=> [126.602, 280.0]
```
Use your calculator to check to see if this location is correct (see Figure 11.5b).

Do some more experiments on your own. Can you use calls to advance_robot and turn_robot to have the robot travel a circular path around the flag in the middle of the terrain and return to its starting point?

T16. Clear the canvas and reposition the robot at its starting location by calling view_robot again. Since the remaining experiments will want to track the motion of the robot around the flag in the middle of the map we can pass options to view_robot so the screen is initialized with a flag and with tracking turned on:
```
>> view_robot(:flag => [200,200], :track => :on)
=> true
```

T17. Tell the robot to move straight ahead for 5 hours:
```
>> robot.advance(5)
=> 5
```

T18. Call orient to have the robot turn so it's arm it pointing at the center again:
```
>> robot.orient
=> 34.70
```
The return value indicates the robot turned 34.7°.

T19. Move the robot ahead and orient it again:
```
>> robot.advance(5); robot.orient
=> 0.0
```
The 0.0 means that this time the robot decided it was still going the right direction, so the turn angle was 0°.

T20. Enclose the two statements on the previous line in a call to `times`, so they are repeated 18 more times:

```
>> 18.times { robot.advance(5); robot.orient }
=> 18
```

Is the robot progressing on a path that will take it back to its "home base"?

T21. Repeat the command that initializes the system with a flag in the middle of the map:

```
>> view_robot(:flag => [200,200], :track => :on)
=> true
```

T22. Repeat the previous exercise, but this time have the robot travel for only 2 hours instead of 5 before it reorients, and have it make a journey involving 50 segments instead of 20:

```
>> 50.times { robot.advance(2); robot.orient }
=> 50
```

In the first experiment the path was adjusted once every 5 hours, but in the second experiment the robot turned once every 2 hours. Do you see that when the robot is told to adjust its direction more frequently the path looks more like a circle?

♦ What do you suppose would happen if the time interval is increased, instead of decreased? Think about where the robot would go if it moved north for 12 hours, and then reoriented and advanced again. Would the line segments that make up the path still be tangent to the circular path we want the robot to make?

♦ Check your answer to the previous question by repeating Exercises T21 and T22, passing 12 in the call to `robot.advance`.

♦ An even more extreme path is made if the robot travels for 24 hours or more, but there isn't room to show this path on the map. But you can get the same effect if you move the robot closer to the flag before it starts, which will have it make a smaller circle. Type this after initializing the map:

```
>> robot.turn(90); robot.advance(8); robot.turn(-90)
=> -90
```

Now send the robot off on a journey that moves for 12 hours before reorienting. What do you see? Is it a circle?

♦ Suppose the robot leaves a monitoring instrument each time it stops. Describe the pattern defined by the locations of the instruments. Are they all equally distant from the flag? Do the locations define a circle? Two circles?

11.2 The Force of Gravity

To simulate the movement of the planets in the solar system, we are going to use the strategy introduced in the last section, where we simulated the motion of an object for a short period and then corrected its heading. The solar system simulation starts with the current location and heading of each planet. The basic equations of motion predict where planets would be a short time later if they all traveled in a straight line. From these new positions the simulator uses gravitational forces between bodies to calculate adjustments to headings, and we will simulate straight-line motions for another short period of time.

The amount of time the bodies move is called the **time step**. As was the case with the robot experiment, the size of a time step is a critical factor in the accuracy of the simulation.

The more often new trajectories for the planets are computed, the more accurate their simulated movements will be.

One difference between the robot experiment and the planet simulation is that we do not have any preconceived notion of where the planets are supposed to move. The adjustments to a planet's heading will be determined only by the gravitational forces from all the other bodies in the system.

Another difference is that the robot always moves with a constant velocity. We simply assumed that once it receives an `advance` command the robot will immediately start off at 10 m/sec. A real robot, like a car, will need some time to accelerate before it reaches the final velocity. Similarly, planets and other bodies in the solar system change velocity, speeding up the closer they get to the Sun.

The next step in the development of our simulation is to explore situations where velocity changes. When a body changes velocity, it is because some external force is being applied that causes the body to accelerate. The project in this section will show how we can simulate the motion of a body when it is accelerating as a result of gravitational forces. We will set up a very simple system with just two bodies: a watermelon and the Earth. We will create an object for each of these two bodies and then simulate the motion of the watermelon as it falls toward the Earth.

The advantage in using a simple "two-body" system for our initial experiment is that there is an equation that predicts how far the melon will fall in a given amount of time, and we can use this equation to check the accuracy of the simulation. Recall from the discussion in the introduction to this chapter that for a general situation, with $N \geq 3$ bodies, there are no formulas to explain how each body will move, but a small object falling toward the Earth is a special case that can be described precisely by a simple equation (Figure 11.6).

Assuming an object is initially stationary, and that it does not slow down because of wind resistance, the equation that predicts how far it will fall in t seconds is $d = 1/2 \ g \times t^2$. The g in this equation stands for the **acceleration due to gravity**, which is the effect of the gravitational force exerted by the Earth. Since we are measuring distances in meters, the value of g is 9.8 m/sec^2. As an example of how this equation can be applied, if the watermelon is dropped from a balcony of a tall building, the distance it falls in 2.0 seconds is $d = 1/2 \times 9.8 \times 2.0^2$, or 19.6 meters.

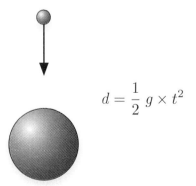

Figure 11.6: *A small object near the surface of the Earth is pulled toward the center of the Earth by the force of gravity. A simple equation predicts how far the object will move in t seconds.*

$$d = \frac{1}{2} \ g \times t^2$$

By rearranging the terms in the equation, we can turn it into a formula that defines t as a function of d:

$$t = \sqrt{2 \times d / g}$$

In this form, the equation tells us how long it will take an object to fall a distance of d meters. For example, if the balcony is 35 meters high, it will take the watermelon $\sqrt{2 \times 35 / 9.8} = 2.67$ seconds to hit the ground. This prediction will be accurate if the watermelon is perfectly stationary when it is let go (*i.e.*, the person holding it doesn't give it a downward push in order to make a bigger splat) and if there is no appreciable air resistance to slow the watermelon down.

The SphereLab module includes a type of object named Body that can be used to represent small bodies like the watermelon, as well as larger bodies like the Earth, Sun, and planets. Each body object has a mass, a position, and a velocity. The methods defined for bodies will do all the necessary calculations for us; all we need to do is create the objects and then call the methods to update their velocities and positions.

A method named `make_system` sets up a computational experiment for a system of bodies. For the watermelon experiment, we pass the symbol `:melon` as an argument to `make_system`:

```
>> b = make_system(:melon)
=> [melon: 3 kg ..., earth: 5.9736e+24 kg ...]
```

The method creates an array containing two body objects, one for the watermelon and one for the Earth. The string printed for each object shows its name and mass, followed by the object's position and velocity. The details of how the position and velocity are represented will be explained in the next section; for now we just need to know that they are included in the representation of each body.

We can monitor the progress of the watermelon experiment by watching the bodies move on the RubyLabs canvas. To initialize the drawing, pass the array of body objects to a method named `view_melon`:

```
>> view_melon(b)
=> true
```

Scientific Notation in Ruby

Since the mass of the Earth (in kilograms) is a very large number, it is usually shown in scientific notation: 5.9736×10^{24} kg.

Unfortunately, Ruby can't use superscripts, so it has an alternative notation. The same number, when printed by Ruby, is `5.9736e+24`.

An e in the middle of a floating point number mean the digits to the right are an exponent, and the value of the number is multiplied by 10 raised to this power. The number to the right of the e can be negative, which will be the case when Ruby prints a very small number.

Figure 11.7: *Each line segment in the drawing represents the motion of a falling melon over a period of one time step. The length of a line segment indicates the distance traveled on that time step.*

The drawing will show the watermelon is initially on the Earth's surface. The first step in the experiment is to raise the melon to a specified height above the ground. A method named `position_melon` sets the melon at a specified height (in meters) above the Earth's surface, *e.g.,*

```
>> position_melon(b, 50)
=> 50
```

The key step in the simulation is performed by a method named `update_melon`, which simulates the motion of the melon for a specified length of time. The argument passed to `update_melon` is the size of the time step. The method will use the force acting on the melon to update its trajectory and compute the new position after the specified amount of time. For example, this expression calculates the predicted height of the melon using a time step of 0.5 seconds:

```
>> update_melon(b, 0.5)
=> 47.54
```

The melon was initially 50 meters above the ground, and the result of this call shows the simulator has determined that after falling for one half second the melon will be 47.54 meters above the ground.

If we repeat the call to `update_melon` we'll see something interesting:

```
>> update_melon(b, 0.5)
=> 42.63
```

It's not surprising that the new height is lower, but notice how much more the melon moved. In the first time step, the melon moved $50 - 47.54 = 2.46$ meters, but in the second time step it moved $47.54 - 42.63 = 4.91$ meters. This is exactly what we expect, since the melon is accelerating. Its velocity is continually increasing as it falls, and the longer it falls the faster it should move.

As we watch the melon's progress on the canvas, we will see a series of dashed lines. The length of a line segment corresponds to the distance the melon moved during one time step. Since the melon is falling faster and faster, it is moving farther on each time step, and we should see the dashes growing longer with each new time step (Figure 11.7).

A method named `drop_melon` will repeatedly call `update_melon` and return as soon as the melon hits the ground. The value returned by `drop_melon` is the amount of time, according to the simulation, required for the melon to drop to the Earth's surface. For example, after initializing the experiment and positioning the melon at 50 meters, a call to `drop_melon` shows it takes 3.0 seconds for the melon to hit the ground:

```
>> drop_melon(b, 0.5)
=> 3.0
```

Note that the return value will always be a multiple of the time step size. One way to interpret this result is that `drop_melon` had to call `update_melon` 6 times before the melon dropped all the way to the ground, because 6 times steps at 0.5 seconds per step adds up to 3.0 seconds.

We can check the accuracy of this result by plugging the melon's initial height into the equation that predicts the time it will take an object to fall this distance:

$$t = \sqrt{2 \times d/g} = \sqrt{2 \times 50/9.8} = 3.19$$

So the simulation was off by 0.19 seconds, which is a sizable error. The project in this section will experiment with shorter time steps to see if they lead to a more accurate simulation.

Tutorial Project

T23. Use the `make_system` method to create the body objects for the "two-body" problem with a watermelon and the Earth:

```
>> b = make_system(:melon)
=> [melon: 3 kg ..., earth: 5.9736e+24 kg ...]
```

T24. The variable `b` is now an array with two objects:

```
>> b[0]
=> melon: 3 kg ...

>> b[1]
=> earth: 5.97e+24 kg ...
```

The output shows the melon has a mass of 3 kilograms (about 6.6 pounds), and the Earth has a mass of 5.97×10^{24} kg (a lot bigger). The other information printed after the mass are the location and velocity vectors (these will be explained in the next section).

T25. Call `view_melon` to make a drawing showing the two bodies in this system:

```
>> view_melon(b)
=> true
```

You should see the melon resting on the surface of the Earth.

T26. Type this expression to raise the melon to a position 100 meters above the Earth's surface:

```
>> position_melon(b, 100)
=> 100
```

You can specify any height between 0 and 100 meters. On the canvas, the circle representing the melon should have moved to the top of the drawing.

T27. When you set up an experiment by calling `position_melon` Ruby adds a method named `height` to the object that represents the melon. To find the current height of the melon:

```
>> b[0].height
=> 100.0
```

T28. Call `update_melon` to compute the melon's new position after a time step of one second:

```
>> update_melon(b, 1.0)
=> 90.18
```

The return value is the new height of the melon. You should also see the melon move toward the ground in the drawing.

T29. Repeat the previous call:

```
>> update_melon(b, 1.0)
=> 70.55
```

Do you see how the melon accelerated during this second time step? The first time step moved it about 10 meters, and the second time step moved it about 20 meters. On the screen, the circle moved twice as far on the second update.

The equation for the distance traveled by a falling object predicts that in one second the melon should fall $1/2\ g\ t^2 = 1/2 \times 9.8 \times 1^2 = 4.9$ meters. But in our simulation the melon fell 9.81 meters in the first time step. The reason for this large error is because the time step was too big. If we move the melon gradually, over smaller time steps, the simulation will be more accurate.

T30. Position the melon 100 meters above the Earth again:

```
>> position_melon(b, 100)
=> 100
```

T31. Call `update_melon` with a time step of 0.1 seconds:

```
>> update_melon(b, 0.1)
=> 99.90
```

As expected, the melon moves a much shorter distance in 1/10 second.

T32. Call the method that updates the position nine more times and get the height of the melon:

```
>> 9.times { update_melon(b, 0.1) }; b[0].height
=> 94.60
```

So in this second simulation, using 10 time steps of 0.1 seconds each, the melon moved $100 - 94.6 = 5.4$ meters. This is much closer to the real-world value of 4.9 meters.

You're now ready to run the complete simulation. The method named `drop_melon` will repeatedly call `update_melon` until the height of the melon reaches or falls below 0 meters.

T33. Reinitialize the drawing and reposition the melon at 50 meters:

```
>> view_melon(b)
=> true

>> position_melon(b, 50)
=> 50
```

T34. Call `drop_melon` to run the complete simulation with a time step size of 0.5 seconds:

```
>> drop_melon(b, 0.5)
=> 3.0
```

The simulator predicts it will take the melon 3.0 seconds to fall from 50 meters.

T35. Repeat the simulation using a smaller time step:

```
>> position_melon(b, 50)
=> 50

>> drop_melon(b, 0.1)
=> 3.2
```

Compare these results with the value predicted by the formula $t = \sqrt{2 \times d/g}$. Is the simulation with the smaller time step more accurate?

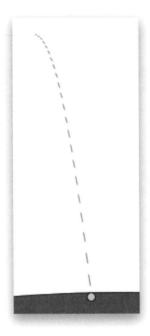

Figure 11.8: *When the melon is given a shove sideways it moves horizontally with a constant velocity while gravity is causing it to accelerate vertically. The resulting path is a parabola that gets steeper the longer the melon falls.*

T36. Use the formula to figure out how long a melon will fall if it is dropped from a seventh floor balcony that is 35 meters above the ground.

T37. Use the simulator to predict how long it will take the melon to fall 35 meters:

```
>> position_melon(b, 35)
=> 35
>> drop_melon(b, 0.1)
=> 2.7
```

We can specify a horizontal motion to cause the melon to move to the right as it is falling, as if the person who drops the melon were to give it a sideways shove when releasing it. Gravity will cause the melon to move with increasing speed toward the ground, but it will move with a constant velocity to the right (Figure 11.8).

T38. Clear the screen, reposition the melon at 100 meters above ground, and assign a velocity of 5 meters/second in the x (horizontal) dimension:

```
>> view_melon(b)
=> true
>> position_melon(b, 100)
=> 100
>> b[0].velocity.x = 5
=> 5
```

T39. Run the simulation using a 0.1 second time step:

```
>> drop_melon(b, 0.1)
=> 4.5
```

T40. How far to the right should the melon move in the amount of time it takes to fall 100 meters?

T41. If the scale of the drawing is such that the vertical distance in the path shown on your screen is 100 meters, is the horizontal distance consistent with your answer to the previous problem?

♦ In the next section we will see that the equations of motion implemented in body objects are based on Newton's law of universal gravitation, which says the force pulling the melon toward Earth is balanced by an equal force that moves the Earth toward the melon.

♦ Make a fresh set of body objects:

```
>> b = make_system(:melon)
=> [melon: 3 kg ..., earth: ...]
```

♦ Type this expression to get the (x, y, z) coordinates of the center of the Earth:

```
>> b[1].position
=> (0,0,0)
```

♦ Raise the melon up to 100 meters and drop it:

```
>> position_melon(b, 100)
=> 100

>> drop_melon(b, 0.1)
=> 4.5
```

♦ Check the Earth's position again:

```
>> b[1].position
=> (0,5.1022e-23,0)
```

The y coordinate in this output shows the Earth moved 5.1×10^{-23} meters.

The Earth was pulled toward the melon, but only by a tiny distance. Look up the diameter of a hydrogen atom in a physics text or on the Internet. How many diameters did the melon cause the Earth to move?

11.3 Force Vectors

The exercise near the end of the previous section showed that an object can be moving in two dimensions at once: an initial sideways shove caused the melon to move horizontally, and gravity caused it to move vertically. In the case of planets, we need to be concerned with motions in three dimensions.

One way to describe these more complex motions is to use **vectors**. Mathematically, a vector is a set of three numbers, corresponding to x, y, and z coordinates. For example, when we assigned a horizontal motion to the melon, the result was a set of three numbers that specified the melon's initial velocity in the x, y, and z dimensions:

```
>> b[0].velocity.x = 5
=> 5

>> b[0].velocity
=> (5,0,0)
```

The 5 in the first position means this body is moving 5 meters per second horizontally (the x dimension), and the 0 in the other two coordinates mean the body initially had no velocity in the y and z dimensions.

The Ruby objects that represent bodies use vectors to keep track of positions, velocities, and accelerations. For this project, we can simply think of these vector as arrows; for example, a velocity vector is an arrow that points in the direction the body will move in the next time step. If you are curious about how the methods carry out the arithmetic operations in terms of vectors, you can read about them in your Lab Manual.

The equation that describes the force that pulls two bodies toward each other is Newton's law of universal gravitation:

$$F \propto \frac{m_1 \times m_2}{d^2}$$

The symbols m_1 and m_2 stand for the masses of the two bodies, and d is the distance between them. This equation describes a general relationship between the gravitational force (F), the size of the bodies, and how far apart they are. When the bodies are larger, *i.e.*, for higher values of m_1 or m_2, the force is larger. Similarly, as the distance grows larger, the force will be smaller, because the value of a fraction shrinks as the denominator grows.

Newton's law is shown pictorially in Figure 11.9. The arrows in the figure represent acceleration vectors. This figure emphasizes the fact that gravity is pushing the bodies in a particular direction, indicated by the arrows that point from each body toward the other one.

An important point to notice is that the length of an arrow corresponds to the magnitude of the acceleration. If the two bodies were the same size, they would each experience the same acceleration. But when they have different masses, the smaller one is accelerated more. This is what we saw at the end of the previous section: the melon moves rapidly toward the Earth, as a result of the large acceleration caused by gravity, while the Earth barely moves toward the melon.

We now come to one of the most important points about the physics behind the motion of the planets: the forces acting on a body are **additive**. What this means is that if we want to figure out how body A will move when it is accelerated by the force of gravity from two other bodies B and C, we can calculate the force moving A toward B, then calculate the force moving A toward C, and then add the two forces together to get the cumulative force.

One reason to use arrows to represent vectors is that it is easy to see how vectors are added together. To add two arrows, simply connect the head of one arrow to the tail of the other. You can imagine using a computer graphics application, and picking up one arrow and dragging it on your canvas until its tail lines up with the head of the other arrow (Figure 11.10). When there are more than two other bodies, the forces can be added one after the other, in any order, because vector addition is like ordinary addition: the sum $x + y + z$ can be computed by adding $x + y$ and then adding z to the result.

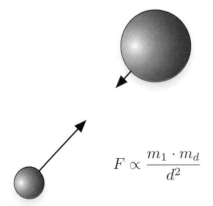

$$F \propto \frac{m_1 \cdot m_d}{d^2}$$

Figure 11.9: *The force attracting two bodies toward each other generates an acceleration that depends on a body's mass. The smaller body will be accelerated more, as indicated by the longer arrow.*

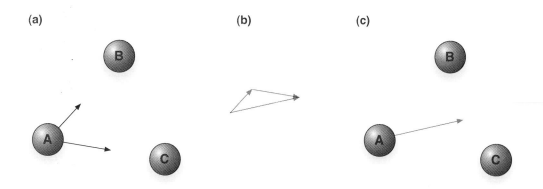

Figure 11.10: *(a) Body A is simultaneously being pulled toward bodies B and C, defined by two different vectors. (b) The two vectors can be added together to compute the cumulative force. (c) Body A will be pulled in the direction defined by the cumulative force.*

To get a sense of how force vectors are added together in an *N*-body simulation, the tutorial exercises in this section will set up an experiment with a set of random bodies. One of the bodies will have a position a little farther from the center of the system than the others. When we start the simulation, all the bodies will be stationary. We will then let the body farthest from the center "fall" toward the others. Just like the melon in the two-body experiment, this body will start slowly, and gradually accelerate as it is drawn toward the other bodies.

Note that this experiment is not intended to be a simulation of a real system of bodies acting under the influence of gravity. In a real system, all the bodies would be in motion. The goal here is simply to illustrate that forces are additive, by focussing on a single body in order to see how it is pulled in a direction that is the sum of the forces acting on it by all the other bodies.

To create the initial data for this experiment, pass the option `:fdemo` in the call to `make_system`:

```
>> b = make_system(:fdemo)
=> [f1: 1e+13 kg (81.472,145.85,0) (0,0,0), ...]
```

When we want this simulation to advance by a single time step, we call a method named `update_one`. The parameters are the body we want to move, an array with all the other bodies, and the time step size. In this data set, the first body in the list is the "falling" body, and the remaining objects represent the stationary bodies, so this expression will simulate the motion of the falling body for a time step of one second:

```
>> update_one(b[0], b[1..5], 1.0)
=> true
```

The main thing to watch for as we run the experiment for several time steps is that at each step the gravitational force acting on the moving body will be the sum of the forces pulling it toward all the others.

What's interesting about this experiment is that it is impossible to predict where the falling body will end up after any given amount of time. We will be able to predict the general direction for the first few time steps, but after that any future motions will be determined by the gravitational effects exerted at the new position, and these effects need to be recalculated at each time step.

One of the hallmarks of chaotic systems is that very small differences in the starting conditions often lead to significantly different states, even after just a few time steps. To demonstrate this effect, before we start the simulation with the falling body, we will make two copies of the object. Then we will run the simulation again, except the copies of the falling body will start in slightly different locations. It will be apparent after 100 time steps that the trajectories of these bodies are very different and that it is impossible to predict where they will be after the next 100 time steps.

Tutorial Project

T42. Load the data set with the six bodies used in this experiment:

```
>> b = make_system(:fdemo)
=> [f1: 1e+13 kg (81.472,145.85,0) (0,0,0), ... ]
```

T43. Call a method named `view_system` to display the bodies on the canvas:

```
>> view_system(b, :pendown => :track)
=> true
```

You should see a canvas with six circles, similar to the one shown in Figure 11.11. One circle, representing the falling body, will be red, and the other five will be blue. The `:pendown` option tells the canvas to draw a line as the falling body moves.

T44. Save the first body in the list in a variable named `f1`, and make two copies of the object, calling them `f2` and `f3`:

```
>> f1 = b[0]
=> f1: 1e+13 kg (81.472,145.85,0) (0,0,0)

>> f2 = f1.clone
=> f1: 1e+13 kg (81.472,145.85,0) (0,0,0)

>> f3 = f2.clone
=> f1: 1e+13 kg (81.472,145.85,0) (0,0,0)
```

T45. Run the simulation with the first falling body for 10 time steps:

```
>> 10.times { update_one(f1, b[1..5], 1.0) }
=> 10
```

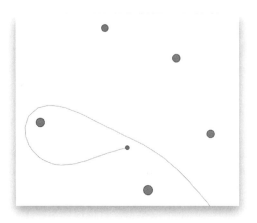

Figure 11.11: *In the "falling planet" simulation one body is allowed to move according to the sum of the gravitational forces acting on it while all the others remain stationary.*

You should see the red circle move slightly on the screen. As was the case with the watermelon simulation, the falling body will move very slowly at first, and then pick up speed. Initially the body will be drawn toward the center of the system. Eventually it will be pulled close to one of the stationary bodies. When that happens, the moving body will speed up the closer it gets to the stationary body. Eventually you will see the falling body get so close to a stationary body that it speeds up fast enough to be flung out of the system in a "slingshot" effect.

T46. Repeat the previous expression to move the falling body for 10 more time steps:
```
>> 10.times { update_one(f1, b[1..5], 1.0) }
=> 10
```

T47. Can you guess where the body will be after the next 10 time steps? Keep repeating the previous expression and watch the progress of the falling body. After about 75 time steps this body will be flung from the system (at which point its trajectory is very predictable).

T48. This expression asks Ruby to print the current location of the first clone:
```
>> f2.position
=> (81.472,145.85,0)
```
The three numbers are the x, y, and z coordinates of the body.

T49. Move this second body slightly to the right by adding 1 to its x coordinate:
```
>> f2.position.x += 1
=> 82.4717
```

T50. Run the simulation for 10 time steps. This expression is the same as in previous calls to update_one, except it uses body f2 instead of f1:
```
>> 10.times { update_one(f2, b[1..5], 1.0) }
=> 10
```

T51. The track for f2 starts out almost identically with the track for f1. But now run the simulation for 50 time steps, and watch what happens:
```
>> 50.times { update_one(f2, b[1..5], 1.0) }
=> 50
```
Can you guess where this body will be after 50 more time steps? Make your guess, and then run the simulation.

T52. Prepare for the next simulation by moving body f3 (the second clone) just a little bit farther to the right than the starting position for f2:
```
>> f3.position.x += 2
=> 83.4717
```

T53. Run the experiment with f3 for 50 time steps:
```
>> 50.times { update_one(f3, b[1..5], 1.0) }
=> 50
```
Again the trajectory is very similar for the first few time steps, and then the path starts to diverge.

As long as f2 or f3 is on the screen you can keep repeating the calls to update_one. When they are ejected from the system do they head out in the same general direction as f1?

If you would like to try some more experiments, the SphereLab module has methods for creating sets of random bodies, using parameters that will make systems similar to the one in the fdemo data set. There are also methods that will simulate the motions of all the bodies, not just one "falling planet." To learn how to set up and run your own experiments refer to the documentation in the Lab Manual.

11.4 *N*-Body Simulation of the Solar System

We're now ready to put all the pieces together to carry out a simulation of the movement of the planets in the solar system.

What we did in the last section for one body—calculate all the force vectors influencing the body, add the vectors together to determine the cumulative force, and then move the body in the direction determined by that force—needs to be done for all the bodies in the system at each time step. The Ruby method that implements this algorithm is shown in Figure 11.12.

The calculation of the gravitational forces between two bodies is carried out by a method named `Body.interaction`, which is called from line 6 of this method. An important detail is that this method computes both accelerations at the same time. Since the force pulling body *j* toward body *i* is the same as the force pulling *i* to *j*, the method only needs to do the force calculation once. The force is then used to update two different vectors: the force pulling *i* toward *j*, and the force pulling *j* toward *i*.

If the loops in this algorithm looks familiar, it's because they have same overall structure as the nested loops in the insertion sort algorithm from Chapter 4. When an array of size *n* is passed to this method, the total number of calls to `Body.interaction` will be $n \times (n-1)/2$. As a result, the algorithm does $\mathcal{O}(n^2)$ force calculations on each time step.

For the solar system simulation, we need an array of body objects for the Sun and each of the nine planets. To make this array, we can pass the `:solarsystem` option in the call to `make_system`:

```
>> b = make_system(:solarsystem)
=> [sun: 1.99e+30 kg ... pluto: 1.31e+22 kg ...]
```

```ruby
# Compute the new positions for objects in array bodies after a time step of size dt

1:    def step_system(bodies, dt)
2:      nb = bodies.length
3:
4:      for i in 0..(nb-1)        # compute all pairwise interactions
5:        for j in (i+1)..(nb-1)
6:          Body.interaction( bodies[i], bodies[j] )
7:        end
8:      end
9:
10:     bodies.each do |b|
11:       b.move(dt)              # apply the accumulated forces
12:       b.clear_force           # reset force to 0 for next round
13:     end
14:   end
```

Figure 11.12: *The Ruby implementation of the N-body simulation algorithm. The method uses nested loops to iterate over all pairs of bodies to do $\mathcal{O}(n^2)$ force calculations on each time step.*

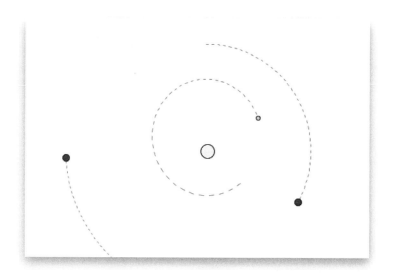

Figure 11.13: *A snapshot of the RubyLabs canvas during the solar system simulation. The simulation uses all 10 bodies, but only the Sun and inner planets are shown on the canvas.*

The parameters for each body are from a solar system database and represent the positions and headings of each body on January 1, 1970. The positions have all been adjusted so the Sun is at the center of the system, *i.e.*, its *x*, *y*, and *z* coordinates are (0,0,0).

To monitor the simulation, we can simply pass the array to `view_system`. But if we do, we'll discover a slight problem: Pluto and the outer planets are so far from the Sun that the inner planets are all crowded together in the middle of the display. A better strategy for viewing the simulation is to pass only the first five bodies in the call to `view_system`. The simulation will still use all 10 bodies, but the drawing will show only the Sun and the planets Mercury, Venus, Earth, and Mars (Figure 11.13).

To run the simulation we need to choose a time step size. To start with, we can use the number of seconds in one Earth day. The fastest moving planet, Mercury, makes a complete orbit in 88 days, so a time step of one day should lead to an ellipse made from 88 line segments, which will be reasonably accurate for our purposes. There are $24 \times 60 \times 60 = 86,400$ seconds in one 24-hour period. If we want to be a little more precise, we should use 86,459 seconds as our time step size, since there are 365.25 days in a year.

As you run this simulation, watch to see if the planets travel in realistic orbits. Are the paths ellipses? If you run the simulation for 365 time steps, does Earth end up more or less where it started, *i.e.*, does it make one complete orbit? Do the planets move faster when they are closer to the Sun?

Tutorial Project

T54. Call `make_system` to create an array of body objects for the Sun and planets:

```
>> b = make_system(:solarsystem)
=> [sun: 1.99e+30 kg ... pluto: 1.31e+22 kg ...]
```

T55. There should be 10 objects in the array:
```
>> b.length
=> 10
```

T56. The first object represents the Sun:
```
>> b[0]
=> sun: 1.99e+30 kg (0,0,0) (0,0,0)
```
As you can see, its position and velocity vectors are both (0,0,0). The Sun starts out at the center of the system, and it is not moving.

T57. Call `view_system` to plot the positions of all the bodies:
```
>> view_system(b)
=> true
```
The distances are scaled so all the bodies will fit on the canvas. As a result, the inner planets are all scrunched together in the center of the canvas.

T58. It will be easier to watch the orbits if you tell `view_system` to track only the first five bodies (Sun, Mercury, Venus, Earth, Mars):
```
>> view_system( b[0..4], :dash => 1 )
=> true
```
The `:dash` option tells the system to mark the progress of each body with a dashed line.

T59. Run the simulation for 10 time steps, using a step size of 1 day:
```
>> 10.times { update_system(b, 86459) }
=> 10
```
You should see the planets move on the canvas. The length of a dash corresponds to the amount of movement in one time step. Mercury is moving much faster than Earth and Mars, as indicated by the longer dashes.

T60. Type this expression to continue the simulation for another 355 days, so the total simulation runs for 365 time steps:
```
>> 355.times { update_system(b, 86459) }
=> 355
```
Did you see what you expected?

When evaluating the previous expression, Ruby updated the drawing as quickly as it could, and the motion may not have been very smooth. If you add a call to `sleep`, Ruby will update the drawing and then pause, and the motion will be smoother.

T61. Type these two expressions to clear the screen and run the simulation with a 1/10 second pause between each update:
```
>> view_system( b[0..4], :dash => 1 )
=> true
>> 365.times { update_system(b, 86459) ; sleep(0.1) }
=> 365
```

The next set of exercises will use a RubyLabs probe to count the number of force calculations in each time step. From Figure 11.12 (or by calling `Source.listing(:step_system)`) you should see that the call to the method that computes forces is from line 6 of the method that does the computations for one time step.

T62. Attach a counting probe to line 6 of `step_system`:
```
>> Source.probe("step_system", 6, :count)
=> true
```

T63. Use the `count` method to count the number of times line 6 is executed in a single time step:

```
>> count { update_system(b, 86400) }
=> 45
```

The discussion of the *N*-body algorithm claimed there would be $n \times (n-1)/2$ force calculations. Is that what you got?

T64. Count the number of force calculations made for a simulation that runs for 100 days. Remember that Ruby programs run much more slowly when a counting probe is attached, so you might have to be patient:

```
>> count { 100.times { update_system(b, 86400) } }
=> 4500
```

Do you see how the number of calculations per time step depends on the number of bodies? And that the total number of calculations is simply the product of the number of calculations per time step times the number of time steps?

As explained at the beginning of the chapter, the reason the planets appear to move in nice, orderly elliptical orbits is that the Sun is much more massive than the other bodies. A small object like a comet will be greatly affected by Jupiter and Saturn, the two biggest planets, but the other planets are both too big and too far away from Jupiter for it to have much of an effect on them.

For the next part of the project, we'll make one of the planets much larger and see what effect that has on the motion of the other planets.

T65. Reinitialize the simulation by making a new list of bodies and a new canvas:

```
>> b = make_system(:solarsystem)
=> [sun: 1.99e+30 kg ... pluto: 1.31e+22 kg ... ]

>> view_system( b[0..4], :dash => 1 )
=> true
```

T66. Type this expression to have Ruby print the parameters for the Sun:

```
>> b[0]
=> sun: 1.99e+30 kg (0,0,0) (0,0,0)
```

The string printed right after the name "sun" is Ruby's way of printing 1.99×10^{30}.

T67. Type this expression to see the parameters for Mars:

```
>> b[4]
=> mars: 6.42e+23 kg ...
```

T68. A method named `mass` will return the mass of a body object:

```
>> b[4].mass
=> 6.4185e+23
```

As you can see, Mars has a mass of 6.4×10^{23} kg, roughly 3×10^6 times smaller than the Sun.

T69. Type this expression to multiply the mass of Mars by 1×10^6:

```
>> b[4].mass *= 1e6
=> 6.4185e+29
```

The circle representing Mars on the canvas does not change, but the planet's effect on the other bodies will be apparent.

T70. Run the simulation for 365 time steps:

```
>> 365.times { update_system(b, 86459); sleep(0.1) }
=> 365
```

Can you explain why each of the planets moved the way they did?

♦ Try some more experiments on your own. Start each experiment by making a fresh array of bodies and reinitializing the display. Some ideas:

- Vary the time step size, and see what effect that has on the accuracy of the simulation. What happens if you increase the time step to the number of seconds in three days? In 30 days?

- View different planets. For example, passing b[0..5] to view_system will include Jupiter in the display.

- Can you figure out how to view only the Sun and the outer planets (Jupiter, Saturn, Uranus, Neptune, and Pluto) during a simulation? Hint: make a new array that has only these six bodies.

- How many steps (with a time step size of 86,459) do you have to run to have Jupiter make a complete orbit?

- Multiply the mass of Mars by different values, e.g., by 10^5 instead of 10^6.

- Increase the mass of different planets, including Jupiter or Saturn. The goal is to have two or more bodies that have roughly the same mass as the Sun.

11.5 Summary

The important concept from the field of computer science introduced in this chapter is the idea of computer simulation. We set up computational experiments in which a set of Ruby objects served as "models" of various real-world objects, and then we used equations of motion to move the objects around in their simulated world.

The main goal for these experiments was to learn about issues involved with computer simulation. One issue we looked at was the tradeoff between the size of a time step and the accuracy of the simulation. Using a smaller time step, so positions are updated more frequently, leads to a more accurate simulation, but it comes at the cost of a higher number of calculations. The idea was introduced by simulating the movements of a hypothetical robot explorer, and we also saw how the size of a time step affected simulations of a falling object and the motions of the planets.

The key ideas used to compute the motions of bodies in the solar system simulation are the fact that bodies move according to the force of gravity and that forces are additive. To determine which direction a body is being pulled, we can compute the force that is pulling it toward each of the other bodies and then compute the sum of the individual forces. Forces are represented in a program in the form of vectors, which can be visualized as arrows pointing in the direction of the force. Computing the sum of forces is equivalent to connecting the arrows, drawing the tail of one arrow next to the head of another.

Several factors besides the size of a time step can have an impact on the accuracy and reliability of a computer simulation. The first, of course, is making sure the software is implemented correctly. But even assuming there are no bugs in the application, there are many other ways in which a simulation can give a misleading result. In our solar system simulation, we assumed the mass of each body remains constant, which is a reasonable assumption for planets. But comets lose a small amount of their mass each time they approach the Sun, so an accurate simulation that includes comets will need some way to update their mass. Asteroids have a very irregular shape, and a typical asteroid is very "lumpy," so its mass is not evenly distributed. That means equations that treat the asteroid as a single uniform object may not give the best results.

Concepts and Terminology Introduced in This Chapter

N-body problem	The problem of trying to predict the motions of a group of three or more celestial bodies; no exact solutions are possible, so positions must be computed
acceleration	A change in velocity; for the N-body problem, acceleration is caused by the force of gravity
computer simulation	A method for solving problems like the N-body problem where computational objects represent real-world objects
time step	An amount of time simulated objects move before adjusting their trajectories
vector	In an N-body simulation, a vector is a set of three points, used to represent the (x, y, z) coordinates of a body's location or its heading
accuracy	In a computer simulation, accuracy is determined by how closely the computed results match real-world measurements

---

Accuracy is the central issue in simulation. For the falling watermelon or solar system, it's easy to check accuracy by comparing the program's results with what happens in the real world. For example, our solar system data was taken from observations of the actual positions of the planets on January 1, 1970. To see how accurate our simulation was, we could go to the same database to retrieve the recorded locations of the planets on January 1, 1971, and compare them to our simulated positions. Once we are convinced the program is working correctly, we can start adding other simulated objects, or use the model to predict future locations.

For other applications, it may be much more difficult to determine the accuracy of a simulation. One important factor is deciding which attributes of a real system need to be included in a computer model. For example, to simulate fuel consumption for a car, attributes like the color of the car can obviously be omitted, but other factors, like the size of the tires, and whether they are inflated properly, will have an effect. But adding additional attributes makes the software much more complicated, and with added complexity there is also a greater chance for programming mistakes.

In spite of the difficulty in designing and testing simulations to make sure they are as accurate as possible, modeling and simulation is an important area in applied computer science. One of the main reasons scientists and people in other fields use computer simulations is that they provide an effective way to understand the workings of a complex system.

Exercises

1. Problem 2 on page 15 listed several fields where computers play in important role. Choose one of these areas and do some research to find out how computer simulation is used in this area.

2. Do you think the Sun moves during the solar system simulation? How could you find out?

3. What would you type in IRB to tell the robot of Section 11.1 to travel in a path that would draw a square?

4. Can you figure out how to get the robot to draw a five-pointed star? Hint: each leg of the star will be the same length, and the angle the robot turns will be the same at the end of each leg.

5. Suppose the watermelon simulation is set up so the melon is positioned 75 meters above the ground.

    a) Use the equation $t = \sqrt{2d/g}$ to calculate how many seconds the melon will fall before it hits the ground.

    b) Here is the result of a call to `drop_melon` which simulates the fall with a 1/4 second time step:
    ```
    >> drop_melon(b, 0.25)
    => 4.0
    ```
    How far off is the computer simulation?

6. Copy the picture of five bodies in Figure 11.14 to a piece of paper, or, if you have a drawing application, make a new document and add five circles of the same size and location as the ones in the figure. Sketch vectors that represent the forces acting on each body, similar to those shown in Figure 11.9 on page 296.

7. Update the drawing you made for the previous exercise to show the cumulative force acting on each body, *i.e.*, for each body draw a new vector that is the sum of the forces pulling that body toward the others.

8. Suppose the solar system data is updated to include the locations and headings of 15 moons, so there are now a total of 25 bodies in the data set. How many force calculations would be made on each time step?

9. In the solar system simulation with 25 bodies, how many force calculations would be made if the program runs for 350 time steps?

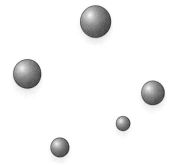

Figure 11.14: *In this drawing, the mass of a body is proportional to its size.*

10. The experiments at the end of Section 11.3 were based on bodies named `f1`, `f2`, and `f3`. The three were identical, except `f2` and `f3` started out in a slightly different position:

    ```
    >> f2.position.x += 1
    => 82.4717
    >> f3.position.x += 2
    => 83.4717
    ```

    What do you suppose would happen if `f3` started out half the distance between `f1` and `f2`? In other words, what if the x coordinate for `f3` was defined by this expression?

    ```
    >> f3.position.x += 0.5
    => 81.9717
    ```

    Would its final position be halfway between the final positions of `f1` and `f2`? Explain why or why not.

11. If you want to use IRB to explore your answer to the previous question, set up some more experiments, using even smaller differences, *e.g.*, start `f3` at 0.5, 0.25, or 0.125 units to the right of `f1`. Are the results still chaotic?

Chapter 12

The Traveling Salesman

A genetic algorithm for a computationally demanding problem

Imagine a situation where a group of friends or relatives have come to visit, and you want to take them on a tour to show them your campus or your home town. For a tour with only a few destinations, the most efficient route may be obvious. But if the tour has more than four or five stops you might need to do some planning. A simple strategy, like moving from a location to the one closest to it, might be all you need (Figure 12.1).

As more and more destinations are added to the tour, however, the simple strategy of going to the nearest location that hasn't been seen already begins to break down. Figure 12.2 shows a larger map, with 25 destinations, and a tour that starts at the location at the top of the map. After visiting the first four locations, the strategy of moving on to the closest point would select the location shown by the dashed line, but it turns out this choice would not lead to the shortest possible tour.

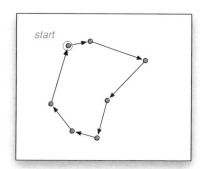

Figure 12.1: *A simple strategy for planning the shortest tour with only a few sites is to start at a random location, and then after each site move on to the nearest location not yet visited.*

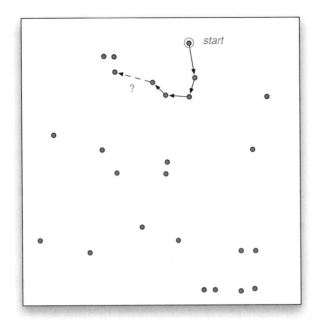

Figure 12.2: *With a larger set of destinations the strategy of moving to the closest point does not always lead to the shortest tour. Before looking at the solution (which is shown in Figure 12.13 at the end of the chapter) can you try to figure out on your own what the shortest tour is? Hint: the best tour does not include the line segment drawn with a dashed line in this picture.*

It's not likely you will ever be called upon to devise a tour that includes 25 destinations. But the same basic idea, of planning the optimal route that connects several different points, is an important part of several real-life situations, and mathematicians and computer scientists have devoted considerable effort to developing algorithms to solve this problem. Some familiar examples are a courier service that wants to plan the most efficient route for their delivery trucks at the start of each day, or a school district that needs to design the shortest route for a set of school buses that pick up and drop off students. A problem that has the same essential structure is figuring out how to connect circuits on the surface of a computer chip. In this case, the "destinations" are electronic components, and the goal is to figure out how to best place metal pathways on the chip to connect the components.

In computer science this problem is known as the **Traveling Salesman Problem**, or **TSP** for short. The input to an algorithm that solves the TSP is a map with a set of cities, where the distance between each pair of cities is defined in a table. The goal is to create the shortest tour that includes all the cities. A tour can start at any city, but it must visit every other city exactly once and then return to the starting point.

In applications of the TSP there are several ways to define the places to visit and the cost of moving from one place to another. For transportation problems, like finding the best route for a delivery truck, the "cities" are simply addresses where a package needs to be dropped off, and the cost can either be the driving distance between points or the expected amount of time it will take to drive between two addresses. For the computer chip layout problem, the "cities" are the components on the chip, and one of the factors in the the overall cost of the design is the total distance along the metal layers that make up the connections between the components.

A simple algorithm guaranteed to find the lowest cost tour is to systematically check every tour. If we define a type of object to represent a map and an iterator to create each tour based on cities on the map, we can use the linear search algorithm presented in Chapter 4 to scan all tours and record the one with the lowest cost. The map objects we will use for projects in this chapter have a method named `each_tour` that creates each possible tour of that map. If a tour is represented by an object named `t`, a call to `t.cost` will compute the cost of the tour. This expression is all we need to find the optimal tour of a group of cities defined in a map object named `m`:

```
>> m.each_tour { |t| best = t if t.cost < best.cost }
```

But before we start applying this iterator to a map with 25 cities it would be a good idea to do a few back of the envelope calculations to estimate how long it might take to find the optimal tour. For reasons that will be explained in Section 12.2, there are $(n-1)!/2$ different ways to visit n cities and then return to the starting point. For a small map with five cities, there are only 12 different itineraries that visit each city exactly once, so the simple linear search would be a reasonable strategy. But if the number of cities doubles, so there are 10 cities on the map, all of a sudden there are over 180,000 different tours. For a map like the one in Figure 12.2, with 25 cities, there are 3×10^{23} different tours. To put this in perspective, even if your computer could somehow compute the cost of one billion tours every second, it would take almost 1,000,000 years to iterate over all possible tours of 25 cities!

In previous chapters, we've referred to iterators as methods that "walk through" a string or an array. In this case, however, the method named `each_tour` does not actually make an array of tours. It uses an algorithm that generates one tour at a time, in such a way that it is guaranteed to eventually make every possible tour. Since the tours are not kept in a list, using the term "linear search" might be misleading, so computer scientists prefer the term **exhaustive search** to describe the algorithm that examines every possible alternative.

The Traveling Salesman Problem is another example of a problem that is at the boundary between what is computable and what is not computable. Like finding a winning move in chess (a problem described in Chapter 1), searching for the best possible tour of a set of cities seems like a very straightforward computation. The computation is relatively simple when there are only a few cities, but is beyond the limits of what is possible for any machine when the map has more than a dozen or so cities.

Factorial Function

The notation $n!$, pronounced "n factorial," stands for the product of the integers from 1 to n.

The factorial function grows even faster than the exponential function introduced in Chapter 5. For every value of $n > 3$,

$$n! > 2^n$$

As you can see from this table, $n!$ is a very large number, even when n is as small as 30.

n	$n!$
3	$3 \times 2 \times 1 = 6$
4	$4 \times 3 \times 2 \times 1 = 24$
5	$5 \times 4 \times 3 \times 2 \times 1 = 120$
10	$10 \times 9 \times \ldots \times 1 = 3{,}628{,}800$
15	$15 \times 14 \times \ldots \times 1 \approx 1.3 \times 10^{12}$
30	$30 \times 29 \times \ldots \times 1 \approx 2.7 \times 10^{32}$

This chapter introduces a new type of algorithm, called an **evolutionary algorithm**, to tackle problems like the traveling salesman where an exhaustive search is simply too expensive. The algorithm begins by making an initial set of random tours and evaluating the cost of each one. A random tour is one that starts at an arbitrary city, and then moves on to random destinations not yet visited, until arriving back at the starting point. While we can't expect any one random tour to be a very good solution, some of them will be much shorter than the others. Evolution comes into the picture when we select the "strongest" of the initial tours and use them as the basis for forming new tours. The idea is to select the most efficient tours, throw out the rest, and then develop new tours that are similar to the "survivors." Usually the new tour is the same as the one it is based on, except for a small "mutation" that changes the tour, perhaps by visiting two of the cities in the opposite order. If we keep iterating this process, of tossing out the most inefficient tours and replacing them with slight modifications of the better tours, eventually a good solution—and in many cases the best possible solution—will emerge.

The project in this chapter will explore how an evolutionary approach can be used to solve the Traveling Salesman Problem. In the first section we will see how to implement a map as an object in Ruby and do some initial experiments that emphasize the point that an exhaustive search of all tours is not feasible. We will then see how to create random tours and how to make slight modifications to them so that eventually an optimal tour evolves from a primordial soup of random tours.

12.1 Maps and Tours

In the days when road maps were printed on paper, one could usually find a table of driving distances between major cities. The entry in row x, column y, would have the distance between cities x and y. Since the distance from x to y is the same as the distance from y to x the tables were often presented with a triangular layout, as shown in the example in Figure 12.3.

In mathematical terms, a rectangular table of numbers like the one shown in the figure is called a **matrix**. Driving distances are a special type of matrix, called a **symmetric** matrix, since the distances are the same for each direction.

TSPLab, the RubyLabs module we will use for this project, includes a type of object named Map that defines the distances between cities. When we create a map object, we can either pass it the name of a file that contains distances between real locations, or we can make up a random map when we want to test our algorithm on maps with more cities.

Here is how to create a matrix to represent the distances between the cities shown in Figure 12.3:

```
>> m = Map.new(:ireland)
=> #<TSPLab::Map [dublin,cork,limerick,galway,belfast]>
```

This assignment statement creates a variable named m and makes it a reference to a map object. The string printed by IRB contains the list of cities that are on the map. As in previous projects, if the argument passed to Map.new is a symbol (a string that starts with a colon) Ruby looks for a data file that is included as part of the RubyLabs software package. You can pass a file name instead of a symbol, and the method will look in your project directory for

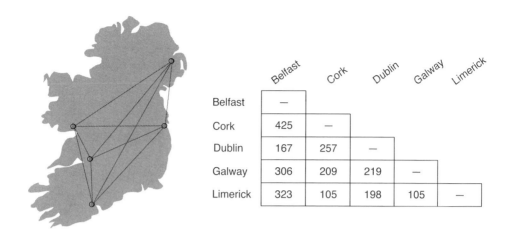

Figure 12.3: *Matrix of driving distances in Ireland.*

a file with that name. Later in the chapter we will see examples where an integer is passed to Map.new in order to make a random map with a large number of cities.

We can use Ruby's index operator to look up the distance between a pair of cities. Just as a[i] refers to the value stored at location i in an array a, the expression m[x,y] means "the distance between cities x and y on map m." For example, to find the distance from Dublin to Galway, two of the cities on our test map named m, the expression is

```
>> m[:dublin, :galway]
=> 219.0
```

A method named make_tour will create an object that represents a tour between two or more cities. One way to create a tour object is to put the names of the cities, in order, in an array and pass the array as an argument to a method named make_tour. For example, an object that represents a three-city tour from Dublin to Galway to Limerick and back is made by this expression:

```
>> t = m.make_tour( [:dublin, :galway, :limerick] )
=> #<TSPLab::Tour [:dublin, :galway, :limerick] (522.00)>
```

Note that this example has only three cities, so it's not a potential solution to the problem of finding the shortest tour of all five cities. However, the ability to make shorter tours like this one will be useful for testing methods that operate on tour objects.

The cost of a tour is the sum of all the distances between destinations in the tour. The total cost of a tour is shown in parentheses at the end of the line when a tour is printed on the terminal; the total cost for the three-city tour shown above is 522. Since this map is based on driving distances, the number 522 represents a distance, in kilometers, but in other maps the cost could be based on travel time or ticket price or some other metric. An important note is that the tour is a round trip, so the total cost of the tour shown above includes the final leg from Limerick back to Dublin (which you can verify by adding the distances between these cities shown in the table in Figure 12.3).

If an algorithm needs information about a tour it can call one of several different methods defined for tour objects. Two of the attributes of a tour that we will be interested in are the path (the list of cities visited) and the cost. Using the example tour object from above:

```
>> t.path
=> [:dublin, :galway, :limerick]

>> t.cost
=> 522.0
```

Tutorial Project

Start IRB and load the module that will be used for projects in this chapter:

```
>> include TSPLab
=> Object
```

T1. Create the map that has the driving distances (in kilometers) between five cities in Ireland:

```
>> m = Map.new(:ireland)
=> #<TSPLab::Map [dublin,cork,limerick,galway,belfast]>
```

T2. A method named `display` will print the distance matrix on your terminal:

```
>> m.display
              dublin      cork limerick    galway  belfast
   dublin      0.00
     cork    257.00      0.00
 limerick    198.00    105.00     0.00
   galway    219.00    209.00   105.00      0.00
  belfast    167.00    425.00   323.00    306.00     0.00
=> nil
```

Note: as a result of the way the city names are represented inside the object they may not be printed in the order you see here or in Figure 12.3, but each pair of cities should be in the matrix.

T3. Use the index operator (square brackets) to find the distance from Cork to Dublin:

```
>> m[:cork, :dublin]
=> 257.0
```

T4. You should get the same distance if you reverse the order of the cities:

```
>> m[:dublin, :cork]
=> 257.0
```

T5. Use the index operator to find distances between other pairs of cities. Do the results agree with the values in the matrix that was printed on your terminal?

T6. Make a three-city tour and save it in a variable named `t`:

```
>> t = m.make_tour([:dublin, :belfast, :galway])
=> #<TSPLab::Tour [:dublin, :belfast, :galway] (692.00)>
```

T7. The string printed as the result of the last expression includes the tour length. You can get the total cost of a tour object by calling its `cost` method:

```
>> t.cost
=> 692.0
```

T8. Verify this value is correct by summing the individual distances of each leg of the trip:

```
>> m[:dublin, :belfast] + m[:belfast, :galway] + m[:galway, :dublin]
=> 692.0
```

T9. Can you type an expression that makes a tour of the same three cities, but starting in Belfast instead of Dublin?

T10. Can you figure out the expression that makes a tour of the same three cities, but goes in the opposite direction (*i.e.*, goes from Dublin to Galway to Belfast)?

Try some more experiments on your own, including some tests with tours of four or five cities. Does it matter which city is the start of the tour, or is the cost the same as long as you visit all the cities in the same order? Do you get the same length tour if the cities are visited in the opposite order?

12.2 Exhaustive Search

The goal for the project in this section is to emphasize just how many tours are possible, even for a small map with as few as 10 cities.

In mathematical terms, the different tours are **permutations**. It's easier to see why tours are permutations if we represent a tour by a string of letters, using just the first initial of each city name. For example, the tour that starts in Dublin and goes to Cork, Galway, Limerick, and Belfast can be described by the string "DCGLB." A better tour, which goes to Limerick after Cork, is represented by the string "DCLGB." A tour that visits the cities in alphabetical order is "BCDGL." Each of these tours have all five letters, but the letters appear in a different order, *i.e.*, they are different permutations of the string "BCDGL" (Figure 12.4).

One way to count the possible permutations is to start by considering the choices for the first city. In the case of the 5-city Ireland tour, there are five choices, corresponding to the strings that start with each of the five letters. There are then four choices for the next city. Since there are four choices for each different starting place, there are 20 different ways to start a tour: the four that start with B ("BC," "BD," "BG," "BL"), the four that start with C ("CB," "CD," "CG," "CL"), and so on. Each of the two-city tours can continue in three different ways, since there are still three cities left to visit, which means there are 60 ways to make a tour that includes three of the five cities.

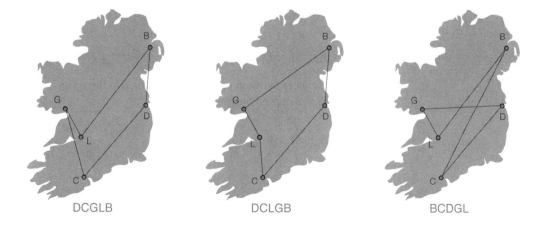

Figure 12.4: *Different tours correspond to permutations of a string made from the first letter of each city name. Tours can start in any city, and can travel in either direction.*

Continuing with this line of reasoning, it should be clear that the total number of strings made from five different letters is $5 \times 4 \times 3 \times 2 \times 1$, or 5!, which is 120. In the general case, for any string with n different letters, there are $n!$ different permutations.

To experiment with permutations, the TSPLab module defines a new iterator named `each_permutation`. When it is called, it will make all possible permutations of a string or array. Like the `each` iterator we've used before, `each_permutation` expects us to call it with a Ruby expression enclosed in braces. The iterator will evaluate the expression once for each different permutation of the string or array.

Here is an example that shows how to print all combinations of letters in a string using `each_permutation`:

```
>> s = "ABC"
=> "ABC"
>> s.each_permutation { |t| puts t }
ABC
ACB
BAC
BCA
CAB
CBA
=> nil
```

The iterator makes a permutation, stores it in the variable `t`, and then evaluates the expression in the body of the block. Since there are 3! = 6 possible permutations of the three letters, the `puts` statement is executed six times, and the six different permutations are printed on the terminal. If we repeat this experiment using a string with five letters, `each_permutation` will generate the 120 strings described above:

```
>> s = "ABCDE"
=> "ABCDE"
>> s.each_permutation { |t| puts t }
ABCDE
ABCED
. . .
```

To relate permutations to tours, first consider that valid tours can start in any city. All that matters is that a tour visits every city and returns to the starting place. Given a permutation, we can "rotate" it so it starts at a different letter but otherwise has all the letters in the same relative order (Figure 12.5). For example, "ABCDE" follows the same roads and has the same cost as "BCDEA," it just starts with B instead of A. Since there are n ways to start a tour, only $n!/n = (n-1)!$ permutations correspond to different tours.

Note also that our hypothetical traveling salesman can travel in either direction on the tour. For the TSP, the tour "ABCDE" is the same as "EDCBA." They both follow the same "roads," but the cities are visited in the opposite order. Since the costs of these tours are the same, we will consider them to be the same solution.

Putting this all together—the fact that $(n-1)!$ permutations correspond to unique paths and that tours can go in either direction—we end up with the formula given in the introduction: an algorithm that searches for the minimal cost tour of a set of n cities potentially has to consider $(n-1)! / 2$ ways of ordering all n cities.

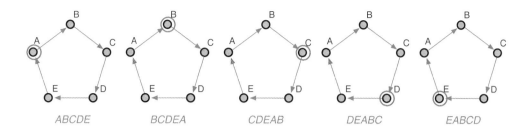

Figure 12.5: *These five permutations all represent the same tour. The starting position has rotated to a different city in the strings that represent the tour, but the path is the same in each case.*

If we have a map object we can call an iterator named `each_tour` to generate tours. The main difference between this iterator and `each_permutation` is that `each_tour` generates only those permutations that correspond to different tours. If the map has n cities, the iterator makes only $(n-1)!/2$ orderings, corresponding to the different paths between the cities. Another difference is that `each_tour` makes tour objects and passes these objects to the block of code. Here is an example, using the map of Ireland:

```
>> m.each_tour { |t| puts t }
#<TSPLab::Tour [:dublin, :cork, ... :belfast] (940.00)>
#<TSPLab::Tour [:dublin, :cork, ... :galway] (1210.00)>
...
```

In all, this iterator passes 12 different tour objects to the block, and the `puts` statement is evaluated 12 times, because for a map with 5 cities there are $(5-1)!/2 = 12$ possible tours.

With this new iterator it is very easy to do an exhaustive search for the minimal cost tour. Simply start by picking any tour to use as an initial value for the best tour found so far. If we call the `make_tour` method without any arguments it will just make a tour with the cities in alphabetical order:

```
>> best = m.make_tour
=> #<TSPLab::Tour [:belfast, :cork, ... :limerick] (1329.00)>
```

Now use `each_tour` to generate every possible tour. Each time we get a new tour, check to see if that tour is better than the best tour seen so far, and if so make it the best tour:

```
>> m.each_tour { |t| best = t if t.cost < best.cost }
```

After the call to `each_tour` returns, just ask Ruby to print the tour object stored in `best`, and you will see the lowest cost tour of the five cities:

```
>> best
=> ... [:dublin, :cork, :limerick, :galway, :belfast] (940.00)>
```

Of course this strategy won't work if we have a map with more than a few cities since there will be far too many tours to check them all. After experimenting with permutations and the `each_permutation` and `each_tour` iterators we'll start looking into the evolutionary algorithm that will work on larger maps.

Tutorial Project

T11. Make a string for the permutation experiments:

```
>> s = "1234"
=> "1234"
```

T12. Before calling the iterator that generates all permutations of this string, can you predict how many permutations are possible?

T13. Use the `each_permutation` iterator to print every possible ordering of the letters in the string:

```
>> s.each_permutation { |t| puts t }
1234
1243
1324
. . .
```

Do you see how this list is organized? The first group of lines all start with 1, then there are lines starting with 2, and so on. The same sort of pattern occurs within a group, also. Did you notice that the six lines in the group that start with 1 are all the permutations of the digits 2, 3, and 4?

T14. The TSPLab module includes a method named `factorial` that will compute the factorial of a number. To compute 4!:

```
>> factorial(4)
=> 24
```

Did you get 4! = 24 output lines from the previous expression?

T15. Another way to count permutations is to save the result of a call to `each_permutation`, instead of using it as an iterator. This expression saves all the permutations of s in a variable named a:

```
>> a = s.each_permutation
=> ["1234", "1243", ... "4321"]
```

T16. To find out how many permutations were created ask Ruby for the number of strings in a:

```
>> a.length
=> 24
```

By looking at the pattern of the strings generated by `each_permutation` you should be convinced there are $n!$ different ways to arrange the letters in a string with n characters.

T17. Make a map object with the driving distances between cities in Ireland:

```
>> m = Map.new(:ireland)
=> #<TSPLab::Map [dublin,cork,limerick,galway,belfast]>
```

T18. Call a method named `size` to get the number of cities in a map:

```
>> m.size
=> 5
```

T19. A method named `ntours` will compute $(n-1)!/2$, the number of tours in a map of a specified size. To compute the number of tours in the 5-city map:

```
>> ntours(m.size)
=> 12
```

T20. Let's check to see if this number agrees with the formula. Since there are $n = 5$ cities, we should have $(5-1)! / 2$ tours:

```
>> factorial(5-1)/2
=> 12
```

T21. This expression will use the `each_tour` iterator to print all 12 tours:
```
>> m.each_tour { |t| puts t }
#<TSPLab::Tour [:dublin, :cork, ... :belfast] (940.00)>
...
#<TSPLab::Tour [:dublin, :belfast, ... :limerick] (985.00)>
```
It's not as easy to see a pattern in these tours as it was to see a pattern in the permutations of strings. The `each_permutation` and `each_tour` methods use different algorithms for reordering the items in a list, but you should see all 12 tour objects.

T22. Make a tour of these five cities, and save the tour object in a variable named `best`. When `make_tour` is called without an argument the cities are in alphabetical order:
```
>> best = m.make_tour
=> #<TSPLab::Tour [:belfast, :cork, ... :limerick] (1329.00)>
```

T23. Use `each_tour` to generate all tours again, but this time compare the cost of each tour to the best cost seen so far, and update the best tour each time a new lower cost tour is found:
```
>> m.each_tour { |t| best = t if t.cost < best.cost }
=> nil
```

T24. Print the best tour and its cost:
```
>> best.path
=> [:dublin, :cork, :limerick, :galway, :belfast]

>> best.cost
=> 940.0
```

If you want to try some more experiments to get a sense of how big a number $n!$ can be, even for strings with as few as 15 letters, try the following optional project, otherwise you can move on to the next section.

♦ Make a new test string with 15 characters:
```
>> s = "ABCDEFGHIJKLMNO"
=> "ABCDEFGHIJKLMNO"
```

♦ Find out how many permutations will be printed if you use `each_permutation` as an iterator:
```
>> factorial(s.length)
=> 1307674368000
```
In scientific notation, that number is 1.3×10^{12}, or about one trillion.

♦ You can ask Ruby to print all the permutations of the new 15-letter string, but be ready to interrupt the computation by typing ^C:
```
>> s.each_permutation { |t| puts t }
ABCDEFGHIJKLMNO
ABCDEFGHIJKLMON
ABCDEFGHIJKLNMO
...
^C IRB::Abort: abort then interrupt
```
As Ruby was printing these strings did you see how the computation was progressing? The letters on the right side of the string were changing places, but the further to the left you looked, the slower the letters were changing.

♦ Redefine `s`, so it is a string with 7 characters:
```
>> s = "ABCDEFG"
=> "ABCDEFG"
```

♦ Enclose the expression that prints the permutations of s in a call to `time`. This statement will ask Ruby to measure how long it takes to generate 7! = 5040 permutations:

```
>> time { s.each_permutation { |t| puts t } }
ABCDEFG
ABCDEGF
ABCDFEG
...
GFEDCBA
=> 0.184829
```

The output shows it took around 0.18 seconds to create and print 5040 permutations (the execution time for your computer will probably by different). At this rate, how long would it take your computer to make and print all permutations of the 15-letter string?

12.3 Random Search

It should be clear by now that for all but the smallest sets of cities it is not going to be possible to find the shortest path connecting all the cities by evaluating every possible tour. The strategy used by the evolutionary algorithm outlined in the introduction is to start with a set of random tours and then apply small adjustments until the optimal tour is found. To prepare for experiments with the evolutionary algorithm, this section will explain in more detail what it means to make a random tour of the cities on a map.

One explanation of a random tour is to imagine the traveling salesman making up his itinerary as he goes. At the start of the day, when he needs to pick the next city to visit, instead of using a strategy such as going to the nearest city, he might just flip a coin and go to a random city he hasn't visited yet.

We can use this same basic idea in a search algorithm by making tours that are a random permutation of the list of cities. The method named `permute!`, first described in Chapter 9, rearranges the items in an array or string and puts them in a random order (note that the exclamation mark is part of the method name). If we want to make a random tour, all we need to do is get a list of city names and then ask Ruby to generate a random permutation. Here is an example, using an object m that contains a map of Ireland:

```
>> a = m.labels
=> [:dublin, :cork, :limerick, :galway, :belfast]

>> permute!(a)
=> [:cork, :belfast, :dublin, :galway, :limerick]
```

As a convenience, the `make_tour` method will carry out these steps for us. All we need to do is pass the symbol `:random` when we call `make_tour` and it will create a random tour:

```
>> t = m.make_tour(:random)
=> #<TSPLab::Tour [:limerick, :belfast, ... :dublin] (1293.00)>
```

The algorithm we will look at in this section tries to find the optimal tour simply by generating lots of random tours and selecting the one that has the lowest cost. We don't really expect the algorithm to find the best tour, or even a reasonably good tour. But this method does establish a baseline. When we start testing the evolutionary algorithm in the next section, we will want to know how well it performs, and one way to do this is to compare the tours it produces with the results of random searches.

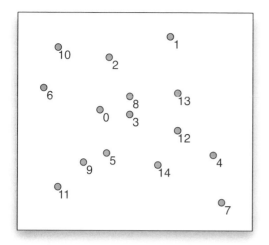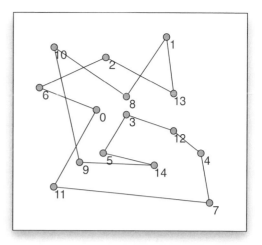

Figure 12.6: *A map of 15 cities placed at random locations by a calling* `Map.new(15)`, *and a tour made by calling* `make_tour(:random)`.

In a random search algorithm we take a "random sample" of all the possible tours and keep track of the best one we find. The Ruby expression to implement this algorithm is very similar to the iteration that does an exhaustive search. The difference is that instead of using the `each_tour` iterator to examine all tours, we just use the `times` method to select a specified number of random tours. For example, to make 100 random tours and save the lowest cost tour in a variable named `best`, the expression is

```
>> 100.times { t = m.make_tour(:random);
               best = t if t.cost < best.cost }
=> 100
```

To test this algorithm and the evolutionary algorithm in the next section we need to make bigger maps. What we want are maps that have enough cities so there are too many tours to be able to find the best one by checking all possible orders, but small enough to be able to display a map and understand at a glance whether a tour is efficient or not. The technique for making larger maps is the same one used in earlier chapters to make large arrays of numbers for testing searching and sorting algorithms. If, instead of passing a file name to `Map.new`, we give it an integer n, we will get back a new map with n cities placed at random locations. For example, this assignment statement will make a map with 15 random cities and save it in the variable named `m`:

```
>> m = Map.new(15)
=> #<TSPLab::Map [0,1,2,3,4,5,6,7,8,9,10,11,12,13,14]>
```

Note that the "names" of the cities are just the numbers from 0 to 14.

A method named `view_map` displays a map on the RubyLabs canvas by drawing a small gray circle for each city (Figure 12.6). For these maps, the x coordinate is the distance from the left edge of the map, and the y coordinate is the distance from the top of the map, *i.e.*, the origin is in the upper left corner of the map. The distance between a pair of cities is the geometric distance defined by their map coordinates.

After displaying the map, we can call `view_tour` to see the connections between the cities defined by a particular tour. Each time we call `view_tour`, any tour that was displayed previously will be erased, and roads used in the new tour will be drawn on the map.

The iteration that makes a series of random tours and saves the lowest cost tour has been implemented in a method named `rsearch`. When we call `rsearch`, we will pass it a map object and tell it the number of random tours to make. It will update the map on the screen each time it finds a new tour with a lower cost, and when it is done it will return the tour object corresponding to the best tour it found.

Tutorial Project

T25. Start by making a map with 15 randomly placed cities:

```
>> m = Map.new(15)
=> #<TSPLab::Map [0,1,2,3,4,5,6,7,8,9,10,11,12,13,14]>
```

T26. Display the map in a RubyLabs canvas window:

```
>> view_map(m)
=> true
```

T27. A method named `coords` will give the (x, y) coordinates of a specified city. Type these expressions to see the locations of the first two cities in your map:

```
>> m.coords(0)
=> [111, 50]

>> m.coords(1)
=> [214, 373]
```

As is always the case for experiments based on random numbers, the actual values you get will be different than the ones shown here. Locate cities 0 and 1 on your canvas, and verify the coordinates that were printed in your terminal window are accurate.

T28. Get the distance between cities 0 and 1:

```
>> m[0,1]
=> 339.02
```

Does the value you got seem accurate? Is it the Euclidean distance, defined by the equation $\sqrt{(x_0 - x_1)^2 + (y_0 - y_1)^2}$?

T29. To see how a random tour is created, first make an array containing the names of the cities:

```
>> a = m.labels
=> [0, 1, 2, 3, 4, 5, 6, 7, 8, 9, 10, 11, 12, 13, 14]
```

T30. Call `permute!` a few times:

```
>> permute!(a)
=> [5, 8, 7, 4, 0, 10, 11, 2, 14, 12, 9, 1, 3, 6, 13]

>> permute!(a)
=> [12, 6, 2, 4, 7, 10, 8, 5, 13, 11, 3, 9, 14, 0, 1]
```

Do you see how each call to `permute!` makes a new random permutation?

T31. Make a random tour of the 15 cities on your map m:

```
>> t = m.make_tour(:random)
=> #<TSPLab::Tour [12, 1, 9, 3, 14, 5, 6, ...] (3257.60)>
```

The `make_tour` method calls `permute!` whenever we ask for a random tour.

T32. Tell Ruby to draw this tour on the map:

```
>> view_tour(t)
=> true
```

The `view_tour` method should have connected the cities that were already on the canvas, using the path defined by the tour object `t`.

T33. This expression tells Ruby to make 10 random tours, displaying each one for one second:

```
>> 10.times { t = m.make_tour(:random); view_tour(t); sleep(1) }
=> 10
```

You should see a series of very different-looking tours, with no relationship between any tour and the one that follows it.

T34. Initialize a variable named `best` with a random tour:

```
>> best = m.make_tour(:random)
=> #<TSPLab::Tour [12, 3, 2, 11, 0, 9, 7, ...] (3377.35)>
```

Note the cost of this tour, which is shown in parentheses at the end of the line.

T35. Type this expression to make 100 random tours, updating `best` whenever a new tour has a lower cost (you can type the entire expression on one line; it has been split into two lines here so it fits the margins of the book):

```
>> 100.times { t = m.make_tour(:random);
                best = t if t.cost < best.cost }
=> 100
```

T36. Print out the best tour:

```
>> best
=> #<TSPLab::Tour [0, 6, 14, 8, 1, 11, 2, ...] (2256.23)>
```

Did the iteration find a tour with a lower cost than the original random guess?

We're now ready to try the `rsearch` method, which collects the steps in the previous exercises into a single method.

T37. Type this expression to find a tour of the cities of map `m` using 100 random samples:

```
>> rsearch(m, 100)
=> #<TSPLab::Tour [3, 5, 9, 7, 2, 11, 12, ...] (2264.95)>
```

The result is the tour object that had the lowest cost of the 100 random tours made by `rsearch`.

T38. Repeat the experiment, but change the 100 to 1000 so the search looks at 1000 random tours:

```
>> rsearch(m, 1000)
=> #<TSPLab::Tour [3, 4, 11, 2, 9, 10, 5, ...] (2033.24)>
```

What you should see is that it's pretty easy for `rsearch` to find a better tour early on, but as the search continues it becomes harder and harder to find a better tour, as indicated by the longer intervals between updates on the canvas.

You should also begin to notice something about the better tours as they are displayed. In general, the fewer the number of roads that cross over other roads, the lower the cost. In fact, the optimal tour will be a loop where no road crosses over any other road in the tour.

Try running the `rsearch` algorithm a few more times, looking at as many as 10^5 tours (written as `100000` in Ruby) or even 10^6 tours (written as `1000000`). You can always type ^C if you get tired of waiting for the method to finish. Can you see how the best tour on the screen generally improves? Also, can you see that improved tours are harder to find the longer the search goes on?

12.4 Point Mutations

The random search algorithm of the previous section is not likely to find the best tour on a map of 15 cities. There are $14!/2 \approx 3 \times 10^{10}$ different tours, so even if we let `rsearch` make 10^6 random tours the probability of its picking the best tour is about $1/30,000$.

It may seem silly to even try to find the best tour just by making random guesses, but the reason we went through the exercise is that it sets the stage for the algorithm described in this chapter, which also works by generating lots of random tours. The difference between the random search algorithm and the evolutionary algorithm is that the latter tries to improve upon the tours it finds. Rather than simply record the fact that it has found a good tour, and going back to guessing, this new algorithm will make minor changes to the good tours to try to make them even better.

It is the idea of making a series of modifications to tours that gives rise to the term "evolutionary algorithm." The approach we are going to look at is a particular kind of evolutionary algorithm known as a **genetic algorithm**. The name comes from the fact that making a series of small changes to tours is reminiscent of the way changes in genes are passed from one generation to the next, as in a culture of yeast growing in a Petri dish.

In our experiments, an array of city names will serve as the "DNA" that defines a single tour. The main step of the algorithm weeds out the most expensive tours—the less fit organisms—and replaces them with new tours that are based on minor modifications of the least expensive tours.

The technique for making a slight change to a tour is called a **point mutation**, based on the terminology used in molecular biology to describe the smallest possible change to the DNA in a real gene. In the TSP, a point mutation corresponds to selecting a city and then exchanging its place with the city that follows it. As an example, suppose a tour named `t` visits five cities in this order:

```
>> t.path
=> [:belfast, :dublin, :limerick, :cork, :galway]
```

Biologically Inspired Algorithms

The genetic algorithm described in this chapter is an example of what computer scientists call a *biologically inspired algorithm*.

The goal for these algorithms is not to simulate real biological systems, but simply to use data structures and operations that are similar to those found in nature to solve computational problems.

Biology	Computation
organism	tour object
DNA	sequence of city names
population	array of tour objects
natural selection	lower cost tours survive
point mutation	rearrange order of cities
crossover	combine subtours

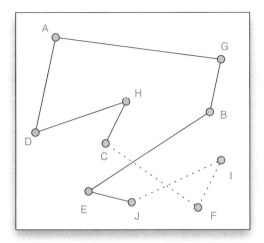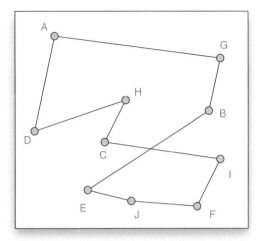

Figure 12.7: *The tour on the left has a small twist, where the road from C to F crosses the road from I to J. The tour can be improved by changing the segment CFIJ to a new segment CIFJ, as shown on the right. The change is the result of a single mutation that switches the position of the letter F with the letter that follows it, transforming the string "...CFIJ..." into a new string "...CIFJ..."*

The city in location 2 of this array is Limerick (remember that the first item in an array is at location 0). A point mutation at this location would exchange Limerick with the city that follows it. The method that makes this change is named `mutate!`, and this is how we would call the method to make a mutation at location 2 in the array:

```
>> t.mutate!(2)
=> #<TSPLab::Tour [...] (940.00)>

>> t.path
=> [:belfast, :dublin, :cork, :limerick, :galway]
```

Notice how Cork and Limerick have traded places in the tour.

Another example of a point mutation is shown in Figure 12.7. On this map, the city names are single letters, and the tour on the left includes a segment that visits cities C, F, I, and J, in that order. One can tell at a glance by looking at this map that the tour is not optimal, because the "road" that connects C to F intersects the road from I to J. A more efficient tour, shown on the map on the right, visits these cities in the order C, I, F, J, and since these paths do not intersect each other the total length of this segment of the tour is shorter. The change in the array that represents this tour, from `[C,F,I,J]` to `[C,I,F,J]`, is a single point mutation that swapped F with the city that followed it.

Figure 12.7 shows the general case of what we hope to achieve with point mutations. At any stage during the search for an optimal tour, there will be tours that have one or more segments like the one shown on the left, where the tour crosses over itself. If we imagine a tour as being a string or rubber band, the change that improves the tour is like untwisting or removing a "kink" from the tour. The optimal tour will be a single loop that has no kinks.

Tutorial Project

T39. Make a map object for the five cities in Ireland:

```
>> m = Map.new(:ireland)
=> #<TSPLab::Map [dublin,cork,limerick,galway,belfast]>
```

T40. Make a new tour with the cities in this order:

```
>> t = m.make_tour( [:dublin, :cork, :limerick, :belfast, :galway] )
=> #<TSPLab::Tour [:dublin, :cork, ... :galway] (1210.00)>
```

Note the cost of this tour is 1210 km.

T41. Draw the map and tour on the canvas:

```
>> view_map(m)
=> true

>> view_tour(t)
=> true
```

You should see a map with five circles, and a tour that looks like a rubber band with a single "twist" in it.

T42. Note that Belfast is in location 3 in the array that represents the tour:

```
>> t.path
=> [:dublin, :cork, :limerick, :belfast, :galway]
```

T43. If we exchange the order that this tour visits Belfast and Galway we will remove the twist. Call t.mutate!(3) to exchange the cities at locations 3 and 4:

```
>> t.mutate!(3)
=> #<TSPLab::Tour [:dublin, ... :galway, :belfast] (940.00)>
```

T44. Print the new path:

```
>> t.path
=> [:belfast, :dublin, :cork, :limerick, :galway]
```

Do you see how the cities at locations 3 and 4 changed places? And that the cost was reduced from 1210 to 940?

T45. Call view_tour again to update the canvas to show the mutated tour:

```
>> view_tour(t)
=> true
```

Notice how the mutation improved the tour, since it is now a simple loop where no line segment intersects any other segment.

Try some more experiments on your own, adding a point mutation at selected places in the array, until you are sure you understand what the mutate! method does. Note that some mutations will make the cost higher; not every mutation is beneficial.

12.5 The Genetic Algorithm

The general outline of the complete genetic algorithm is shown in the box at the top of the next page. The main data structure is an array of tour objects; we will call this array the "population," and we will refer to the size of the population as a number p. After initializing the population with random tours, the main steps of the algorithm remove the higher cost tours and rebuild the population using slightly modified copies of the tours that remain (the "survivors").

> **A Genetic Algorithm for the Traveling Salesman Problem**
>
> 1. Create an array of p objects, each representing a different tour.
>
> 2. Iterate over the array to remove tours with higher costs.
>
> 3. Add new tours to the array, until there are once again p objects. New objects are slightly modified copies of objects left in the array after Step 2.
>
> 4. Halt if any tour in the array is optimal or close to optimal, otherwise continue at Step 2.

The initialization step is straightforward. To create an array of tour objects, the algorithm just makes p calls to `make_tour(:random)` to get a collection of random tours. Each tour object will have its own path, *i.e.*, its own array of city names. Most of the paths will undoubtedly be wild zigzag tours with several roads crossing each other, but we expect some of the tours to have fewer twists than others.

In Step 2, when the program iterates over the population to remove some of the tours, an important question is how to decide which tours to delete from the array. One approach would be to remove a fixed percentage of tours. For example, suppose $p = 10$ and the goal is to remove half the tours. We could implement this operation by sorting the array by cost and then deleting items $p/2$ through $p - 1$, *i.e.*, removing items 5 through 9. But a more subtle approach is one that mimics what happens in nature. The goal is to maintain the "genetic diversity" of the population by allowing some of the less fit tours to survive while removing some of the stronger (less expensive) tours. The idea is that even if a tour is very expensive, after a few generations one of its descendants might be radically different, and that descendant might be a better solution than one derived from the best tours in the founding population.

The technique used in the RubyLabs implementation is to sort the array by cost, so the lower cost tours are at the front of the array. Then, when iterating over the array, the tour in location i is deleted with probability i/p. For example, with $p = 10$ the first tour is always kept because $0/10 = 0$. The probability of removing the second tour is $1/10$, and so on (Figure 12.8). As a result, the best tour in any generation is always preserved, but the others are deleted with a probability that depends on their relative fitness.

The operation in the third step is implemented by the `make_tour` method. Instead of passing `:random` as the argument, we can pass the symbol `:mutate` and a reference to an existing tour. Here is an example, assuming m is a map object and t1 is a tour that survived the round of cuts:

```
>> t2 = m.make_tour(:mutate, t1)
```

This expression makes a copy of the existing tour t1, applies a point mutation at a random location, and stores the result in the variable t2. Note that t1 is not modified. To use the analogy from biology, t1 is the parent of t2, and the descendant t2 is slightly different as a result of the mutation in its "DNA."

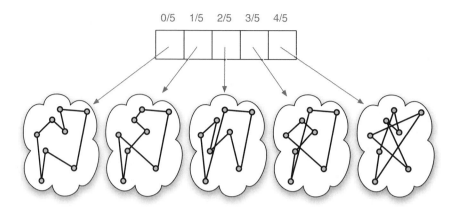

Figure 12.8: *The tour objects in the population array are sorted according to cost, with the lowest cost tours at the front of the array. The probability of removing a tour is based on its position in the array, e.g. the probability of removing tour 3 is* $3/5 = 0.6$.

After rebuilding the population, the algorithm must decide whether to continue and begin another round of selection and rebuilding. That poses a difficult question for our genetic algorithm, one we have conveniently ignored up until this point. How long should we let the algorithm run? How many generations should we let a population evolve before the algorithm halts? There are several different ways to solve this problem, but we will postpone this discussion for later. For our first experiments we will simply pick some number of generations ahead of time and tell the algorithm to execute that many iterations of its main loop.

The RubyLabs implementation of the genetic algorithm is a method named `esearch`, which stands for for *evolutionary search*. The top-level `esearch` method has several "helper methods" to carry out the individual steps in the algorithm:

- The initial population is created by a method named `init_population`, which repeatedly calls `make_tour` to create a set of random tours.

- The step that removes tours as a function of their fitness is implemented by a method named `select_survivors`.

- To process of creating new offspring by copying survivors and adding mutations to implemented by a method called `rebuild_population`.

As we work through the tutorial project for this section, we will be call each of these methods individually, to get a sense of how they work, and then call `esearch` to run complete experiments that search for the optimal tour in maps of varying sizes.

If a map has been displayed on the RubyLabs canvas, a call to `init_population` will draw a histogram, or bar chart, using one bar to represent each tour (Figure 12.9). The height of a bar is proportional to the cost of the corresponding tour. There can be up to 50 bars in the histogram. When a population has 50 or fewer tours there is one bar per object, but when there are more than 50 tours each bar is the average of two or more tours, *e.g.*, if there are 100 individuals each bar represents the average cost of two tours.

Figure 12.9: *The methods that implement the genetic algorithm update the canvas to show the progress of the search for the best tour. The lowest cost tour found so far is displayed on the map, and a histogram shows the cost of each tour in the population. In this snapshot, gray bars correspond to tours that have been deleted; they will be replaced on the next call to* rebuild_population.

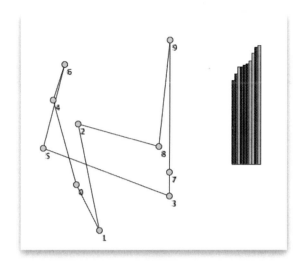

The project will set up some experiments with the genetic algorithm so we can compare tours found by esearch with tours found by calling the random search method rsearch for the same map. A natural question is, how many tours will be created by the esearch method? If it makes just as many tours as rsearch, and the results aren't that much better, then we might as well not bother, and just stick with a random search.

What we will see is that not only does esearch produce much better solutions, it does so by looking at far fewer tours. On each iteration, esearch replaces about half of the population: the tours at the front of the array have a low probability of being deleted, but this is balanced by the fact that tours at the end of the array have a higher probability of being removed. The number of tours deleted by the esearch method is thus approximately one half of the population on each iteration, and these tours are replaced with new tours by the rebuild_population method. At the end of the search, given a population of size p evolving over g generations, we can expect the number of tour objects made by the genetic algorithm to be roughly $(p/2) \times g$.

The genetic algorithm gives noticeably better results than the random search, but there is still room for improvement. With the strategy of making a series of small adjustments to tours, the small "kinks" in the paths will be smoothed out, but there will be many situations where evolution won't make any progress unless it is able to make more dramatic changes. We'll come back to this idea in the next section, where we introduce a second type of mutation that makes much larger changes to a tour.

Tutorial Project

T46. Make a new tour with 20 cities:

```
>> m = Map.new(20)
=> #<TSPLab::Map [0,1,2,3,4,5,6,7,8,9,10,...]>
```

T47. Use the ntours method to compute the number of possible tours:

```
>> ntours(20)
=> 60822550204416000
```

That's about 6×10^{16}, which is 10 quadrillion (or 10,000 trillion!).

T48. Recall from the last section that `make_tour` can be used to make a random tour. Type this expression to make a tour and save it in a variable named `t`:

```
>> t = m.make_tour(:random)
=> #<TSPLab::Tour [11, 17, 5, 7, 6, 12, 10, 9, ...] (4115.27)>
```

You will see a different result, of course, but you should see a path with the numbers from 0 to 19 in some random order. The cost of the tour is shown in parentheses at the end of the line.

T49. Call `make_tour` again, but this time have it create a new tour that is a copy of `t` with a mutation added at a random location:

```
>> m.make_tour(:mutate, t)
=> #<TSPLab::Tour [11, 17, 5, 7, 12, 6, 10, 9, ...] (4120.13)>
```

In this example the mutation exchanged cities 6 and 12. Can you spot the change in the tours you got in your own IRB session?

T50. The mutation in the example shown above also increased the cost of the tour, from 4115.27 km to 4120.13 km. Did the mutation in your own IRB session increase or decrease the cost?

T51. Type this expression to make 10 copies of `t`, printing each new tour in the terminal window:

```
>> 10.times { puts m.make_tour(:mutate, t) }
#<TSPLab::Tour [11, 17, 5, 7, 6, 12, 10, 9, ...] (4179.23)>
#<TSPLab::Tour [11, 5, 17, 7, 6, 12, 10, 9, ...] (4169.07)>
...
```

How many of these "descendants" of `t` have a "beneficial mutation" that gives them a lower cost than `t`?

T52. Display the map on the canvas:

```
>> view_map(m)
=> true
```

T53. Call `init_population` to make an array of 10 random tours based on the map `m`, saving the array in a variable named `pop`:

```
>> pop = init_population( m, 10 )
=> [ ... ]
```

In your canvas window you should see a histogram with 10 bars, one for each tour.

T54. It will be easier to see the tours if you print the array with `puts` (since `puts` prints each item in an array on a separate line):

```
>> puts pop
#<TSPLab::Tour [4, 8, 0, 9, 14, 12, 1, 10, ...] (3263.49)>
#<TSPLab::Tour [14, 12, 7, 10, 18, 3, 2, ...] (3627.96)>
...
#<TSPLab::Tour [14, 13, 1, 12, 2, 8, 5, ...] (4467.65)>
=> nil
```

You should see one tour object on each line. Notice they are sorted, with the lowest cost tours at the front of the array. Do the heights of the bars on the canvas correspond to the costs of these tours?

T55. The `select_survivors` method will delete some of the tours from the array, with a higher probability of keeping the lower cost tours at the front of the array:

```
>> select_survivors( pop )
=> [...]
```

The histogram should now have some gray bars, to indicate the tours that will be deleted from the population, and there should be more gray bars toward the right side, where the most expensive tours are.

T56. Call `rebuild_population` to add new tours to replace the ones marked for deletion by `select_survivors`:

```
>> rebuild_population( pop, 10 )
=> [...]
```

There should now be 10 blue bars in the histogram again. The ones left from the previous generation are shifted left, and the new ones will be on the right.

An important point is that the new tours are all copies of tours from the previous generation, and the new tours all differ from their "parents" by a single mutation.

T57. Print the population array:

```
>> puts pop
#<TSPLab::Tour [4, 8, 0, 9, 14, 12, 1, 10, ...] (3263.49)>
#<TSPLab::Tour [18, 4, 1, 7, 6, 0, 11, 10, ...] (3588.86)>
...
```

Can you find the new tours? Can you can tell which of the earlier tours they are derived from? Are any of these new tours improvements?

T58. This expression will repeat the calls you typed above 10 times, telling Ruby to pause for a half second after each method call (you should type this entire expression on a single line):

```
>> 10.times { select_survivors(pop); sleep(0.5);
                rebuild_population(pop, 10); sleep(0.5) }
=> 10
```

By watching the histogram, do you see how tours are being deleted, survivors being shifted left, and empty places being refilled with new random tours based on the survivors?

The method named `evolve` will do what you did in the last exercise (except without the pauses).

T59. Call `evolve` to tell it to continue the process of selecting survivors and rebuilding the array. This expression tells `evolve` to run for 10 generations:

```
>> evolve(pop, 10)
=> #<TSPLab::Tour [8, 19, 14, 18, 15, 3, 17, ...] (2284.48)>
```

The return value is the lowest cost tour in the population.

T60. Repeat the previous expression a few more times, perhaps running for 100 generations or more. Is the best tour continually getting better?

We're now ready to run some experiments with `esearch`. At the start of each experiment, `esearch` calls `init_population` to make a set of random tours, and then repeatedly call `evolve`. As we will see, we can also pass it options to tell it to vary the population size and control a number of other parameters.

T61. Type this expression to have `esearch` run for 100 generations to find a tour for map `m`:

```
>> esearch(m, 100)
=> #<TSPLab::Tour [2, 17, 11, 10, 19, 4, 18, ...] (2465.38)>
```

Notice that `esearch` also displays some text on the canvas to tell you how the search is progressing.

T62. Call `esearch` a few more times, and record the costs of the tours you get.

One of the pieces of data displayed on the canvas is the number of tours created by `esearch`. If the population size is $p = 10$ and the algorithm runs for $g = 100$ generations, `esearch` should make roughly $(p/2) \times g = 500$ tours. Is that about how many you saw next to `#tours` on the canvas?

Next let's see how `rsearch` will do if it checks the same number of tours.

T63. Use the random search algorithm to find the best tour out of 500 random samples:
```
>> rsearch(m, 500)
=> #<TSPLab::Tour [13, 19, 1, 17, 5, 2, 3, 4, ...] (2782.08)>
```

T64. Repeat the previous expression a few times, and record the costs of the tours you get from `rsearch`.

What do you conclude from these experiments? Is `esearch` generally producing better tours than `rsearch` when both are generating the same number of tours?

The number of tours in a population has a big effect on the search. When there are more tours to choose from, the evolutionary algorithm has a better chance of finding a high-quality tour.

T65. To increase the population size, pass an option named `:popsize` in the call to `esearch`. Run the genetic algorithm for 200 generations on a population of size 50:
```
>> evolve(m, 200, :popsize => 50)
=> #<TSPLab::Tour [19, 16, 2, 6, 8, 9, 18, 4, 1, ...] (1656.45)>
```

T66. Repeat the previous expression a few more times. Does a larger population size give better results?

T67. Do a few more experiments on your own, running for up to 500 generations and using a population size of up to 100. Remember that if the algorithm is not making progress you can always halt it by typing ^C.

How would you characterize the tours made by `esearch`? In most cases, they should appear less jumbled than the random tours found by `rsearch`, but they will still have a few large "twists" or "kinks" where one segment of the path crosses over another one (Figure 12.10)

♦ Try some more experiments on your own, perhaps with a map of 30 or 40 cities, or with a larger population size or running for more generations. In most cases the genetic algorithm will make lots of improvements at first, and then make fewer and fewer improvements as the search continues, and eventually seem to get stuck on a tour with one or more large twists.

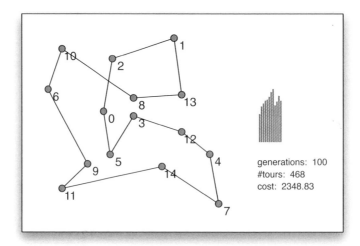

Figure 12.10: *A tour produced with point mutations only. These tours typically have one or two large loops that are hard to "untwist" by exchanging a city with the one next to it on the tour.*

12.6 Crossovers

Most of your experiments with `esearch` probably ended with a tour that looks like the one in Figure 12.10, where there are several large regions that look like they might be part of the optimal tour along with other areas where there are twists or "kinks" in the path. The tour in Figure 12.10 would be improved if a mutation exchanged cities 2 and 8, making the tour one large loop with no roads that cross each other. But there are other cities in the path between 2 and 8. The `mutate!` method cannot exchange 2 and 8 because they are not right next to each other in this tour.

The reason it is so difficult for `esearch` to work out this sort of wrinkle is that smaller changes that are a step in the right direction, such as exchanging 2 with 1 to make the loop one path shorter, often increase the overall cost, so any tour that has the change doesn't survive long enough to pass its "genes" to succeeding generations.

One approach to solving this problem is to use a second type of mutation, one that makes larger changes in the "DNA" of a tour. A common type of mutation, and one that is implemented in TSPLab, is based on another process that occurs naturally in real DNA in real cells. In genetics, a **crossover** mutation occurs when two chromosomes (which are very long strands of DNA) break apart. When the pieces are brought back together, to form whole chromosomes again, there is often some mixing, and as a result new chromosomes can have a mixture of parts from the original chromosomes. This is the source of genetic recombination, and it is the reason we all have a combination of traits from both of our parents.

In a genetic algorithm for the TSP, a crossover is basically a "cut-and-paste" operation. An array of city names from one tour is appended to an array of city names from a second tour, resulting in a third tour that has large pieces from each of the original tours (Figure 12.11). We won't go into the details of how this operation works; for the tutorial project, it is sufficient to know that it is implemented by the same `make_tour` method that creates random tours and tours based on point mutations. If you are interested in how this operation is implemented, you can find a description in the TSPLab documentation section of the Lab Manual.

To make a new tour that combines large pieces of the paths of two other tours, the expression is

```
>> t3 = m.make_tour(:cross, t1, t2)
```

Here `t1` and `t2` are references to two existing tour objects. The return value is a new tour object that has a path made from large subpaths from `t1` and `t2`.

Having a second type of mutation to use when making new tours raises a new question, however: when the `rebuild_population` method creates new tour objects, how does it decide whether to make the new tour with a point mutation or with a crossover? How does it know whether to call `make_tour` with the old `:mutate` parameter or with the new `:cross` parameter? The answer is that `rebuild_population` basically just "flips a coin." Each time it goes to make a new tour, the method uses a random number generator to decide which argument to pass to `make_tour` (Figure 12.12).

The coin flipping analogy is a little misleading, since the odds are not always 50-50 for each kind of mutation. The probability of each type of mutation is defined by a parameter called the **distribution**. By default, if we call `esearch` without any extra parameters, it

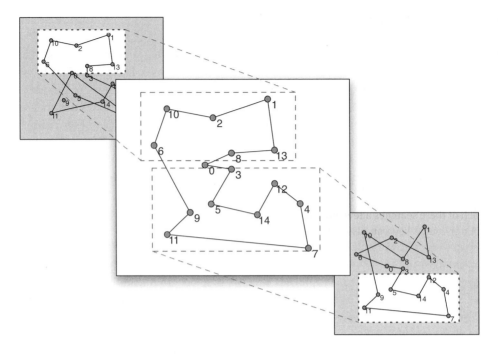

Figure 12.11: *In a crossover mutation, portions of two existing tours (shown in gray) are spliced together to form a new tour. The new tour often combines the best portions of its "parents."*

uses a distribution that forces all the mutations to be point mutations. If we want to include crossovers, the easiest way is to use a distribution named :mixed, as in this example:

```
>> esearch(m, 100, :distribution => :mixed)
```

There are several different sorts of distributions defined in TSPLab, and it is possible to define new ones, *e.g.*, to tell rebuild_population to make 90% point mutations and 10% crossovers (see the Lab Manual for details).

The project for this section will carry out some experiments on maps with 20 or more cities, using various combinations of the different types of mutations. What we will see is that both types of mutations are necessary. Without crossovers, as we have seen, the evolutionary search algorithm cannot make changes it needs to get "unstuck" from solutions that have large twists. On the other hand, using only crossovers will lead to wild changes almost like those seen in a random search, and the algorithm won't be able to "fine-tune" any solutions that are almost optimal. With a combination of both types of mutations, some very low cost tours, and perhaps even the optimal tour, will eventually emerge.

Tutorial Project

T68. If you still have the map with 20 cities from the previous section you can use it for this
project, otherwise make a new map and display it:

```
>> m = Map.new(20)
=> #<TSPLab::Map [0,1,2,3,4,5,6,7,...]>

>> view_map(m)
=> true
```

T69. Call `esearch` with these parameters to see what kind of tour it can find using point mutations only:

```
>> esearch(m, 500, :popsize => 50)
=> #<TSPLab::Tour [6, 15, 9, 11, 1, 19, 14, ...] (1969.10)>
```

It's possible this search may lead to a tour with no twists, but usually there will be one or
two places where roads intersect each other.

T70. Repeat the previous expression, but this time tell `esearch` to use a combination of point
mutations and crossovers:

```
>> esearch(m, 500, :popsize => 50, :distribution => :mixed)
=> #<TSPLab::Tour [2, 3, 13, 4, 0, 8, 12, 7, ...] (1522.62)>
```

Was the algorithm able to find a better tour?

Try each of these searches a few more times, to convince yourself that using crossovers helps `esearch`
find better tours. You can also try each search (with and without the `:distribution` parameter)
on different maps of size 20. Some maps may be laid out in such a way that only point mutations are
required, and some may have regions that are difficult even when both types of mutations are used.

T71. At this point you might be wondering, if crossovers are so effective, why even have point
mutations? Type this expression to tell `esearch` to use only crossover mutations:

```
>> esearch(m, 500, :popsize => 50, :distribution => :all_cross)
=> #<TSPLab::Tour [13, 3, 2, 5, 17, 16, 11, 1, ...] (1606.77)>
```

T72. Repeat the previous expression a few more times. Do the tours made using only crossovers
have anything in common?

```
# Add new Tour objects to array a until there are n objects in a.  The variable named
# pcross is a number between 0 and 1 that represents the probability of making
# the new tour as a crossover of two existing tours.

def rebuild_population( a, n )
  prev = a.length

  while a.length < n
    r = rand
    if r > pcross
      mom = a[ rand(prev) ]
      kid = map.make_tour( :mutate, mom )
    else
      mom = a[ rand(prev) ]
      dad = a[ rand(prev) ]
      kid = map.make_tour( :cross, mom, dad )
    end
    a << kid
  end

end
```

Figure 12.12: *Outline of the Ruby code for the* `rebuild_population` *method. The actual code can
be seen by calling* `Source.listing`.

When `esearch` is not able to use any point mutations it usually finds a tour that has the best overall "shape," but there will be several places where a point mutation that exchanges two cities right next to each other would make a better tour. But such small changes are not likely when the only way to make a new tour is to "cut and paste" from two different tours.

If you would like to try some more experiments on larger maps here are some suggestions.

♦ Make a map with 30 cities and draw it on the canvas:

```
>> m = Map.new(30)
=> #<TSPLab::Map [0,1,2,3,4,5,6,7,8,9,10,...,28,29]>

>> view_map(m)
=> true
```

♦ Call `esearch` using the same parameters you used for the map with 20 cities:

```
>> esearch(m, 500, :popsize => 50, :distribution => :mixed)
=> #<TSPLab::Tour [15, 18, 23, 6, 17, 20, 24, ...] (2113.99)>
```

♦ One way to see if `esearch` can find a better tour is to let it run longer. Repeat the previous expression, but increase the number of generations to 1000:

```
>> esearch(m, 1000, :popsize => 50, :distribution => :mixed)
=> #<TSPLab::Tour [4, 11, 28, 2, 29, 14, 10, ...] (2103.86)>
```

♦ Another possibility for getting a better tour is to use a larger population size. Type this expression to use the original number of generations (500) but twice the population size:

```
>> esearch(m, 500, :popsize => 100, :distribution => :mixed)
=> #<TSPLab::Tour [26, 8, 2, 5, 15, 29, 23, ...] (2043.61)>
```

♦ If you ever see a search that seems like it would produce an answer if only it could run for a few more generations, you can tell `esearch` to continue the previous search. Just repeat the expression that started the search, but increase the number of generations, and pass an option named `:continue`. For example, to continue the search from the previous exercise for another 100 generations (so the total is 600 generations) the command is:

```
>> esearch(m, 600, :continue => true)
=> #<TSPLab::Tour [23, 19, 27, 15, 7, 5, 20, ...] (2030.86)>
```

All the other parameters (population size, *etc.*) will have the same value as in the previous search.

♦ Do some more experiments on your own. Try increasing the number of generations up to 2500, and the population size up to 300. In general, which strategy is more effective, letting the search run for more generations or increasing the population size?

♦ If you want to see how well this method works on larger maps try making maps of up to 100 cities, and using population sizes of up to 1000. The canvas only has room for 400 cities, so with more than 100 cities the map will be very crowded. The population size can be as big as you want, but obviously the larger the population size the longer it will take to make each new generation.

12.7 Summary

This chapter introduced an important problem in computer science known as the Traveling Salesman Problem. Although it sounds like a simple puzzle or brain-teaser, it is actually a very challenging problem faced by professionals in a wide variety of different areas. Traveling to a set of cities on a tour that visits each city exactly once before returning to the starting point is the same basic problem as driving a delivery truck to drop off packages for

---

Concepts and Terminology Introduced in This Chapter

Traveling Salesman Problem	An optimization problem where the goal is to find the lowest cost tour that visits each city on a map exactly once
random search	An algorithm that looks at random solutions; for the TSP, the algorithm evaluates random tours
genetic algorithm	A biologically inspired algorithm that begins with random tours and then tries to "evolve" better tours through a series of mutations
point mutation	A small change in DNA, or, in the TSP, in the order in which two cities are visited
crossover	A process that combines DNA from two sources, or, in the TSP, makes a new tour by combining large parts of existing tours

---

customers, laying out components on a circuit board, and a variety of other problems in transportation and manufacturing.

What makes this problem so difficult is the huge number of alternatives to consider. There are $(n-1)!\,/\,2$ different itineraries for visiting n cities. This equation grows very quickly with increasing n, so that for as few as 20 cities there are far too many alternatives for an algorithm to examine each possible tour.

Instead of doing an exhaustive search that considers every tour, we tried two approaches that examine random tours. A method named `rsearch` simply makes random tours, hoping to find a reasonably good one. Building on that idea, a method named `esearch` implements an evolutionary search that tries to improve on any good tours it finds.

The idea behind a genetic algorithm is to mimic how a population of real organisms changes over time. The algorithm starts with a collection of random tours, and over a series of iterations throws out the more costly tours and replaces them with slight variations of the better tours. Under the right conditions, involving a mixture of the different kinds of "mutations" that modify tours and a large enough "population" to work with, a very good tour eventually emerges.

One of the curious things about the Traveling Salesman Problem is that it is possible to calculate the *cost* of the optimal tour without knowing the actual *path* for the best tour. There are algorithms that use distances between cities to compute a lower bound on the cost of a tour. One might think that the algorithm to compute the cost of the best tour could also be used to find the tour itself, but unfortunately that's not the case. Some more advanced algorithms for the TSP do the same sort of calculations used to find the lower bound, but they still need to carry out some form of search to actually find a tour that has that lowest cost.

Knowing the lower bound for the cost of a tour allows different strategies for halting the algorithm. The simple technique used in the lab projects was to specify a fixed number of

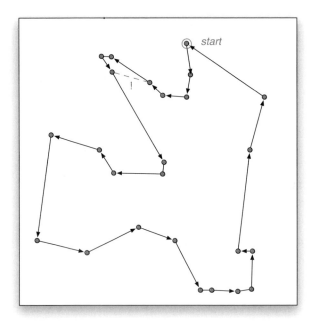

Figure 12.13: *A tour of the 25 cities from the map in Figure 12.2. This tour was made by a call to* `esearch`, *using a combination of point mutations and crossovers.*

generations ahead of time, and then just see what the method would produce in that number of generations. But if we know the cost of the optimal tour, we can tell the algorithm to run until it finds a tour with this cost.

In many cases, such as a courier company planning the routes for a delivery truck, it may not be necessary to have the very best tour. The company may be happy to know the driver will follow a tour that is almost as good as the best possible tour. In this case, they might be satisfied with a computer program that runs for half a minute to produce a tour that is within 5% of the lowest cost, instead of a program that runs for five hours to find the absolute best tour.

A resource for learning more about the Traveling Salesman Problem, including the history of the problem, real-world applications based on the TSP, and information about computing lower bounds, is a web site maintained by a research group at the Georgia Institute of Technology.[1] This site also has a Java applet that will allow you to try to solve the TSP yourself. If you open this applet with your web browser, it will display a set of cities, like the one shown earlier in Figure 12.2, and let you click on the cities to try to make the best tour.

If you took the challenge at the beginning of the chapter, to find the optimal tour of the map in Figure 12.2, you can compare your solution to the one found by `esearch`, which is shown here in Figure 12.13. After visiting four cities, it's tempting to move on to the closest city, as shown by the dashed line, but that would be a mistake. The best tour has to skip over this city, and then visit it on the way toward the cities at the bottom of the map. How did your solution compare to the one found by Ruby?

---

[1]`http://www.tsp.gatech.edu`

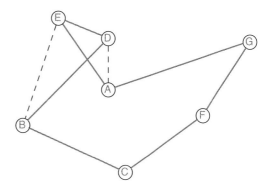

Figure 12.14: *This map with 7 cities is used in problems 1 to 6. If you want to check your answers to these problems you can use IRB, and make a map with these cities by calling* `Map.new(:test 7)`.

Exercises

The map in Figure 12.14 is used in the first set of problems. The distances between the cities on the map are defined by this matrix:

```
         A       B       C       D       E       F       G
A     0.00
B   151.54    0.00
C   202.99  187.45    0.00
D    82.02  199.41  284.85    0.00
E   142.56  185.95  325.58   89.55    0.00
F   167.36  302.42  242.33  202.43  290.11    0.00
G   253.62  404.88  383.51  239.02  323.36  144.51    0.00
```

1. What is the distance between these pairs of cities?

   A—E   F—C   B—E

2. What is the length of the three-city tour A—D—E?

3. What is the length of the seven-city tour shown by the solid lines in Figure 12.14?

4. The variable t contains the tour shown by the solid lines in the figure. Its path is:
   ```
   >> t.path
   => [:C, :F, :G, :A, :E, :D, :B]
   ```
   What parameter would you pass in a call to t.mutate? to "straighten out" the tour so it uses the connections shown by the dotted lines?

5. What is the cost of the path that follows the dotted lines in Figure 12.14? That is, find the cost of the path
   ```
   [:C, :F, :G, :A, :D, :E, :B]
   ```

6. Draw the map that would result from calling t.mutate!(1).

7. Suppose you have a tour named t, and you call t.mutate!(2) to exchange the cities in locations 2 and 3 of the path. What would happen if you called t.mutate!(2) a second time?

8. How many possible tours are there for a map with 16 cities?

9. What proportion of the possible tours of 16 cities would rsearch consider if it looked at 10^6 random tours?

10. Approximately how many tours of a map with 16 cities would be evaluated when esearch is called with the following sets of parameters?

    a) population size = 25, number of generations = 50

    b) population size = 50, number of generations = 100

    c) population size = 100, number of generations = 1000

11. In the previous problem, will the number of tours evaluated by `esearch` change if a different size map is used? Or is the number of tours created independent of the map size?

12. ♦ A formula for estimating the number of tours created by a genetic algorithm was given on page 331. It turns out this formula overestimates the number slightly. In each generation, the best tour is never thrown out. Devise a formula that is a more accurate prediction of the number of tours that will be created, and compare the number of tours made in your experiments with this new formula.

Appendix A

Answers to Selected Exercises

Introduction

1.3: Which of these methods for finding a book in a library are algorithms?

- Ask the first person you see: Not an algorithm, since the person probably won't know where the book is, and may or may not want to help you look in the catalog.

- Find a librarian: Maybe an algorithm. If the library always has a librarian on duty, and the librarian is always in a designated location (*e.g.*, a reference desk) you know how to find, and the librarian on duty is always willing to look in the catalog and give you directions, then this method would be considered an algorithm. Theoretical computer scientists often analyze properties of algorithms where key steps are carried out by "oracles" that always give the right answer.

- Wait by the book return: Not an algorithm. Even if you are very patient, this method will never terminate unless someone else checks out and then returns your book.

- Use an electronic catalog: Probably the most effective algorithm for this task.

- Work systematically, shelf by shelf: This is an algorithm, but not the best way to solve this problem. Note that you would have to search every shelf to learn the book has been checked out.

- Pick shelves at random: The key word here is "random." The project in Chapter 9 explores algorithms that generate sequences of numbers that appear to be random (they are called pseudorandom number generators), and if you use one of these algorithms you can set it up so that it eventually chooses every shelf. If you use some other method for selecting shelves (*e.g.*, rolling dice to get shelf numbers) then many people would argue this method is not an algorithm because there is a chance it could go on forever.

- Ten friends each search a designated area: This is an example of a "parallel algorithm." Parallel algorithms are used on systems that have more than one processor, and are especially useful in large scientific problems where a calculation is divided into smaller parts for each processor to work on.

1.6: An algorithm for making a table of squares of numbers:

- Input: a number n. Result: a table that shows the values of i^2 for every number i between 1 and n.

- Start by making a table with n rows and three columns.

- Write the numbers between 1 and n in the first column.

- Put a 1 in the second and third columns of the first row.

- Fill in the second column from top to bottom, setting each cell to a value 2 greater than the value in the cell above.

- Fill in the third column from top to bottom, setting each cell to the sum of the number above and the number to the left.

The Ruby Workbench

2.5: Ruby should handle this expression with no problem. The equation for converting from Fahrenheit to Celsius works for negative temperatures as well as positive temperatures.

2.6: To find the freezing point of water call `celsius(32)`. To find the boiling point call `celsius(212)`.

The Sieve of Eratosthenes

3.3: Ruby removes every string with an even number of letters.

3.4: $\sqrt{1000} = 31.62$, so as soon as the first number in the worksheet is greater than 31 the algorithm can stop.

3.7: To find the number of primes less than some number n call `sieve(n-1)` to make a list and use the `length` method to find the number of items in the list. This can be done with a single Ruby expression. For example, to count the number of primes less than 1000:

```
>> sieve(999).length
=> 168
```

A Journey of a Thousand Miles

4.6: (a) Ruby will print every string in the array named `languages`. (b) Ruby will print strings shorter than 5 characters in length. (c) Ruby will print "found it" if the string `"fortran"` is in the array; since it isn't there, nothing is printed on the terminal.

4.9: The linear search methods stop after they find the first occurrence of the item they are looking for. Since there is a 7 in the third location in this array the search stops after doing three comparisons.

4.16: Insertion sort will make the fewest comparisons when the list is already sorted. It only makes one comparison on each iteration, and it does $n - 1$ iterations to sort an array with n items, so it would make 19 comparisons for an array of 20 items and 49 comparisons for an array of 50 items.

4.17: Iteration sort makes the most comparisons if the array is sorted, but in the opposite order. The algorithm needs to make a comparison for each dot shown in Figure 4.9. For an array with 20 items, `isort` will make $20 \times 19/2 = 190$ comparisons, and for an array with 50 items, $50 \times 49/2 = 1225$ comparisons.

Divide and Conquer

5.6: If the item being sought is in the first location in the array, `search` will find it on the first comparison, but `bsearch` will have to "zero in" on the item and do four comparisons.

5.12: If the array is already sorted, the merge operation needs to do half as many comparisons as it would in the worst case. To see why, suppose it is merging two groups of size four. The algorithm does four comparisons, each time moving something from the left group to the output. But then it can copy all four items from the right group without doing any comparisons. In the worst case, there would be seven comparisons to merge two groups of four items.

When Words Collide

6.6: A 3-digit decimal number has a value between 100 and 999, so an "average" three-digit number would be around 500, or $10^3/2$. By analogy, a rough estimate for the numeric value of an n-digit radix-26 string would be $26^n/2$.

6.8: A hash table with one row would put all the items in one long bucket in row 0. Since the bucket would not be sorted, looking for an item in the table would be a linear search (plus the wasted cost of computing the value of the hash function, which would be 0 for every string).

Bit by Bit

7.1: A 6-bit code has $2^6 = 64$ different patterns, so there is room for $64 - 50 = 14$ additional locations before the code has to be extended to 7 bits.

7.2: Encoding 192 countries would require $\lceil \log_2 192 \rceil = 8$ bits.

7.3: If 9 bits are used to encode team names, $2^9 = 512$ different names could be encoded.

7.7: Codes b, f, and g have an odd number of 1 bits.

7.14: If each letter has the (roughly) the same probability, there will be the same number of bits in each code, and the Huffman code will not be any shorter than a regular code. To see why, consider what happens when the first interior node is created: its value is the sum of the two letters, and when it is added to the queue it will go at the end, after the last letter. The same will be true of every new interior node, so the final tree will be "balanced."

The War of the Words

8.1: To multiply 2×8, the algorithm makes 8 iterations.

8.3: Multiplying $0 \times n$ means adding 0 n times. It doesn't matter how many iterations are made, the answer is always 0.

8.4: The way the Redcode program is written, the loop always does at least one addition before testing to see if the counter has reached 0. That means x will be added to `acc` one time, and the answer will be wrong.

8.5: Having $x < 0$ is no problem, but if $y < 0$ the program will give the wrong answer (it's the same bug that gives an error when $y = 0$, described above).

8.6: If $y = 0$ the subtract instruction sets x to $x - 0$, *i.e.*, it doesn't modify x. That means the program will be stuck in an infinite loop! Does this exercise give you any insight into why calculators give an error message (or ∞) when you try to divide something by 0?

8.7: Since the Dwarf occupies only four words in memory, and the first bomb goes in the word right after the end of the program, the bombs will not hit the program's instructions when they wrap around.

8.8: If bombs go in every other word, then yes, the Dwarf will bomb itself after the addresses wrap around. Since there are 4096 memory locations in the RubyLabs version of MARS, the Dwarf can run 2048 iterations before it bombs itself.

Now for Something Completely Different

9.1: An interval of 5 hours leads to a schedule with all 12 times. To make a schedule that includes all 12 numbers, the interval and 12 must be relatively prime, *i.e.*, there is no other number that divides both evenly. For a 12-hour clock, the only numbers that fit this criterion are 5 and 7.

9.6: Any interval that is relatively prime with 24, so any interval that is a prime number will work, *e.g.*, 7, 11, 13, 17.

9.9: To see if an array a contains every number between 0 and m-1:

```
>> a.uniq.sort == Array(0..(m-1))
```

Ask Dr. Ruby

Here are some reasons Eliza would not be able to respond to the statements given in the exercises for Section 10.7 (page 275).

```
H: I remember when I lived in the country and there were
   cows and chickens everywhere.
C: Why do you mention cows just now?
```

This sentence is too complex. Eliza would probably apply a rule for the word "remember," which has a high priority, and make a suspicious response like "Do you often think of when you lived in the country and there were cows and chickens everywhere?" instead of simply "Do you often think of cows?"

```
H: They were always looking at me.
C: The cows were looking at you?
```

This response depends on the previous sentence. The program would need some sort of memory to record the fact that the first sentence was about cows, and then be able to figure out the word "they" in the second sentence refers to the "cows" in the previous sentence.

H: Our parents got mad when we tipped the cows.
C: Did you really push over a cow?

This response requires the program to have some real-world knowledge and to be able to associate words with their meanings. It would need to know that, in this context, "tipping" is the same as "pushing," and that when something is tipped far enough it will fall over. The context is important, too, since the correct meaning is the one in "he tipped the cup" but not in "he tipped the waiter."

H: The one with the big horns was the scariest.
C: Are you referring to a bull?

For a computer to give this response, it would have to remember that the conversation was about cows, so "horns" refers to the things on the bull's head, not musical instruments, and it would require some real-world knowledge that cows (females) often don't have horns.

H: Yeah, that guy was pretty aggressive.
C: Did the horns seem dangerous?

This response is based on context (a conversation about cows and horns) and on the meaning of the word "horns" – that they are sharp, pointy, and, when on the head of a live animal, potentially a threat.

The Music of the Spheres

11.2: Every body is influenced by the gravitational pull of the others, so yes, the Sun moves. Since the Sun is the first body in the array, ask Ruby to print that object's position:

```
>> b[0].position
=> (-7.3353e+07,-4.4832e+07,1.4706e+06)
```

11.3: Repeat the advance and turn commands four times, *e.g.*,

```
>> 4.times { robot.advance(10); robot.turn(90) }
=> 4
```

11.8: With 25 bodies, the simulation does $(25 \times 24)/2 = 300$ force calculations on each time step.

11.10: If f3 starts out at a position halfway between f1 and f2, its track will initially stay between the tracks for the other two bodies, but at some point f3 will veer away and head off on its own track. One of the important attributes of chaotic systems is that small changes in the starting conditions may lead to very big changes later on, so it is impossible to predict where the body will end up.

The Traveling Salesman

12.2:   The cost of the tour A–D–E is the cost of the path from A to D, plus the cost of the path from D to E, plus the cost of the path from E back to A: $82.02 + 89.55 + 142.56 = 314.13$.

12.4:   The current path is

```
[:C, :F, :G, :A, :E, :D, :B]
```

The path that follows the dotted lines is

```
[:C, :F, :G, :A, :D, :E, :B]
```

The difference in these paths is the two cities in locations 4 and 5 of the array, so the call to `mutate!` is

```
>> t.mutate!(4)
=> #<TSPLab::Tour [:C, :F, :G, :A, :D, :E, :B] (1185.43)>
```

12.6:   If the current path is

```
[:C, :F, :G, :A, :E, :D, :B]
```

a call to `t.mutate!(1)` would produce this path:

```
[:C, :G, :F, :A, :E, :D, :B]
```

The new tour is

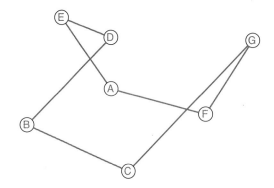

12.7:   A second mutation at the same location would simply change the path back to what it was before the first mutation.

12.9:   There are $15!/2 = 653,837,184,000$ possible tours of a map with 16 cities. This is approximately 6.5×10^{11}. Dividing by 10^6 means the random search would look at 1 out of every 650,000 tours. We can use IRB to get the exact percentage:

```
>> 1000000.0 / ntours(16)
=> 1.52943274636396e-06
```

If you convert this to a percentage, it's .000153% of the total number of tours.

12.10:   The evolutionary search will replace roughly half the tours on each generation, so just multiply the population size by the number of generations and divide by 2.

Appendix B

Ruby Reference

This appendix is a quick reference guide to numbers, strings, arrays, and other types of objects used in projects throughout the book. The types of data described here are those that are defined already as part of the Ruby programming language. The types of object created specifically for the projects in this book – RandomArray, Body, Tour, *etc.* – are described in the Lab Manual (since they may have been updated since the book was published). The *Explorations in Computing* web site also has an on-line reference section for the RubyLabs software:

```
http://www.cs.uoregon.edu/eic
```

Objects and Classes

In Ruby (and other object-oriented programming languages) each piece of data is called an *object*. There are many different kinds of objects, from simple pieces of data like numbers and characters, to more complicated data structures like arrays, which are lists of objects. Each of these categories is called a *class*, and we say each individual object belongs to a class, or is an *instance* of a class.

Here is a simple example. We can define a variable named n to have the value 42 with this assignment statement:

```
>> n = 7 * 6
=> 42
```

The class that represents integers in Ruby is named Fixnum. The statements "n is an integer," "n belongs to the Fixnum class," and "n is an instance of Fixnum" are three different ways of saying what type of object n refers to.

A second example defines a string:

```
>> s = "Hello"
=> "Hello"
```

The name of this class is String, so we would say "s is an instance of the String class," or simply "s is a String."

If you ever forget what an object is, or are curious to find out what the name of its class is, you can call a method named `class`:

```
>> n.class
=> Fixnum
>> s.class
=> String
```

Every class in Ruby has a name that starts with an uppercase letter. Most classes have a method named `new` that can be called to create a new object of that class. For example, another way to make a new String object is

```
>> t = String.new
=> ""
```

As you can see by the output from IRB, a String made by calling `new` has no characters in it; this method for making a string might be used by a program that needs to initialize a string and later add new text as it works its way through an algorithm.

Object-oriented languages allow programmers to define their own new classes. The Body class used in the solar system simulation, the Tour class used in the traveling salesman, and several other classes are all defined in the RubyLabs module. Defining new types of data is beyond the scope of this book, but if you're curious you can find more information in any book on Ruby, and you can also look at the source files for the RubyLabs gem that you installed when you set up your system to run the tutorial projects.

Modules

A module is a software package that contains the definition of several related classes and methods. Ruby's math library is an example of a module, and the RubyLabs software used for the tutorial projects in this book is also organized as a group of modules.

Normally, to use a function in the math library, we need to put the module name in front of the function name. For example, to ask Ruby to compute $\sqrt{5}$, we would type:

```
>> Math.sqrt(25)
=> 5.0
```

If we are going to be using a lot of math library functions, we can tell Ruby to get the entire library and include it as part of IRB, so all of the library's functions are available for the remainder of that IRB session:

```
>> include Math
=> Object
```

Then we can just call a math function directly:

```
>> sqrt(36)
=> 6.0
```

If you installed the RubyLabs software and configured your system according to the guidelines in the Lab Manual, the Math module and RubyLabs are automatically included at the start of each new IRB session. Then you simply have to type an `include` command to specify which particular part of the RubyLabs package you want to use. For example, at the start of an IRB session to work on the traveling salesman project, you would type

```
>> include TSPLab
=> Object
```

Variables

Variables in Ruby are simply labels for objects. A variable is defined in an *assignment statement*, which has the name of a variable, an equal sign, and an expression:

```
>> x = 3
=> 3
>> y = x * 2
=> 6
```

Note that the second statement above has a variable name in an expression on the right side. To evaluate this expression, Ruby looks to see if x has been defined, and if so, it uses the value of the object x refers to, which is how y got the value 6.

Variable names in Ruby always begin with a lowercase letter. After the initial letter, names can have any combination of upper and lowercase letters, digits, or an underscore character.

Something that seems strange at first is to see the same name on the left and right side of an assignment:

```
>> x = x + 1
=> 4
```

Ruby treats this just like it does any assignment: it first evaluates the expression on the right side, using the current value of x to produce the number 4, and then it updates x to refer to this new object. Updating a variable by adding a value to it is such a common operation there is a special symbol for it:

```
>> x += 1
=> 5
```

The notation n += m is shorthand for n = n + m, where the value on the right side of the operator is added to the variable named on the left side.

Ruby does not care what type of object a variable refers to. We can have a variable refer to any type of object, and we can even have it refer to different types of objects at different times. After telling Ruby we want x to be a number in the previous example, we can later tell Ruby we want x to refer to a string:

```
>> x.class
=> Fixnum
>> x = "aloha"
=> "aloha"
>> x.class
=> String
```

Changing the kind of object referred to by a variable is not something you will see in the text, but it does help illustrate that a variable is nothing more than a label for an object, and that we are allowed to modify a variable by updating it to refer to a new object.

Methods

In an object-oriented programming language, objects are the individual pieces of data, and methods are operations that use the data. Each different kind of object has methods that are defined specifically for that type of object. A large part of learning how to use a language is learning what sorts of things can be done with the different kinds of objects.

Most methods have names, and the operation defined by the method is performed when we use the name in an expression. A simple example is a method named `length`, which counts the number of characters in a string. If `s` is a String object, we can use the `length` method to count the number of characters:

```
>> s = "abcde"
=> "abcde"
>> s.length
=> 5
```

The notation shown above is the most common way to use a method, where we write the name of an object, a period, and the name of the method.

Some other terminology associated with methods:

- When Ruby evaluates an expression that has a method name, we say Ruby *calls* the method.

- A value computed by a method is *returned* to be used in the expression.

- When we call a method, we often pass it a set of *arguments* to use in its calculation; arguments are simply objects that have been created already.

An example of a method that expects us to pass it an argument is `include?`, which searches through a string. For example, given the definition of `s` above, if we want to know whether or not `s` contains the substring `"bc"` we would type this expression:

```
>> s.include?("bc")
=> true
```

Note the question mark is part of the method name. Ruby methods that return `true` or `false` often have question marks at the end of their name.

Often methods are called without putting an object name in front of the method name. This is especially true of methods that we write ourselves, to test an algorithm. For example, the project in Chapter 2 shows how to define a new method to convert temperatures from Fahrenheit to Celsius. To call this method, we just write its name, followed by the temperature value we want to convert:

```
>> celsius(100)
=> 37
```

A few methods are associated with symbols that can be used in expressions. An example is a method used often in projects in this book to attach a new item to the end of an array:

```
>> a = [2, 3, 5]
=> [2, 3, 5]
>> a << 7
=> [2, 3, 5, 7]
```

The "name" of this method is <<, and we can use this symbol in an expression, just like we use + and other arithmetic operators in expressions with numbers.

It turns out there really is a method with this name, and if you want, you can use the normal method-calling syntax to call it. To add the number 11 to the end of a you could type

```
>> a.<<(11)
=> [2, 3, 5, 7, 11]
```

That expression is not very readable, compared to the example above that appended 7 to the list, but it does show that the operation that attaches items to the end of an array is a method, just like other methods in Ruby. In an object-oriented language, every piece of data is an object, and every operation is a method.

Numbers

By default, every number we use in an expression in Ruby is an integer. An integer in Ruby is called a "Fixnum," as shown by this example:

```
>> n = 20
=> 20
>> n.class
=> Fixnum
```

If we want to make a real number, we need to include a decimal point:

```
>> n = 20.0
=> 20.0
>> n.class
=> Float
```

The word "Float" is short for "floating point," which refers to the fact that these numbers are stored internally in a type of scientific notation.

An important thing to remember is that when the operands in an expression are integers, the result is an integer. This can lead to unexpected results, especially in a division operation:

```
>> 7 / 10
=> 0
>> 100 / 40
=> 2
```

Arithmetic Operators in Ruby

+	addition	*	multiplication	**	exponentiation
–	subtraction	/	division	%	modulo

When Ruby divides one integer by another, the result is truncated, *i.e.*, Ruby just keeps the quotient and throws away the remainder.

Sometimes we want the remainder and don't care about the quotient. These are places where the modulo operator comes into play. The meaning of x % y is "the remainder after dividing *x* by *y*."

If either or both of the operands is a Float, the result will be a Float. We can make a number a Float by adding ".0" to the end:

```
>> 100.0 / 40.0
=> 2.5
```

If an integer is stored in a variable, we can call a method named to_f to convert it to a Float:

```
>> x = 100
=> 100

>> x / 6
=> 16

>> x.to_f / 6
=> 16.6666666666667
```

Ruby has a third type of number, named Bignum. This kind of number is used when the result of a numeric operation is very large. For example, we can compute 30! using a method named fact defined for the traveling salesman project:

```
>> n = fact(30)
=> 265252859812191058636308480000000

>> n.class
=> Bignum
```

For the projects in this book we don't have to worry about whether an object is a Fixnum or a Bignum; Ruby switches over to using Bignum when it needs to, and switches back to Fixnum when it can:

```
>> m = n / fact(29)
=> 30

>> m.class
=> Fixnum
```

Strings

Strings are pieces of text. The easiest way to make a String object in an IRB session is to simply type a sequence of characters enclosed in double quotes:

```
>> s1 = "green"
=> "green"

>> s2 = "fields"
=> "fields"
```

Some operations on strings are associated with symbols. For example, the + operator, when applied to String objects, tells Ruby to append one string to the end of another:

```
>> s1 + s2
=> "greenfields"

>> s2 + " of " + s1
=> "fields of green"
```

Ruby has several more operations for string objects. Most of the operations have English or English-sounding names, and what they do is self-explanatory:

```
>> s2.length
=> 6
>> s2.reverse
=> "sdleif"

>> s2.upcase
=> "FIELDS"
```

An important detail is that methods usually do not modify a string. Instead, methods like `reverse` and `upcase` make a new String object by copying the existing string, and then performing their operation on the copy.

A few methods do change the string. These almost all have names that end with an exclamation mark:

```
>> s2.reverse!
=> "sdleif"

>> s2
=> "sdleif"
```

Note that as far as Ruby is concerned, `reverse` and `reverse!` are different methods, with two different names. The exclamation mark is not a special operator that tells Ruby it can modify the string, it's an extra character added to the name to make a new method name.

We can access individual characters in a string using the index operator, which consists of an integer between two square brackets. The first character is at location 0, and the last character in an n-letter string is in location $n - 1$. In Ruby 1.8.7 and earlier, the value returned by the index operator is the numeric code for the character at the specified position:

```
>> s1[0]
=> 103
>> s1[1]
=> 114
```

Numeric codes are the topic of Chapter 7. You can also find more information by searching for "ASCII" at Wikipedia or another Internet reference.

Symbols

A Symbol in Ruby is a special type of string used when a program needs an object to represent a simple name. Instead of putting quotes around the characters in the string, we just write a colon and then the name:

```
>> color = :green
=> :green
>> hash_function = :h0
=> :h0
```

As always, after storing an object in a variable, we can call a method named `class` to ask Ruby what type of object is referred to by the variable:

```
>> color.class
=> Symbol
```

If a Symbol is basically a string, why have a new kind of object? Why not just use String objects? The first assignment above could just as easily have been written this way:

```
>> color = "green"
=> "green"
```

The answer is that Symbol objects and String objects have different properties, so they behave differently in a program. In general, a String object is used when the object might change. Strings have methods like `reverse`, which reverses the order of the characters, or `upcase`, which capitalizes all the letters in the string. Strings also have a method named `length`, which counts the number of characters. We can access the individual letters, *e.g.*, by calling `color[0]` to get the code for the first letter in a string, or by calling `each_byte` to iterate over all the characters.

A Symbol object is used when we just want to identify something or give a name to a value. The RubyLabs method that creates a new TestArray object expects a parameter that specifies what kind of items to include in the array:

```
>> a = TestArray.new(10, :cars)
=> ["chevrolet", "oldsmobile", ... "bentley"]
```

The previous example made an array of car names, but if we want an array of colors:

```
>> a = TestArray.new(10, :colors)
=> ["black", "coral", ... "bisque"]
```

The symbol tells the method what type of string to include in the output. The method won't do anything with the Symbol – it doesn't need to count the letters or do anything other than just make sure it refers to one of the known types of data.

Internally, Symbols are neat and tidy little packages, so programmers often prefer to use Symbol objects instead of String objects when all they need is a name.

```
i = 1
while i <= n
  puts i ** 2
  i += 1
end
```

```
for i in 1..n
  puts i**2
end
```

Figure B.1: *Two different ways to print the values of i^2 for numbers ranging from 1 to 10.*

Ranges

A Range in Ruby is written as pair of numbers separated by two periods.

One place to use a range is when asking Ruby to make a list of numbers:

```
>> a = Array(1..10)
=> [1, 2, 3, 4, 5, 6, 7, 8, 9, 10]
```

As you might expect, the notation n..m means "numbers between n and m." In this case Ruby uses the range as a specification for the items to include in an array.

Another place ranges are used in the text is in asking for a substring or a part of an array. In this case, the range is written between the square brackets of an index operator:

```
>> s = "abracadabra"
=> "abracadabra"
>> s[3..6]
=> "acad"
>> a = [2, 3, 5, 7, 11, 13, 17]
=> [2, 3, 5, 7, 11, 13, 17]
>> a[2..4]
=> [5, 7, 11]
```

Ranges can also be used to control iterations. A concise way to write a loop that iterates over every number in a specified range is shown in Figure B.1. The statement on the left uses a while statement. Before the while loop begins, we need to initialize the control variable i, and inside the loop we need to increment the variable so the terminating condition is eventually false. The statement on the right takes care of these details for us and is a convenient way to write a loop that executes a specific number of times (see page 359).

Arrays

An array is a *container*, meaning it is an object that holds references to other objects. The easiest way to make an array in Ruby is to simply make a list of objects that go in the array, separated by commas and enclosed in square brackets:

```
>> a = [53, 20, 28, 56, 47]
=> [53, 20, 28, 56, 47]
```

Most operations on arrays are performed by calling a method. Many of the method names defined for strings are also defined for arrays. For example, we can ask Ruby to count the number of items, or reverse the items:

```
>> a.length
=> 5
>> a.reverse
=> [47, 56, 28, 20, 53]
```

As is the case with Strings (and most other classes), calling a method like `reverse` does not change the object, but instead makes a copy of the object, so `a.reverse` returns a copy of `a` but with the items in the opposite order.

In another similarity with strings, some array operations come in two "flavors." There is a version that returns a modified copy of the array, and a version (usually with an exclamation mark in its name) that modifies the array:

```
>> a.reverse => [53, 20, 28, 56, 47]
>> a
=> [53, 20, 28, 56, 47]
```

Here are some other examples of array operations:

```
>> a[0]
=> 53
>> a.sort
=> [20, 28, 47, 53, 56]
>> a.insert(1,99)
=> [53, 99, 20, 28, 56, 47]
```

The first example uses the index operator to access a single item in the array, the second returns a sorted copy of the array, and the third adds a new item to the array at the specified location (note that this method *does* modify the array).

Arrays can contain references to any type of object – including other arrays.

```
>> fruits = ["apple", "lime", "kiwi"]
=> ["apple", "lime", "kiwi"]
>> matrix = [ [0,1], [2,3] ]
=> [[0, 1], [2, 3]]
```

Besides listing all the items we want to put in an array, we can make an array by calling the method named `new`. By default, a call to `Array.new` makes an empty array, but we can also ask for an array of a particular size, or an array of a specified size initialized with a certain value:

```
>> Array.new
=> []
>> Array.new(10)
=> [nil, nil, nil, nil, nil, nil, nil, nil, nil, nil]
>> Array.new(10,0)
=> [0, 0, 0, 0, 0, 0, 0, 0, 0, 0]
```

A special case is making an array with a range of values – note the word `new` is not used in this expression:

```
>> Array(2..10)
=> [2, 3, 4, 5, 6, 7, 8, 9, 10]
```

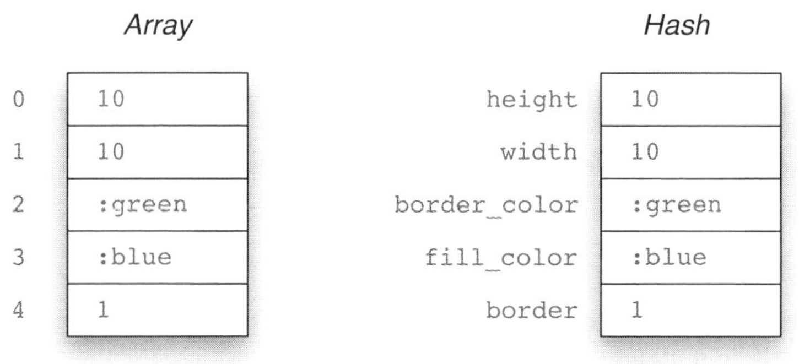

Array   *Hash*

0	10	height	10
1	10	width	10
2	:green	border_color	:green
3	:blue	fill_color	:blue
4	1	border	1

Figure B.2: *Hashes and arrays are both containers, meaning they are collections of references to other objects. Items in an array are accessed by their location in the array, but items in a hash are accessed by their name.*

Hashes

A type of object known as a "hash" is a container that holds references to other objects. In Ruby, a hash is much like an array. To access an object in a hash, use the same index operator as for arrays: write the name of the hash, followed by a specification of an item in square brackets. The difference between an array and a hash is in how the items are identified. In an array, items are identified by their location in the array, but in a hash items are identified by name.

As an example of how (and why) to use a hash, suppose a program needs to represent the attributes of a rectangle that will be displayed on a canvas. The attributes are the height, width, border color, interior color, and the width of the border. It would be possible to put all these values in an array:

```
>> box = [10, 10, :green, :blue, 1]
=> [10, 10, :green, :blue, 1]
```

Then the attributes would be accessed by their location. For example, to fill in the color, the program would call a method named `fill_rectangle`, passing it the attribute stored in the array:

```
>> fill_rectangle( box[3] )
```

There are several problems with this approach. It's hard to understand, from looking at this expression, what `box[3]` means. It also forces the programmer to remember the order in which attributes are stored in the array, and it makes it easy for errors to creep into the program, *e.g.*, if the programmer forgets where the fill color is stored and calls `fill_rectangle(box[2])` by mistake.

A better solution is to use a hash to represent the attributes, so the attributes could be referred to by their names (Figure B.2). This is how to define the hash:

```
>> box = { :height => 10, :width => 10, :border_color => :green,
           :fill_color => :blue, :border => 1 }
=> {:border_color=>:green, :fill_color=>:blue, :width=>10,
           :border=>1, :height=>10}
```

When written out in this form, the hash is a set of name-value pairs, separated by commas, and enclosed in curly braces. Each item in the hash has an attribute name on the left side, an arrow (written with an equal sign and a greater-than sign, with no spaces between), and the attribute value on the right side of the arrow. Ruby doesn't care if we put spaces between the items, or spread the definition out over several lines.

Note that in this example the attribute names (and some of the values, as well) are Ruby symbols. This is one of the main uses for Symbol objects in Ruby, but a hash can use strings, numbers, or almost any kind of object for attribute names.

Once the hash object is created, we refer to attributes by their name, not their location. For example, to get the height of the box:

```
>> box[:height]
=> 10
```

To pass the fill color to the method that draws the box:

```
>> fill_rectangle( box[:fill_color] )
```

This version of the program is much easier to write, and less error-prone. Ruby takes care of figuring out where the attributes are stored, and we just have to remember the attribute names.

A typical application will first create an empty hash, and then add items to it as the program runs. Two ways to make a hash are to write a pair of braces with nothing in between:

```
>> circle = { }
=> {}
```

Or we can call a method named `Hash.new`, which creates a new, empty, hash object:

```
>> circle = Hash.new
=> {}
```

To add an item to a hash, just use an assignment statement, *e.g.*,

```
>> circle[:radius] = 10
=> 10
>> circle[:fill_color] = :red
=> :red
```

Now if we ask Ruby to print the hash object, it will show the two new name-value pairs:

```
>> circle
=> {:radius=>10, :fill_color=>:red}
```

There are several methods in Ruby for working with hash objects. A method named `keys` will return an array of all the attribute names:

```
>> box.keys
=> [:border_color, :fill_color, :width, :border, :height]
```

Note that several methods defined for arrays don't make sense for hashes, so they are not allowed for Hash objects. For example, since the items in a hash are not in any particular order, there are no methods for sorting or reversing a hash.

Iterators

An iterator is a method that repeats an operation several times.

Most of the iterators used in projects in this book are methods that "traverse" an array in order to perform an operation with every item in the array. The simplest example is `each`, which tells Ruby to execute the statements between braces for every item in an array:

```
>> a = ["He", "Ne", "Ar", "Kr", "Xe", "Rn"]
=> ["He", "Ne", "Ar", "Kr", "Xe", "Rn"]
>> a.each { |s| puts s }
He
Ne
...
Rn
=> ["He", "Ne", "Ar", "Kr", "Xe", "Rn"]
```

The items in the array are fetched, in order, and assigned one at a time to the variable named between vertical bar characters. After storing a value in the variable Ruby executes the statement(s) between the braces.

A method named `delete_if` selectively removes items from an array. The statement between the braces is expected to be a Ruby expression that evaluates to either `true` or `false`. If the expression is `true` for any item in the array, that item is removed from the array. Using the array from the previous example, this call to `delete_if` removes every string that would follow `"Mg"` in an alphabetical listing of elements:

```
>> a.delete_if { |s| s > "Mg" }
=> ["He", "Ar", "Kr"]
```

To iterate over the characters in a string, use `each_byte`:

```
>> s = "Curie"
=> "Curie"
>> s.each_byte { |i| puts i }
67
117
114
105
101
=> "Curie"
```

In Ruby 1.8.7 and earlier, the `each_byte` iterator works with the ASCII codes of the characters in the string.

Many of the projects also used iterators that do not operate on arrays. The simplest is named `times`:

```
>> 5.times { puts "yeah" }
yeah
yeah
yeah
yeah
yeah
=> 5
```

If the code between the braces includes a variable name, the variable will be set to increasing values on each successive iteration. The values start at 0:

```
>> n = 5
=> 5
>> n.times { |i| puts 5 * i }
0
5
10
15
20
=> 5
```

To perform the same sort of iteration, but on a decreasing set of values, call downto:

```
>> 10.downto(5) { |i| puts 5*i }
50
45
40
35
30
25
=> 10
```

The starting value is the value of the object before the period, the ending value is the argument passed to downto, and the variable named between vertical bars takes on every value in this range.

Finally, there is something called a "for statement" (shown earlier in Figure B.1) that is not technically an iterator, but is a convenient notation that is translated into an iterator. This statement asks Ruby to execute the statements between the line with the word for and the line with the word end, setting the variable i to each value in the specified range:

```
for i in 1..10
    ....
end
```

This type of statement is awkward to type in IRB, since we have to use an alternative syntax to mark the beginning and ending of the statements to iterate. If you want to try it, just remember to type do instead of an opening brace and end instead of a closing brace:

```
>> for i in 2..7 do puts 2**i end
4
8
...
128
=> 2..7
```

Conditional Execution

There are situations in a program where we want Ruby to execute a statement only under the right circumstances. The general idea is to have Ruby evaluate a Boolean expression (one that evaluates to `true` or `false`) and then perform an operation only when the expression is `true`.

The simplest form of conditional execution is to attach a *modifier* to the end of a statement. A modifier consists of the word `if` followed by a Boolean expression. Here is a simple example. Assuming `a` is an array of color names, this statement prints every string in `a`:

```
>> a.each { |s| puts s }
almond
antique white
...
```

But suppose we just want to see the names that contain a space in the middle (*e.g.*, "spring green"). If `s` is a string, the expression `s.contains?(" ")` will be `true` if a space occurs anywhere in `s`. We can edit the previous expression by adding a modifier after the call to `puts` so Ruby prints only those color names that contain a space:

```
>> a.each { |s| puts s if s.include?(" ") }
antique white
cadet blue
dodger blue
...
```

Attaching a modifier to an expression is the simplest (and often the easiest to read) technique for writing a conditional operation, but there are many situations where it is necessary to perform two or more operations when a condition is met. The way to do this is to write an *if statement*. Figure B.3 has two examples. An `if` statement is similar in structure to a `while` statement. The first line has the word `if` followed by a Boolean expression. Following this line there can be any number of Ruby statements, followed in turn by a line that just has the keyword `end`. When Ruby sees an `if` statement in a method, it evaluates the expression, and the statements in the body (the statements up to the closing `end`) are executed only if the expression is `true`.

The method named `emphasize` in Figure B.3 expects us to pass it a string. If the string is one of the words "red," "green," or "blue" (the `||` operator means "or" in Ruby) the method converts all the letters to upper case and adds an "S" to the end of the word:

```
>> emphasize("red")
=> "REDS"
```

If the argument passed to `emphasize` is not one of these three words the method doesn't do anything:

```
>> emphasize("white")
=> "white"
```

```ruby
def emphasize(s)
  if s == "red" || s == "green" || s == "blue"
    s.upcase!
    s += "S"
  end
  return s
end

def drink_cup(n)
  if n == 12
    return "tall"
  elsif n == 16
    return "grande"
  elsif n == 20
    return "venti"
  else
    return n.to_s + " ounce"
  end
end
```

Figure B.3: *Examples of* if *statements.*

The other method in Figure B.3 shows that we can add one or more "else clauses" to an if statement. As you might expect, Ruby evaluates the Boolean expressions in order, and as soon as it finds one that evaluates to true the statements following it are executed. The method named drink_cup expects us to pass an integer, and it will return one of the strings in the body of the if statement:

```
>> drink_cup(16)
=> "grande"
>> drink_cup(20)
=> "venti"
```

Note that only one of the groups of statements is executed. If all of the Boolean expressions are false, the last group (the one preceded by the keyword else) is executed:

```
>> drink_cup(8)
=> "8 ounce"
```

Blocks

There are times when we want Ruby to operate on two or more expressions as a single group. A good example is when we iterate over an array. Suppose we want to compute the sum of a set of numbers. This expression tells Ruby to add each item to the sum, and to print the partial sum after each addition:

```ruby
a.each { |x| sum += x; puts sum }
```

In Ruby, a collection of statements, enclosed between braces and separated by semicolons, is known as a *block*. Several methods, defined in a wide variety of Ruby classes, work with blocks. Iterators, like `each` and `delete_if`, are the most common examples. Another one is `sort`. By default, a call to `sort` arranges an array of strings in alphabetical order:

```
>> a = TestArray.new(10, :cars)
=> ["peugeot", "maserati", "pontiac", ... "kia"]
>> a.sort
=> ["chevrolet", "fiat", "honda", "kia", ...]
```

But we can also supply a block of code when we call `sort`, and the method will use the block to compare items. This is how we can sort the array of strings by their length, so the shortest car names are at the front:

```
>> a.sort { |x,y| x.length <=> y.length }
=> ["kia", "fiat", "honda", ... "chevrolet"]
```

In our lab projects we use blocks when monitoring a call to a method or measuring execution time. For example, to ask Ruby to report the amount of time required to sort an array with a method named `isort`, we make a block of code containing the call, and use the block with a method named `time`:

```
>> time { isort(a) }
=> 0.6108
```

Blocks are also used by the `trace` method. Several projects put "software probes" in place, where they can be monitored when a block of code is executed by `trace`. To allow the probes to work with any arbitrary Ruby expression, `trace` allows us to write a block expression. Here is an example from Chapter 5:

```
>> trace { x = a.random(:fail); puts x; bsearch(a, x) }
```

This block chooses a random value that is known to not be in an array, prints the value, and then calls a search method. Since there is a probe in place, we can watch the search in action, and see how it tries to find the missing item.

Index